T0142445

EcoProduction

Environmental Issues in Logistics and Manufacturing

Series Editor

Paulina Golinska-Dawson, Poznań, Poland

The EcoProduction Series is a forum for presenting emerging environmental issues in Logistics and Manufacturing. Its main objective is a multidisciplinary approach to link the scientific activities in various manufacturing and logistics fields with the sustainability research. It encompasses topical monographs and selected conference proceedings, authored or edited by leading experts as well as by promising young scientists. The Series aims to provide the impulse for new ideas by reporting on the state-of-the-art and motivating for the future development of sustainable manufacturing systems, environmentally conscious operations management and reverse or closed loop logistics.

It aims to bring together academic, industry and government personnel from various countries to present and discuss the challenges for implementation of sustainable policy in the field of production and logistics.

More information about this series at http://www.springer.com/series/10152

Adam Kolinski · Davor Dujak ·
Paulina Golinska-Dawson
Editors

Integration of Information Flow for Greening Supply Chain Management

Editors
Adam Kolinski
Poznan School of Logistics
Poznań, Poland

Davor Dujak
Faculty of Economics
Josip Juraj Strossmayer University of Osijek
Osijek, Croatia

Paulina Golinska-Dawson
Faculty of Engineering Management
Poznan University of Technology
Poznań, Poland

ISSN 2193-4614 ISSN 2193-4622 (electronic)
EcoProduction
ISBN 978-3-030-24357-9 ISBN 978-3-030-24355-5 (eBook)
https://doi.org/10.1007/978-3-030-24355-5

This Springer imprint is published by the registered company Springer Nature Switzerland AG
The registered company address is: Gewerbestrasse 11, 6330 Cham, Switzerland

Preface

The information management plays a crucial role in decreasing the time needed for making the right management decisions. The undeniable impact of the information flow in enterprises on the logistic processes efficiency forces the focus on the integration of the information flow in the entire supply chain. Current trends in business practice make it also necessary to consider opportunities related to the information integration of the supply chain in terms of the impact of the environmental aspects at an early implementation stage of the sustainable development concept. Information flow integration provides an opportunity to focus on greening the supply chain. It is currently difficult to identify the impact of information integration on the greening of the supply chain in a wide range of practical applications. This monograph therefore focuses on the potential of information integration solutions for the greening supply chain. Through 20 chapters that positively passed at least two double-blind review processes, 36 authors from 10 countries present their latest scientific results providing a significant contribution to the area of information flow integration in supply chain management.

The scope of the presented monograph is divided into three main parts:

- Application of Information Flow Standards in the Supply Chain,
- Information Systems and Technological Solutions Integrating Information Flows in the Supply Chain,
- Modelling and Simulation of Logistics Processes as a Decision-Making Tool.

The first part concerns the application of information flow standards in supply chains. The central task of integrated supply chains is to ensure access of all of the SC participants to updated, consolidated and coherent information with regard to the currently performed processes. Having access to reliable and valid information anywhere and anytime is a significant factor affecting competitive advantage and efficiency of actions. The application of information and communication standards eliminates errors in the implementation of logistic processes and decreases overall lead time, which in return supports ecological orientation in supply chain management.

The second part focuses on the role of IT systems and technological solutions in the integration of supply chain information flow between individual business partners in the supply chain. An important factor influencing this integration is the application of technological solutions that support the integration of logistic processes in the supply chain, both at the strategic level through the application of integration platforms or data exchange clouds and at the operational level through the application of technologies that improve the physical implementation of logistic processes. The Internet of things and the Industry 4.0 concept play an increasingly important role in efficient supply chain management.

The third part of the problem concerns modelling and simulation of logistic processes as decision-making tools. In the era of dynamically changing market situation, the time spent on information flow and decision-making is one of the key factors influencing the competitiveness of enterprises, also in the environmental aspect. The decision-making process can be effectively supported by simulation tools, which enable a multi-criteria analysis of potential benefits or risks resulting from the implementation of the planned measures. The stage immediately preceding the use of simulation tools is process modelling, without which it is not possible to effectively monitor and control the implementation of activities. As the decision-making process has a key impact on the logistics process efficiency, it is an important element in the process of information integration for the greening supply chains.

Although not all of the received chapters appear in this book, the efforts spent and the work done for this book are very much appreciated.

We would like to thank all reviewers whose names are not listed in the volume due to the confidentiality of the process. Their voluntary service and comments helped the authors to improve the quality of the manuscripts.

Poznań, Poland Adam Kolinski
Osijek, Croatia Davor Dujak
Poznań, Poland Paulina Golinska-Dawson

Contents

Application of Information Flow Standards in the Supply Chain

Using the Standards of Electronic Communication on the Example of Solutions Applied in Creating Logistics Single Window at the Ports of Portugal

Waldemar Osmólski and Marina A. Zhuravskaya

Abstract The following chapter discusses the practical application of the latest technological solutions for the exchange of electronic data between the entities of the logistic ecosystem. One of the most important features taken into account in the construction of system connections is the component architecture, which in this case is based on the SOA (Service Oriented Architecture) model, and the standard of information exchange based on the eDelivery and eFreight models. All these solutions were used in the construction of LSW (Logistics Single Window) in the seaport of Lisbon and were the basis for its application in other ports of Portugal. An extremely important feature is, above all, the interoperability of the issue, which is based on the possibility of easy connection with European port systems and using transoceanic connections based on the solutions used in Brazilian ports. Such a modern approach to communication solutions puts Portuguese ports in a leading position in comparison with other European ports, thus setting the direction for their further development.

Keywords Communication standards · Single Window · e-Delivery · e-Freight

1 Introduction

The world is undergoing changes caused by increasingly common presence of technology in our lives. The introduction of artificial intelligence and self-learning machines changes the way decisions are made and the way people communicate. The process is irreversible and will certainly proceed. In the nearest future, chatbots will replace people, while Blockchain and cryptocurrencies will completely change how the world works. Industrial revolution 4.0 (Prajogo and Olhager 2012) will lead to compulsory automation in production and supply chain covering raw materials,

W. Osmólski (✉)
Institute of Logistics and Warehousing, Estkowskiego 6, 61-755 Poznan, Poland
e-mail: waldemar.osmolski@ilim.poznan.pl

M. A. Zhuravskaya
Ural State University of Railway Transport, Kolmogorova Str., 66, Yekaterinburg, Russia
e-mail: MZhuravskaya@usurt.ru

A. Kolinski et al. (eds.), *Integration of Information Flow for Greening Supply Chain Management*, Environmental Issues in Logistics and Manufacturing, https://doi.org/10.1007/978-3-030-24355-5_1

components and even finished goods. Services, initially only the repeatable ones, will be handled by machines or software. Automatic and self-operating machines will be able to replace humans.

To make sure that these processes take place in an appropriate and undisturbed way, particular attention should be drawn to the correct construction of IT systems' component architecture and efficient data exchange between cooperating economic entities, institutions or public administration bodies. A discussion on contemporary supply chains therefore involves aspects related to the digitisation of processes occurring within them. A supply chain is presently a set of independent, discrete and, to a large degree, autonomous events controlled by actions related to marketing, production or distribution. Digitisation of these processes triggers a continuous change, which creates an integrated ecosystem (Debicki and Kolinski 2018) based on the transparency of processes occurring between entities involved and on the application of communication standards or common structural architecture (Fig. 1).

An immensely important element of this concept is the minimisation of activities that should be carried out as part of transferring necessary information, or the clarity and uniformity of processes occurring in a specific ecosystem. One of the pillars of such type of solutions is using the concept of Single Window (Osmólski et al. 2018), defined as one access window representing an ecosystem for cooperation or a decentralised system based on an existing or planned IT tool, an electronic shared services platform in particular. It allows parties participating in the trading

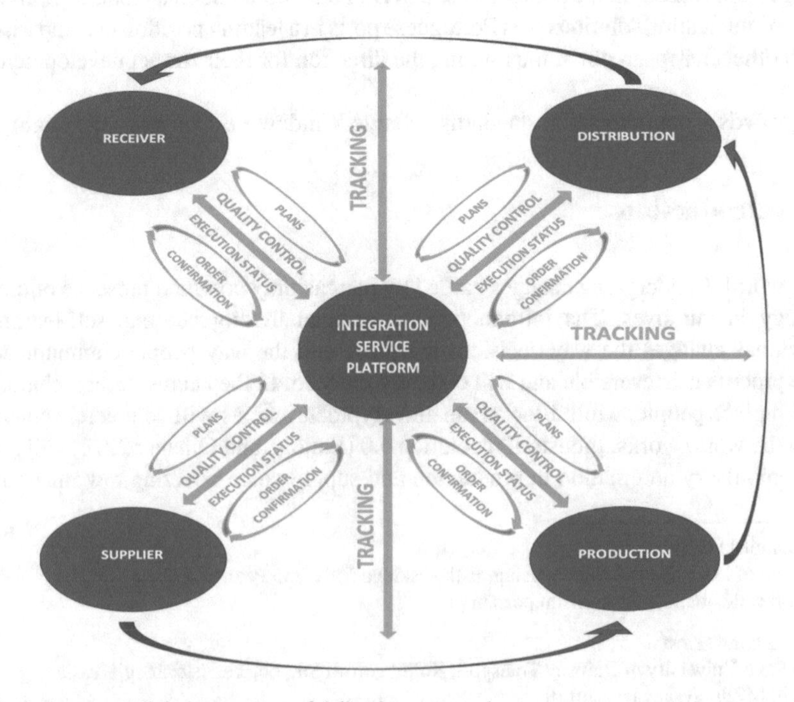

Fig. 1 Integrated digital ecosystem of deliveries, own study

of goods to submit standardised information and documents in one place, and at the same time to meet all legal requirements in terms of form. It should be emphasized that the concept is recognised and promoted by several global organisations dealing with facilitations to trade. They include the United Nations Economic Commission for Europe (UNECE) and its Centre for Trade Facilitation and Electronic Business (UN/CEFACT), the World Customs Organisation (WCO), the United Nations Network of Experts for Paperless Trade and Transport in Asia and the Pacific (UNNExT) (APICS Dictionary 2004), SITPRO Limited in United Kingdom and the Association of Southeast Asian Nations (ASEAN).

The basic value of a specific solution for the ecosystem's individual interested parties is more efficient handling of the procedures related to the trading of goods, such as customs clearance or applying for different permits, by shortening their duration or reducing their cost. Thus, implementing a solution based on the Single Window rule allows entering information to an ecosystem on a one-off basis, without having to repeat the action multiple times at different stages of the decision-making chain. In Portugal, for example, an innovative approach to customs clearance processes has been implemented. It is based on the creation of a national system of comprehensive service points which allow entrepreneurs to submit all required information concerning imports, exports and transit via one electronic gate. The logic of this approach is clear and obvious. Yet, certain intricacies and the specific nature of the entire solution must be taken into account during its implementation.

2 Research Methodology

Research on the effectiveness of logistic processes in terms of integration of information flow in logistic ecosystems functioning in sea ports has been conducted since 2015 within the framework of research and development projects of the Institute of Logistics and Warehousing in Poznań, constituting the basis for the creation of single window solutions. Figure 2 shows the general methodology of research.

Research logic includes a broad literature analysis of solutions to integrate information flows in logistics ecosystems, as well as their application in business practice. Theoretical basis is based on a critical review of literature on: the application of communication standards (Pedersen 2012; Sliwczyński et al. 2012) and modern integration trends in the supply chain (Speier et al. 2008; Prajogo and Olhager 2012; Leuschner et al. 2013; Trojanowska et al. 2017; Dujak and Sajter 2019; Osmólski and Koliński 2018). Within the framework of the above mentioned research work, an analysis of the possibilities of using the analysis of the possibilities of integration of information flow in supply chains was carried out. The article focuses on the description of the solution used in Portuguese ports.

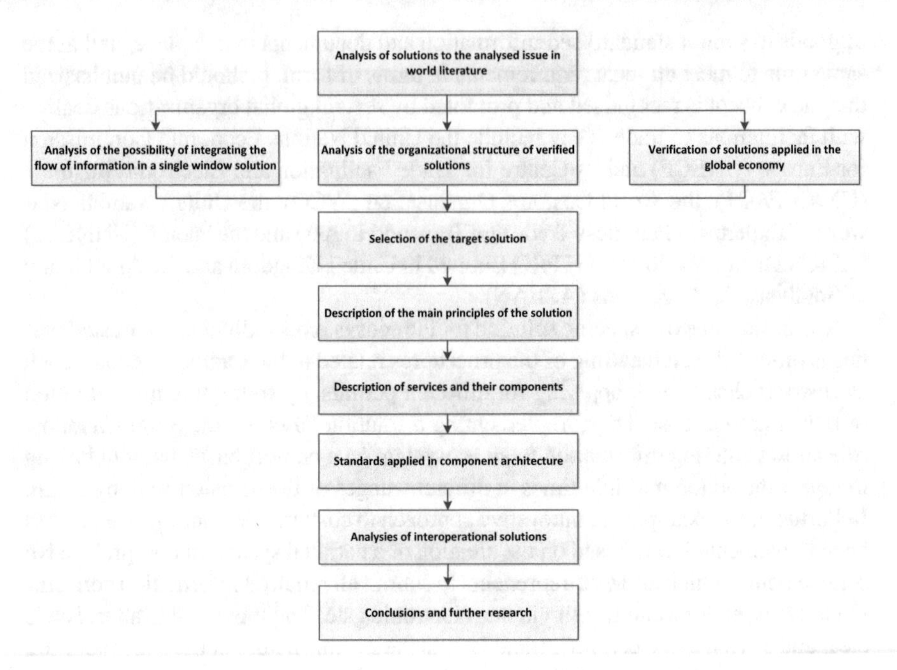

Fig. 2 Research methodology, own study

3 Single Window in Portugal—Main Assumptions

This type of approach is most commonly applied in IT systems operating in certain European ports. A solution which may serve as an example is a structure developed as part of WiderMos project (Trojanowska et al. 2017), implemented in Portuguese ports, referred to as LSW (Logistics Single Window). Processes occurring in the defined ecosystem integrate a number of services, solutions and applications in a common ecosystem covering logistics, reservation, planning and management. All of the parties interested in the process were enthusiastic about such an undertaking and about applying it in daily work. They included forwarders, operators of container terminals, multimodal operators, road carriers, rail carriers and carriers handling air transport.

The main objectives of creating LSW were:

- to promote interoperational solutions for the TSL industry in Portugal,
- to optimise connections and synchronise the flow of information,
- to apply standards both in the structure of the solution, and in the exchange of messages taking place in the process of information relocation.

During the development of a specific solution, the obligation to rely on an open and interdisciplinary structure, which may contribute to information exchange ecosystems that are more complicated in terms of architecture, e.g. the Port Community System (Sliwczynski et al. 2012), was taken into account.

The detailed developmental objectives of LSW included:

- optimisation and synchronisation of multimodal flows,
- increased efficiency and reliability of logistics operations,
- greater transparency of processes occurring in supply chains,
- reduction of costs for interested parties involved.

LSW was created on the basis of results of different European projects concerning the interoperability of logistics, including solutions such as e-Freight (Osmólski and Koliński 2018) or e-Delivery (APICS Dictionary 2004). Owing to the solutions applied, the project was considered one of the most modern IT solutions implemented on a European scale.

4 Services and Components

LSW is nothing else but an electronic catalogue of logistics services defined and described on the basis of TSD (Transport Service Description) standards included in ISO/IEC FDIS 19845 [Information Technology—Universal Business Language Version 2.1 (UBL v2.1)]. This type of services usually covers such areas as:

- transporting goods with the use of various means of transportation,
- warehousing services,
- terminal, e.g. intermodal, services,
- services related to handling customs documents or phytosanitary documents,
- other services related to the flow of goods in a supply chain.

The catalogue of logistics services may cover logistics services provided by companies operating in the area of local functional ecosystems, such as terminals at ports, e.g. intermodal terminals, services related to mass products or mixed products. It may also cover logistics services from beyond the local environment, focusing on mutual relations between different ecosystems which constitute separate functional organisations, also referred to as External Communities. The INTTRA portal, used by large container operators to offer their services in the area intercontinental transport of containers, or GT-Nexus, i.e. a set of services for integrating supply chains, based on cloud solutions, may serve as examples of interoperational (Dujak and Sajter 2019) solutions.

The work of the Portuguese LSW is based on data obtained from sources such as:

- rail carriers,
- road carriers or logistics operators,
- INTTRA, GT Nexus (over 70% of services related to ocean shipping containers),
- CESAR, a platform used by Kombiverkehr and partners of Adria Kombi, CEMAT, HUPAC, NOVATRANS and Rail Cargo Austria to handle rail connections,
- ocean carriers operating in Portugal who are not included in the INTTRA or GT-Nexus catalogue.

According to the federal approach, LSW is also adapted to receive and share service catalogues with other countries. Test with Brazil and Spain are under way. After full implementation, LSW Portugal intends to be the first country in the world to offer a complete electronic catalogue of rail, sea and inland navigation services, shared locally and updated regularly.

The LSW system also includes a routing mechanism which allows logistics operators to search through indexed catalogues and combine services to offer their customers a comprehensive service based on multimodal transport (Fig. 3).

One of the most important features of LSW is the possibility to control and track the status of a dispatched container by all of the users of the ecosystem. The information is generated on the basis of messages received from:

- seaports (by the port's LSW), namely:
  - entry/exit gate,
  - movement of containers from the terminal to the ship and the other way round,
  - arrival or departure of a ship,

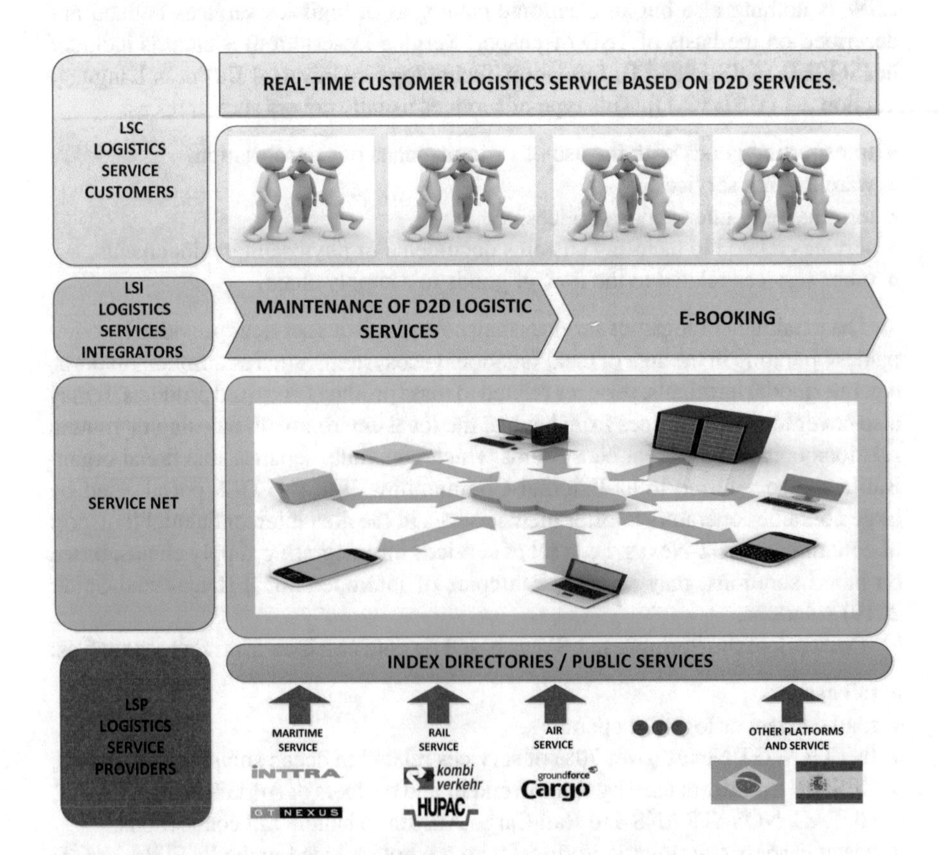

Fig. 3 Structure of a catalogue of logistics services, own study

- external communities:
  - logistics companies,
  - transport companies,
- public offices and institutions:
  - veterinary testing office,
  - the customs office,
- applications and external services:
  - INTTRA or GT-Nexus – allows the tracking of containers,
  - applications used to verify driving time for transport companies,
- other sources, such as EPCIS (Speier et al. 2008) or NEAL-NET (Pedersen 2012).

To be able to handle all of the occurrences listed above, LSW creates a service called the LSW Control Tower, as part of which a specific dispatcher or transport integrator may have access to information on container shipments, status occurrences and analyses—all in one place. The graphical form of the system has been shown in (Fig. 4).

LSW provides simple and available online tools for transport planning and management (Fig. 5). The tools may be used by cargo managers, forwarders and other providers of logistics services. These solutions also take full advantage of available status occurrences and are directly integrated with service catalogues and components of reservation, planning, reservation management or order completion monitoring.

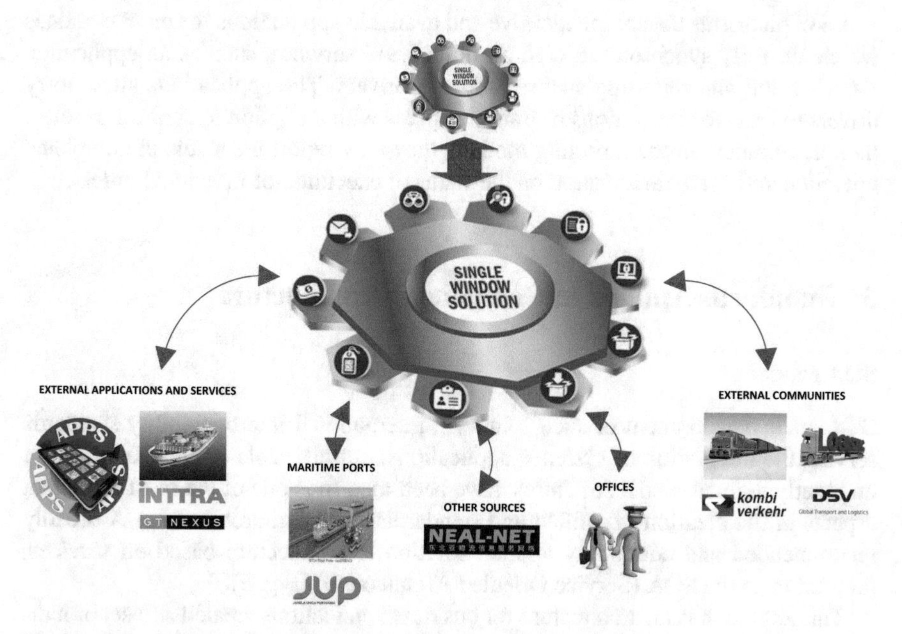

Fig. 4 Information exchange system—LSW Control Tower, own study

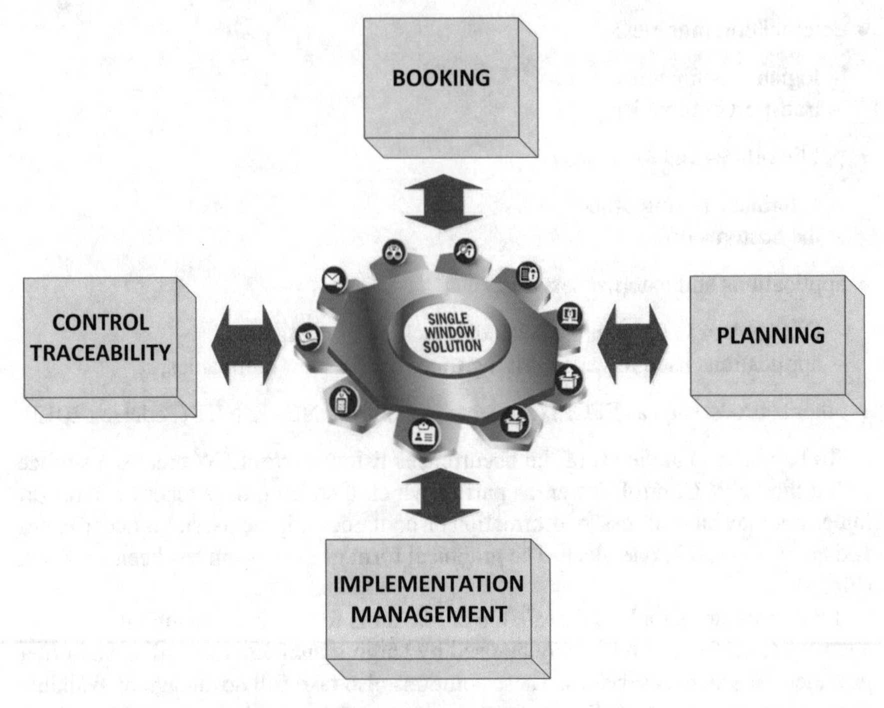

Fig. 5 Transport planning and management, own study

LSW platforms ensure inexpensive and available applications for mobile phones which are fully synchronised with all other LSW services, such as an application for reserving and reporting statuses of lorry drivers. The applications allow lorry drivers to receive reservations or transport plans with an option to confirm or reject their acceptance. In the reporting module, they may report the status of completed operation and share information on the status of execution of individual orders.

5 Standards Applied in Component Architecture

SOA Model

During the development of such solutions, international interoperability standards serving the integration of systems, applications, portals, tools or data bases, based on strictly defined model structures, have been met. In is one of the most important aspects in the creation of unified and standardised system architecture. A broadly recommended and commonly applied solution is architecture based on services, referred to as the SOA (Service Oriented Architecture) (Fig. 6).

This approach is an architecture for business applications created as a set of individual components organised in a way that allows the provision of services operating

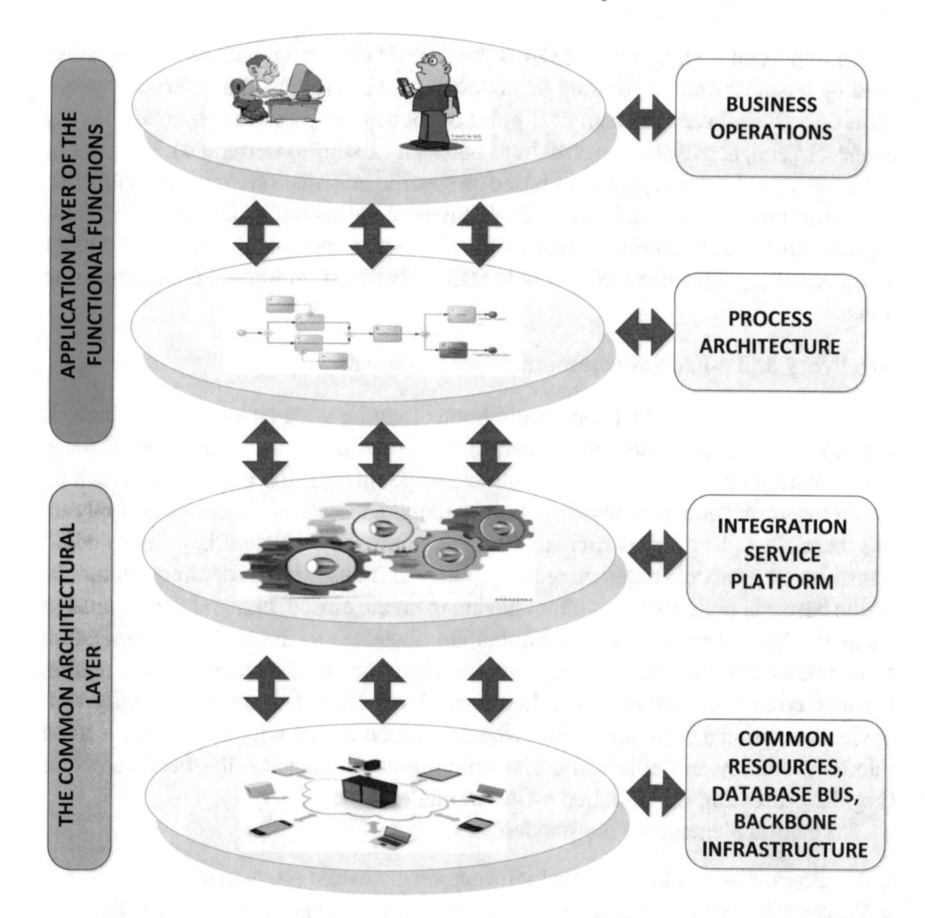

Fig. 6 Service model (SOA), own study

according to specified criteria, supporting the execution of business processes. One might say that the essence of SOA depends on the perspective. From the perspective of business (service recipient), it is a set of services supporting the execution of business processes. From the perspective of IT, on the other hand, it represents infrastructure necessary to provide these services. It should be emphasized that SOA is not a universal approach to information technology. The application which prevails is the application in IT, where we deal with direct support of business processes (which can be provided in the form or services). The basic features characteristic for SOA include:

- use in business applications,
- the architecture of components is a set of 'black boxes',
- the components of the architecture are loosely connected,
- the components are created and cooperate with each other while carrying out business processes.

An important assumption of SOA is the use of existing applications and systems used by business entities, boiling them down to an ecosystem characterised by uniform operation. From the technical point of view, it is necessary to create closely connected and, above all, universal links between existing systems and new systems, e.g. by using integration platforms based on specific information exchange standards. Such an approach also requires the development of the so-called informational architecture, which will combine elements functioning in individual areas of computer systems, taking advantage of available standards based on unified communication units.

e-Delivery and e-Freight communication standards

A standard based on e-Delivery assumptions may be used as an example here. It is a result of works carried out as part of EU projects in the recent years. It helps public institutions and entrepreneurs to exchange information and documents in an electronic form, in a way which is both reliable and safe. As a result of applying this technology, its every participant becomes a node in the network, using standard transmission protocols based on security policy. It is also a basis for direct communication between participants, without having to create closed, bilateral data exchange channels. Such a solution helps eliminating obstacles to the technical integration of computer systems, at the same time accelerating the exchange of information between companies and reducing their operational costs. It is most frequently visible in international communication, where an immense quantity of data needs to be relocated quickly and efficiently. The structure of the standard has been shown in (Fig. 7), presenting the so-called 4-Corner model.

The central elements of the standard are:

- Access Point—ensures standard information exchange protocols,
- Connector—ensures correct interaction between Access Point and the user,
- Certificate—ensures safe exchange of information on the Internet.

Another very important aspect in communication used in the e-Delivery model structure is the e-Freight standard (Debicki and Kolinski 2018). Its main components are:

- e-Freight structure—reference model for transport and logistics, describing processes supporting electronic exchange of messages between parties in all types of transport;
- e-Freight platform—comprehensive software infrastructure reflecting the framework of e-Freight structure, facilitating the development and implementation of e-Freight solutions. The platform has three functions:
  - it provides a repository (a set) which solutions and services can be taken from,
  - it works as a "run-time" environment, which supports operations and interactions between solutions,
  - it ensures an environment for developing additional services and solutions,

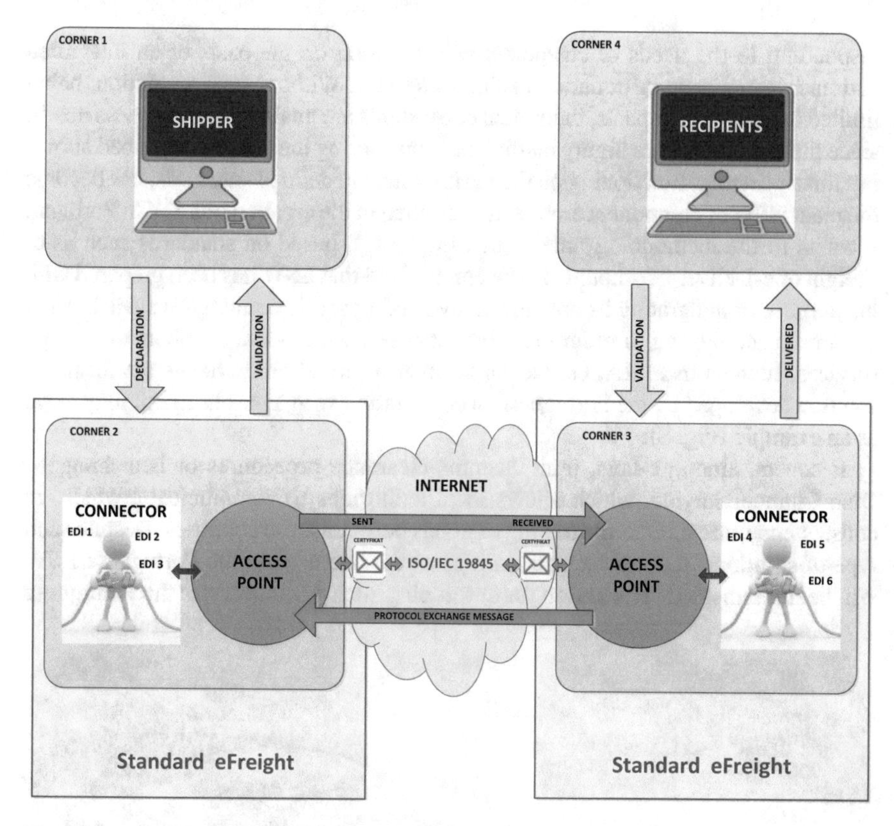

Fig. 7 e-Delivery standard (4-Corner model), own study

- e-Freight services are basic elements of software used as components of e-Freight solutions. e-Freight services represent entirely interoperational software originating directly from the structure of e-Freight,
- e-Freight solutions—A2A & A2B and B2A & B2B applications, systems consisting of elements of software and data exchange channels (e.g. traffic intensity, weather, flows of goods) that significantly affect the area of transport and logistics.

6 Interoperational Solutions

In Europe, there is a considerable number of logistic ecosystems supported by IT services. Some of them are separate organisms operating only within their own impact zone, which is primarily local, regional or functional, without being fully integrated with other ecosystems. One of the most striking facts in this situation is the loss of large potential of values brought by the synergy of actions in the area of logistic services. A solution to this problem would be to implement a broad variety of services responding to the needs of customers who function on the basis of an integrated and interdisciplinary information exchange system. Without such a solution, being limited to their own impacts, individual ecosystems are unable to satisfy these needs, since their structures are highly inefficient. Considering the aspects described above, interdisciplinarity, functioning on the basis of strictly defined standards, has become the main pillar of component architecture applied in the creation of LSW in Portugal. It refers to the methodology mentioned in point 3, based on standards such as e-Freight or e-Delivery. It should also be emphasized that LSW has been prepared with the purpose of integrating its operations, by making service catalogues available and by current monitoring of order execution statuses, with global suppliers of e-cargo services, such as INTTRA, GT-Nexus or other information exchange platforms. A solution developed by the Portuguese in cooperation with a port in Brazil may serve as an example (Fig. 8).

It covers, among others, joint customs clearance procedures or launching the Blue Channel service, which allows local authorities to communicate, which, in consequence, streamlines the trading of goods between these countries. Despite such type of solutions, the same functionalities as the one used in the Portuguese LSW will be implemented. The reason for developing such a solution was the Portuguese

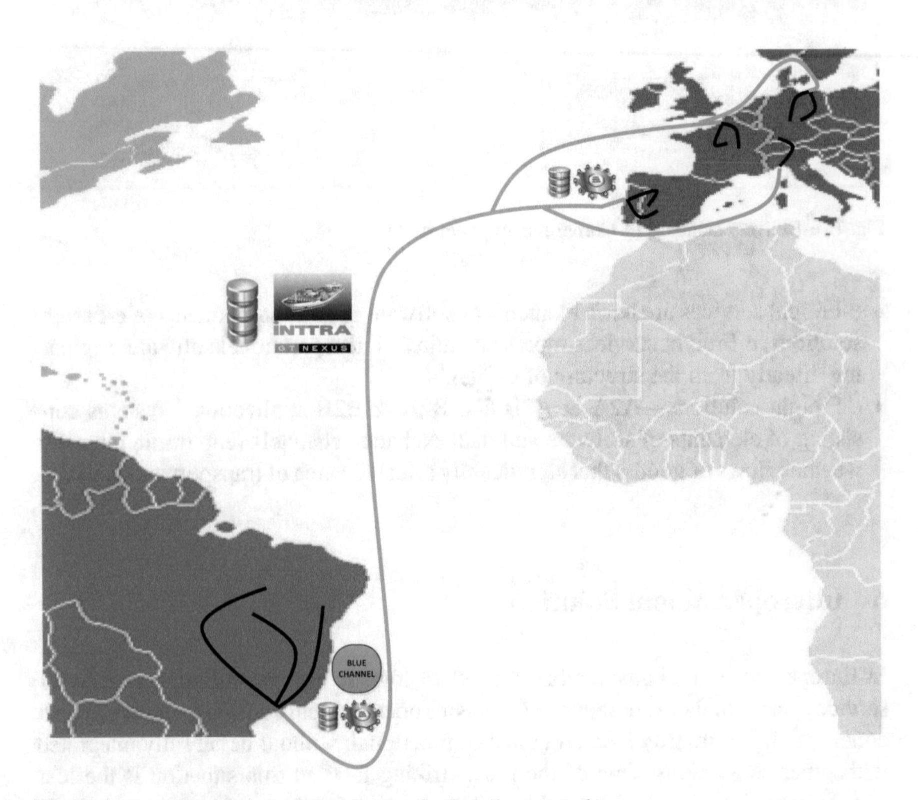

Fig. 8 An interoperational solution Portugal—Brazil, own study

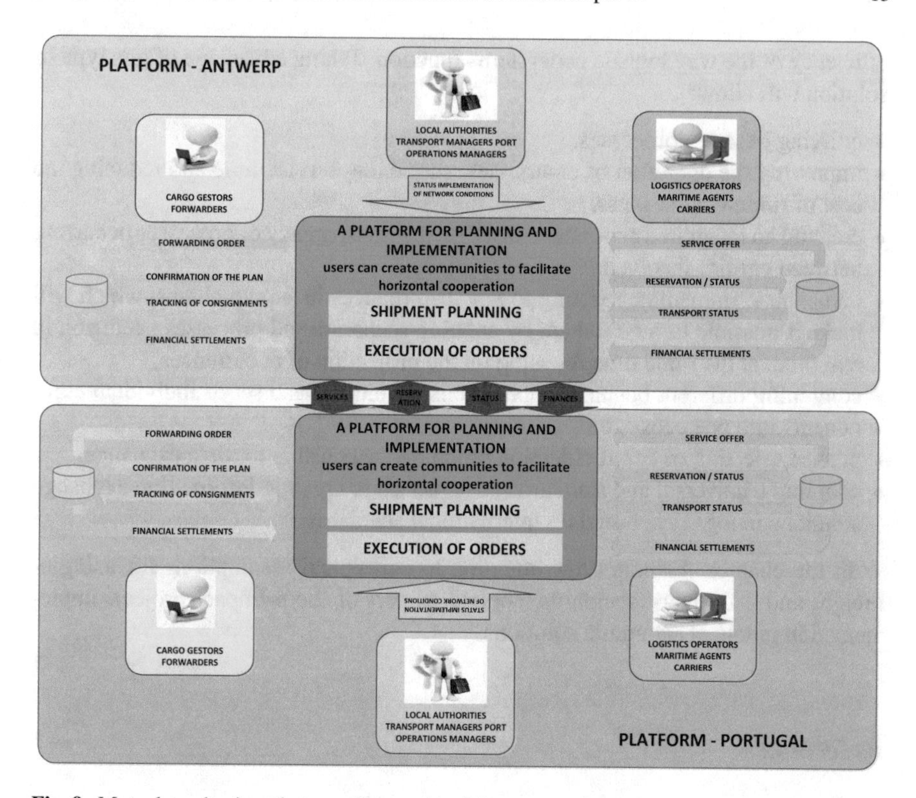

Fig. 9 Mutual mechanisms between Portugal and Antwerp, own study

people's intention to gain competitive advantage in the trading of goods with Brazil, which would make Portugal one of Europe's largest centres of trading with Brazil. The mechanisms of efficient information exchange applied in this solution have been based on existing structures of cooperation developed between Portuguese LSW and the information platform in the Port of Antwerp which has been shown in Fig. 9.

7 Conclusions and Further Research

The aim of this article was to analyse the possibilities of information flow integration, to present the possibilities of using standards in the process of creating component architecture and to present structural solutions on the example of LSW used in Portugal, as well as in the process of interoperable exchange of information between different port ecosystems. Presenting a detailed discussion on the issue has a considerable influence on creating IT structures which are efficient solutions related to information exchange based on standards, which greatly affects the improvement of efficiency of the way logistic ecosystems function. Taking advantage of this type of solution will allow:

- unifying business processes,
- improving the operation of enterprises, shortening service time and reducing the cost of running a business,
- creating an integrated ecosystem based on the transparency of processes occurring between entities involved,
- companies' immediate reacting to any disturbances in supply chains, which will make it possible to model them by creating scenarios and processes occurring in real time, at the same time focusing on the projection of occurrences,
- combining different business applications developed as a set of individual components into one ecosystem,
- precise selecting of communication protocols responsible for data exchange,
- adopting a universal and standardised approach to creating information exchange solutions in logistics, based on international standards.

In the chapter, Authors drew attention to conceptual assumptions for using e-Freight and e-Delivery standards, applied as part of the e-Impact project, implemented in practical economic solutions.

References

APICS Dictionary (2004) 11th edn. American Production and Inventory Control Society, Inc. Falls Church

Debicki T, Kolinski A (2018) Influence Of EDI approach for complexity of information flow in global supply chains. Bus Logistics Mod Manage 18:683–694

Dujak D, Sajter D (2019) Blockchain applications in supply chain. In: Kawa A, Maryniak A (eds) SMART supply network. Springer International Publishing AG

Leuschner R, Rogers D, Charvet FF (2013) A meta-analysis of supply chain integration and firm performance. J Supply Chain Manage 49(2):34–57

Osmólski W, Koliński A (2018) Wykorzystanie technologii Blockchain w obrocie produktami spożywczymi, Przemysł Spożywczy, Wydawnictwo Czasopism i Książek Technicznych Sigma-Not Sp. z o.o., Nr. 8 rok 2018, Warszawa

Osmólski W, Kolinski A, Dujak D (2018) Methodology of implementing e-Freight solutions in terms of information flow efficiency. In: Interdisciplinary management research XIV-IMR 2018, Osijek, pp 306–325

Pedersen JT (2012) One common framework for information and communication systems in transport and logistics: Facilitating interoperability. In: Golinska P, Hajdul M (eds) Sustainable transport. Springer, Berlin, pp 165–196

Prajogo D, Olhager J (2012) Supply chain integration and performance: the effects of long-term relationships, information technology and sharing, and logistics integration. Int J Prod Econ 135(1):514–522

Sliwczynski B, Hajdul M, Golinska P (2012) Standards for transport data exchange in the supply chain–pilot studies. KES international symposium on agent and multi-agent systems: technologies and applications. Springer, Berlin, pp 586–594

Speier C, Mollenkopf D, Stank TP (2008) The role of information integration in facilitating 21(st) century supply chains: a theory-based perspective. Transp J 47(2):21–38

Trojanowska J, Varela MLR, Machado J (2017) The tool supporting decision making process in area of job-shop scheduling. In World conference on information systems and technologies. Springer, Cham, pp 490–498

Reference Model of Information Flow in Business Relations with 4PL Operator

Adam Kolinski, Agata Horzela, Marta Cudzilo and Roman Domanski

Abstract Nowadays, apart from the physical availability of goods in a given link in the logistics chain, great importance is attached to the possession of up-to-date information about the quantity, structure of assortment, or location of goods flowing through the supply chain, which a given link is not (first of all) or was already the owner of. Goods movement monitoring enables a wide range of information from the logistics chain to be obtained and used to deliver faster and at the same time at a lower cost—more effectively reconcile the ubiquitous compromise in logistics between customer service level and its cost. The aim of this chapter is to present studies on the problems and needs of information integration in business relations between the 4PL Operator and the contractors. The chapter uses both secondary sources—analysis of the literature of the subject, as well as primary sources—various analyses of business practice. The authors developed and implemented their own multi-stage methodology of studies covering procurement companies, distribution companies, 4PL operators and their customers. The studies were conducted in years 2017–2018 on a sample of 76 companies from the TSL sector (targeted selection) operating on the Polish market. The study perspective includes a cross-confrontation of positions of both parties to the information flow in the logistics chain: opinions of suppliers and distributors regarding the functioning of 4PL operators and the opinions of 4PL operators regarding the functioning of their customers—suppliers and distributors. The reference model of information flow between a business entity

A. Kolinski (✉)
Poznan School of Logistics, Estkowskiego 6, 61-755 Poznan, Poland
e-mail: adam.kolinski@wsl.com.pl

A. Horzela
GS1 Poland, Estkowskiego 6, 61-755 Poznan, Poland
e-mail: agata.horzela@gs1pl.org

M. Cudzilo
Institute of Logistics and Warehousing, Estkowskiego 6, 61-755 Poznan, Poland
e-mail: marta.cudzilo@ilim.poznan.pl

R. Domanski
Poznan University of Technology, Strzelecka 11, 60-965 Poznan, Poland
e-mail: roman.domanski@put.poznan.pl

A. Kolinski et al. (eds.), *Integration of Information Flow for Greening Supply Chain Management*, Environmental Issues in Logistics and Manufacturing, https://doi.org/10.1007/978-3-030-24355-5_2

and a 4PL operator proposed in this chapter may constitute a standardized method of monitoring of communication in business practice of companies, providing a basis for further analyses and evaluations of the improvement of the efficiency of logistics processes in procurement and distribution chains.

Keywords Fourth party logistics · GS1 · Supply chain integration

1 Introduction

The development of information technology has led to a situation in which man and his creations function under conditions of information redundancy. At any time in the logistics system, huge amounts of data on the information and goods flow are collected by means of various sensors (Domanski and Adamczak 2017). Automatic data collection systems, including systems to identify logistics traders and objects, are helping in this area. The problem of the logistics industry in today's business reality is not the access to different data, but the ability to make the right use of them. In the context of the ubiquitous Internet and mobile technologies, the world is in a situation of an oversupply of ubiquitous and varied data. The lack of precise indications makes it impossible to integrate the information flow in a comprehensive way, taking into account the multidimensional analysis of relations taking place in the supply chain. The research conducted by the Authors responds to the identified research gap.

The easiest way to reduce the diversity of data in the logistic system is to try to standardize and to unify them. Various standards, from the simplest and historically first—such as barcodes—to the most advanced and nowadays common—GS1 communication standards come to the aid in this area. However, even in this situation, the amount of data in the logistics systems of companies is still enormous, difficult to extract the desired knowledge. The solution to this problem is based on Big Data—heuristic algorithms able to provide correct information on the basis of intelligent, selective search of the whole set of data.

The reality of modern business—the needs and expectations of customers—is evolving towards the search for comprehensive solutions. In the TSL (Transport-Spedition-Logistics) market until recently, customers selectively outsourced only one or a few functions to logistics operators. Today there is increasing pressure on the overall system offer—the logistics operator performs the full range of functions of its customer. This results in a natural increase in the complexity of logistics systems of logistics operators and their customers (Lopes-Martínez et al. 2018), both in terms of information flow (communication) and goods flow (logistics). Logistics operators are therefore taking on more and more logistics tasks. New conditions are connected with the occurrence of new, probably not yet occurring, problems. So how to overcome them?

With the help of companies comes the Supply Chain Council (SCC). The result is the Supply Chain Operation Reference Model (SCOR), a solution that describes the flow of information and goods in the logistics chain. The SCOR model, due to its generality, can be treated in the category of guidelines for further work, designing more detailed solutions dedicated to particular categories of enterprises or industries. The main contribution of this chapter is to present to the scientific community the author's reference model of information flow in the TSL industry between a given participant in the logistics chain and the logistics operator operating it. In the opinion of the authors, their model presented in this chapter is mainly characterized by utilitarian character of solutions (detailed algorithms of realization of specific logistic processes) together with allocation of specific communication standards of GS1 system in the supply chain (reduction of information redundancy, reduction of misunderstandings in information exchange).

2 Research Methodology

The development of a reference model of information flow in relations with the 4PL operator in the supply chain is the result of many years of studies by the authors and observation of business practice. The developed research methodology includes both a literature research and an analysis of business practice. Companies were studied in the years 2017 and 2018. Only their comprehensive analysis allows for generation of constructive conclusions regarding the structure of a reference model adequate to the needs and requirements of a modern business practice. Figure 1 presents a detailed methodology applied during the conduct of the studies in this area.

Due to the complexity of the issues raised, it was decided to divide the presentation of the conducted studies into two publications. A detailed analysis of literature on information flow and application of communication standards as well as the presentation of results of the studies on distribution processes in supply chains were described in the publication (Horzela et al. 2018). Theoretical foundations were based on a critical review of literature on: application of communication standards (Pedersen 2012; Sliwczynski et al. 2012; Power and Gruner 2015; Chituc 2017), 4PL operator functions (Ozovaci 2016; Pavlić Skender et al. 2017) and modern integration trends in the supply chain (Speier et al. 2008; Stajniak and Guszczak 2011; Prajogo and Olhager 2012; Kawa 2012; Leuschner et al. 2013; Cyplik et al. 2014; Awasthi and Grzybowska 2014; Wong et al. 2015; Hadas et al. 2015; Kawa and Zdrenka 2016; Trojanowska et al. 2017; Domanski et al. 2018; Dujak and Sajter 2019).

This chapter focuses mainly on the results of the studies of business practice which had a key impact on the structure of the reference model of information flow in relation with the 4PL operator and the structure of the matrix of application of GS1 standards in the supply chain.

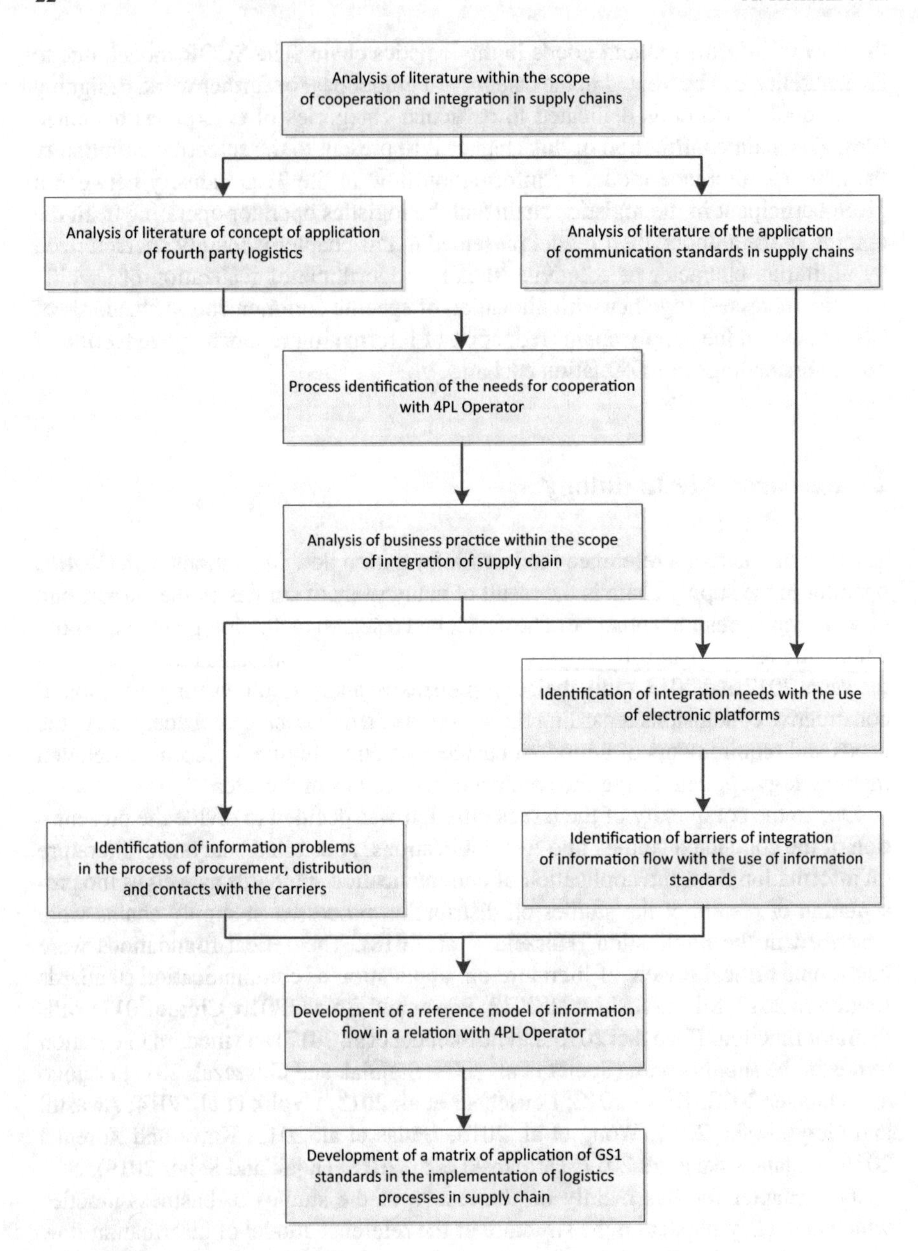

Fig. 1 Research methodology, own study

3 Identification of Integration Problems in the Supply Chain

While monitoring the results obtained in connection with the analysis of needs of the integration portal concept, it was decided to carry out an additional analysis concerning the identification of information and integration needs with the carrier/logistics operator from the point of view of contractors. The study was based on the logic shown in Fig. 2.

The study was conducted in the third and fourth quarter of 2017, in 76 companies from the TSL sector. The survey consisted of 15 single- or multiple-choice questions. The first stage of the study, which initiated the analysis of integration needs in the supply chain, was the analysis of the frequency of misunderstandings between the company employees and the contractors. Misunderstandings should be understood as all business situations in which an improper, incomplete flow of information or its failure has caused the process to be carried out in the opposite way to what was intended. Figure 3 shows the detailed results of this analysis.

It should be noted that most of the analyzed companies declares lack of problems concerning misunderstandings with their contractors, or occasional problems. However, as many as 46% of companies confirm that misunderstandings are frequent or very frequent. On the basis of the analysis of this stage of the study, it is not possible to state unequivocally that the absence of problems related to misunderstandings with contractors is due to information integration in the supply chain.

In the study, it was assumed that the concept of significance (materiality) expresses the degree or strength of conviction/confidence of the respondent as to the appropriateness of using a given function in the effective flow of information in the supply chain. The same mechanism of their evaluation was established, based on a five-step Likert scale, supplemented by a zero level. The position of many research teams presented in the literature of the subject indicates that the analyses based on Likert's scale assumptions are of quantitative nature (Elliott and Woodward 2007; Gamst et al. 2008; Gatignon 2013). Individual assessments meant:

- 0—insignificance of the problem,
- 5—very high significance of the problem.

When identifying information problems in relations with the carrier, more than 61% of the companies (47 out of 76) declared that they used the services of transport companies to supply their business. When analyzing the degree of significance of information problems in procurement processes, it should be stated that only four identified problems are characterized by high significance from the perspective of business practice (the result above 3.0—average level of significance). According to the analysis carried out, it should be concluded that companies identify the following listed basic information problems with the carrier as significant:

- misunderstandings regarding the delivery date (61.70% of companies declaring that they use carriers in the procurement process)—high significance (3.85),

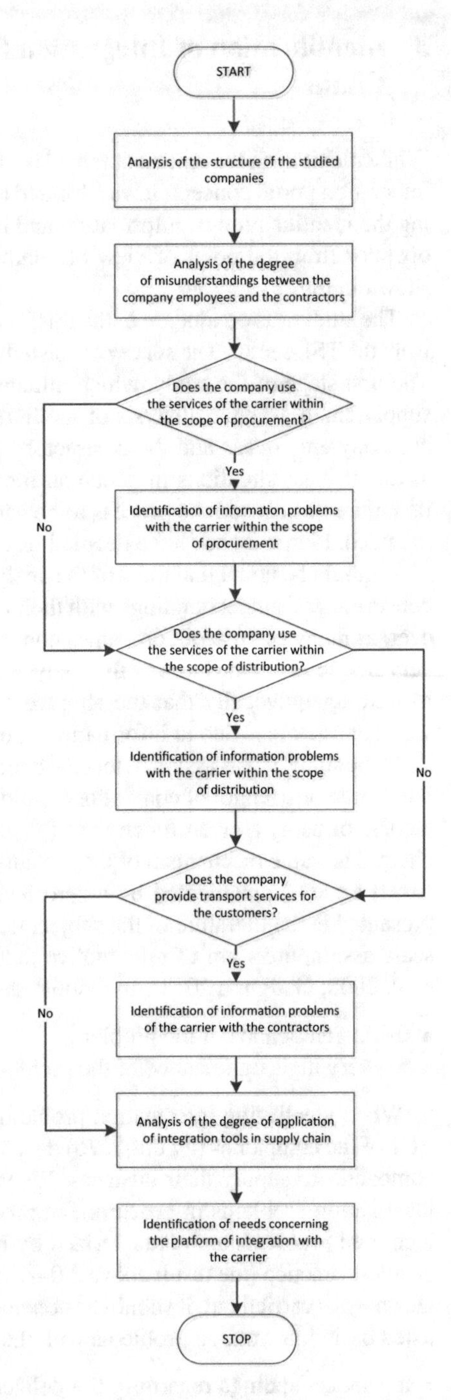

Fig. 2 Algorithm for identification of information problems between the 4PL operator and contractors, own study

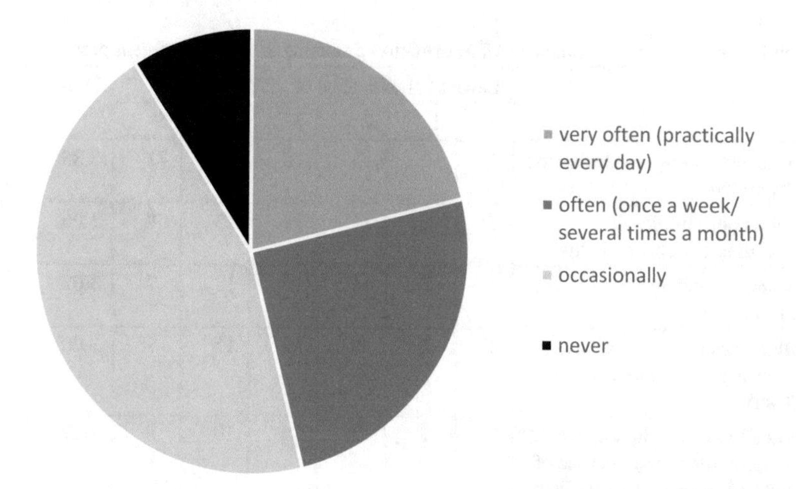

Fig. 3 Frequency of misunderstandings between the 4PL operator and the contractors, own study

- limited possibilities of monitoring the delivery status (55.32%)—average significance (3.06),
- misunderstandings regarding the delivery item (38.30%)—average significance (3.02),
- discrepancies between an advice of delivery and the actual delivery (36.17%)—average significance (3.00).

Table 1 presents detailed results of the distribution of significance for the identified problems of effective information flow in the procurement process.

It is worth noting that the remaining problems, which were reported by more than 36% of the studied companies (limited possibilities of checking the content/completeness of delivery during transport and difficult contact with the carrier in current affairs), were not considered by the same companies to be a significant problem.

When analyzing the degree of significance of information problems in distribution processes, it should be stated that only two identified problems are characterized by high significance from the perspective of business practice (the result above 3.0—average level of significance). According to the analysis carried out, it should be concluded that the companies identify the following listed two main information problems with the carrier as significant:

- misunderstandings regarding the delivery date (70% of companies declaring that they use carriers in the distribution process)—high significance (4.00),
- limited possibilities of monitoring the delivery status (60%)—average significance (3.33).

Table 2 presents detailed results of the distribution of significance for the identified problems of effective information flow in the distribution process.

Table 1 Distribution of significance of information problems in the procurement process

Problem identified	Level of significance of the problem						Mean value
	0	1	2	3	4	5	
Misunderstandings regarding the delivery date	2	2	3	8	11	21	3.85
Limited possibilities of monitoring of delivery status	3	2	11	11	13	7	3.06
Misunderstandings regarding the delivery item	3	3	11	10	13	7	3.02
Differences between the advice of delivery and the actual delivery	2	6	9	10	13	7	3.00
Difficult contact with the carrier in current affairs (e.g. in case of sudden changes in delivery date, breakdown of the means of transport, etc.)	1	7	15	6	10	8	2.87
Limited possibilities of checking the content/completeness of the delivery during transportation	7	4	9	13	10	4	2.57
Difficult flow of information concerning the settlement of the provided transport service (invoicing)	4	10	10	9	11	3	2.47

Source Own study

Table 2 Distribution of significance of information problems in the distribution process

Problem identified	Level of significance of the problem						Mean value
	0	1	2	3	4	5	
Misunderstandings regarding the delivery date	1	0	3	8	10	18	4.00
Limited possibilities of monitoring of delivery status	2	2	7	9	10	10	3.33
Misunderstandings regarding the choice of means of transport	3	2	14	7	7	7	2.85
Difficult contact with the carrier in current affairs (e.g. in case of sudden changes in delivery date, breakdown of the means of transport, etc.)	4	3	11	9	10	3	2.68
Difficult flow of information concerning the settlement of the provided transport service (invoicing)	4	3	12	11	7	3	2.58

Source Own study based on Horzela et al. (2018)

It is worth noting that the problem of difficult contact with the carrier in current affairs (e.g. in case of sudden changes in the delivery date, breakdown of the means of transport, etc.), despite being reported by 47.50% of the studied companies, was not assessed by these companies as particularly significant (average significance value—2.68).

When analyzing the level of significance of information problems in relations with the customers, it should be stated that only two identified problems are characterized by high significance from the perspective of business practice (the result above 3.0—average level of significance). According to the analysis carried out, it should be stated that 4PL operators identify the following listed two basic information problems in customer relations as significant (Horzela et al. 2018):

- misunderstandings regarding the delivery date (73% of companies declaring that they operate as a carrier)—high significance (4.10),
- difficult contact with the customer in current affairs (40%)—average significance (3.10).

Table 3 presents detailed results of the distribution of significance for the identified problems of effective information flow in the 4PL operator's relations with customers. The most frequently used communication methods are e-mail contact and telephone contact (average value of results above 4.0—high level of significance).

Table 3 Distribution of significance of information problems in relations with customers

Problem identified	Level of significance of the problem						Mean value
	0	1	2	3	4	5	
Misunderstandings regarding the delivery date	0	1	2	5	7	15	4.10
Difficult contact with the customer in current affairs (e.g. in case of sudden changes in delivery date, breakdown of the means of transport, etc.)	3	2	6	3	10	6	3.10
Discrepancies between customer advice note and the final order	1	3	8	7	9	2	2.87
Misunderstandings regarding the choice of means of transport	1	4	9	7	4	5	2.80
Misunderstandings regarding the delivery item	0	3	12	7	5	3	2.77
Limited possibilities of monitoring the delivery schedule at the customer's	0	5	11	4	7	3	2.73
Difficult flow of information concerning the settlement of the provided transport service (invoicing)	2	7	11	3	3	4	2.33

Source Own study

It is worth noting that the problem with misunderstandings regarding the choice of means of transport, despite being reported by 46.67% of the studied companies, was not assessed by these companies as particularly significant (average significance value—2.80).

When analyzing the level of significance of the use of methods and tools integrating the flow of information, it should be stated that all the evaluated are characterized by high use in business practice (all had results above 3.0—average level of significance). Table 4 presents detailed results of the distribution of significance for the identified problems of effective information flow in the implementation of transport processes.

The last stage of the studies was identification of companies' needs concerning the development of the concept of the integration portal. This stage of the studies included all companies. As in the penultimate stage, the Likert scale, complemented by a zero level, was used to assess the importance of the project related to the creation of a new integration portal with the carrier. Figure 4 shows the detailed results of significance of introducing a new integration platform.

When analyzing the companies' needs, it should be noted that more than 63% of the companies (first three results of the significance level) do not see the need

Table 4 Distribution of the level of use of information flow methods and tools in the supply chain

Problem identified	Level of significance of the problem						Mean value
	0	1	2	3	4	5	
Information and communication system based on telephone contact	1	0	1	8	33	33	4.25
Information and communication system based on e-mail contact	1	0	4	11	35	25	4.03
Portal for integration of information flow with the customer	8	3	5	16	28	16	3.33
Portal for integration of information flow with the carrier	9	3	5	29	21	9	3.01

Source Own study

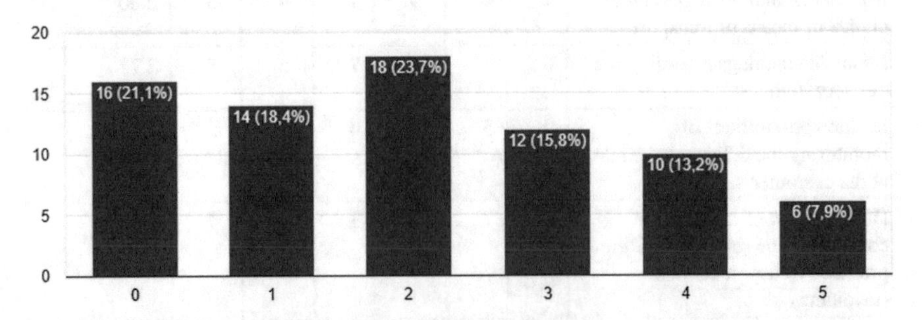

Fig. 4 Identification of needs for the introduction of a new integration platform, own study

to introduce a new platform or consider this need to be of little significance and insignificant. Also, the result based on the weighted[1] mean of the obtained results indicates that there is no market need for an integration portal (the result of the mean value: 2.01). The studies conducted in 2017 were completed with the conclusion that there is no business need for development of a new integration platform, therefore it would be advisable to focus on the integration capabilities of the platforms already in use.

As part of the research work in this area carried out in September 2018, it was decided to analyze the reasons for the lack of business need related to the development and implementation of a new platform aimed at integrating the flow of information within the transport processes of the supply chain. The studies were conducted in the form of consultations and questionnaires in 24 companies, which also participated in the 2017 studies and rated their need for a new integration platform low (on a scale from 0 to 2). The surveyed companies represented both carriers (8) and their contractors (16—of which operators—10, manufacturers—6).

In the course of the consultation, companies' representatives were asked to identify the problem that makes the use of electronic platforms and/or the development of a new integration platform, in their view, an ineffective solution. Additionally, the degree of impact of the indicated problem on the lack of interest in the perspective of developing a new integration platform was identified. Table 5 shows the detailed results of the conducted studies.

It should be noted that all of the problems mentioned above are characterized by a fairly high level of significance in business practice (all of them were ranked

Table 5 Distribution of the level of significance of problems related to effective implementation of electronic platforms

Problem identified	Level of significance of the problem						Mean value
	0	1	2	3	4	5	
Necessity of multiple entering of data (into the system and the platform)	0	0	1	7	7	9	4.00
Application of different communication standards by business partners	0	0	0	0	13	11	4.46
Fear of losing company's sensitive/crucial data	0	0	0	5	9	10	4.21
Low reflection of the economic returns of using such platforms	0	0	3	8	9	4	3.58
Small number of contractors using the same communication platform	0	0	8	8	5	3	3.13

Source Own study

[1]The weighted mean takes into account both the significance level and the frequency of indications of a given significance level by the individual companies studied.

above 3.0—average level of significance). The key problem is the integration inconsistency of the platforms with the IT systems of the companies, which results in the need to enter data several times, data security issues and, above all, the variety of communication standards that are available on the market.

The solution to these problems may be the application of the EPCIS standard. EPCIS may enable the creation of an information service on the Internet, which ensures provision of information on events concerning the movement of goods and cargo, their groupings, quantities or dates. EPCIS standard defines four types of events for the electronic recording of object information: object events, aggregation events, transaction events and transformation events. These events contain information on the unique identity of the object (answer to the question "what?"), time and date (answer to the question "when?"), the location of a specific event (answer to the question "where?") and the reason for which the event took place (answer to the question "why?"). The concept of EPCIS can also be extended in terms of its use as a database, which represents a great potential to implement this standard in existing electronic platforms as well as during implementation of initiatives connected with the development of new integration platforms.

Considering the analysis of identification of information problems and the scope of integration with the 4PL operator, one should remember about the matrix correlation of the obtained results. Table 6 presents a synthetic summary of information problems in the process of procurement, distribution and in relations with customers in terms of processes that are most often the subject of cooperation with the 4PL operator.

Based on Table 6, it should be stated that regardless of the perspective of the analysis (cooperation with the operator in the scope of procurement, distribution, as well as the operator's cooperation with other contractors), when analyzing the possibility of applying GS1 standards in relations with the 4PL operator, all the above issues should be taken into account both in terms of the warehouse and transport process.

When developing the matrix for the proposed application of GS1 standards in 4PL, particular attention should be paid to:

- monitoring of the status of deliveries,
- minimization of misunderstandings regarding the delivery item and the delivery date.

4 Reference Model of Cooperation in 4PL Operators

Model of business cooperation with the 4PL operator should include identification of information flow with contractors. The development of a general model was supported by an exchange of experience with logistics operators. The main objective of the developed model is to present the basic scope of possibilities of cooperation with 4PL operators with regard to the flow of information. A general reference model of information flow with the 4PL operator is shown in Fig. 5.

Table 6 Correlation of information problems and the scope of cooperation with the 4PL operator

Scope of problem identification	Information problem	Scope of cooperation with the 4PL operator	
		Warehousing	Transport
Procurement	Misunderstandings regarding the delivery date	X	X
	Misunderstandings regarding the delivery item		X
	Limited possibilities of monitoring of delivery status	X	X
	Discrepancies between the advice of delivery and the actual delivery	X	
Distribution	Limited possibilities of monitoring of delivery status	X	X
	Misunderstandings regarding the delivery date	X	X
Relation of operator with contractors	Misunderstandings regarding the delivery date	X	X
	Difficult contact with the customer in current affairs (e.g. in case of sudden changes in delivery date, breakdown of the means of transport, etc.)		X

Source Own study

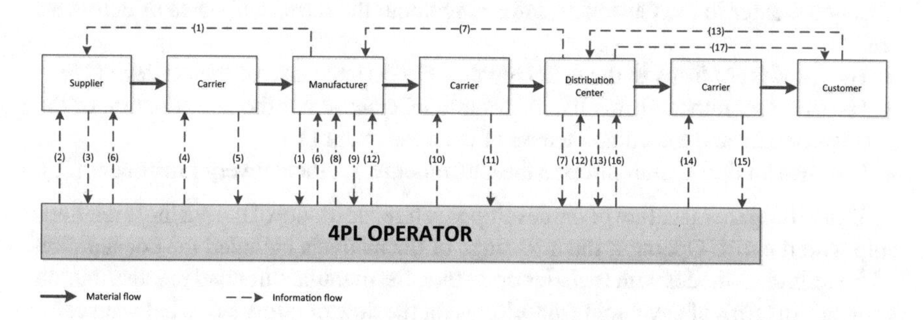

Fig. 5 General reference model of information flow with the 4PL operator, own study

The presented reference model is based on the classic material flow in the supply chain. The reference model shows the model information flow, which is presented from the manufacturer's perspective in the following logical process:

- The Manufacturer reports the need for materials and components to the Supplier (1), specifying the delivery date, and at the same time informs the 4PL Operator about this fact (1),

- The 4PL Operator contacts the Supplier and makes an inquiry regarding the delivery date, completeness of delivery, advice, possible business and operational conditions (2),
- The Supplier accepts and confirms the fulfillment of the order from the Manufacturer, informing the 4PL Operator about it (3),
- The 4PL Operator notifies the Carrier about the transport order from the Supplier to the Manufacturer, informing about the terms and dates of deliveries (4),
- The Carrier confirms to the 4PL Operator the performance of the carriage (5),
- The 4PL Operator informs the Manufacturer about the confirmation of the delivery date and the completeness of the assortment (6),
- The Distribution Center sends the purchase order to the Manufacturer (7) specifying the expected delivery date, at the same time informs the 4PL Operator about this fact (7),
- The 4PL Operator contacts the Manufacturer and makes an inquiry regarding assortment and time compliance of the order (8),
- The Manufacturer confirms the order fulfillment to the 4PL Operator (9),
- The 4PL Operator notifies the Carrier about the transport order from the Manufacturer to the Distribution Center, informing about the terms and dates of deliveries (10),
- The Carrier confirms to the 4PL Operator the performance of the carriage (11),
- The 4PL Operator informs the Distribution Center about the confirmation of the delivery date and the completeness of the assortment (12),
- The Customer places an order at the Distribution Center (13), which informs the 4PL Operator about this fact (13),
- The 4PL Operator notifies the Carrier about the transport order from the Distribution Center to the Customer, informing about the terms and dates of deliveries (14),
- The Carrier confirms to the 4PL Operator the performance of the carriage (15),
- The 4PL Operator informs the Distribution Center about the confirmation of the delivery date and the completeness of the assortment (16),
- The Distribution Center informs the Customer about the delivery fulfillment (17).

Due to the general nature of the developed reference model of the business relationship with the 4PL Operator, the next stage of the analysis included the consultation of the reference model with logistics operators, the manufacturer and the distribution company in terms of developing algorithms for the flow of information between companies and the operator.[2] Figure 6 presents a general algorithm of information flow and logistic processes implementation in business relations with the 4PL Operator.

Due to the nature of cooperation with the 4PL Operator, further analysis of information flow was carried out within the following processes:

- (2) Purchases and procurement, in particular (2.2) procurement transport and (2.3) receipt of delivery to the warehouse,

[2]The presented algorithms of information flow have been developed in consultation with representatives of logistics operators, carriers and manufacturers.

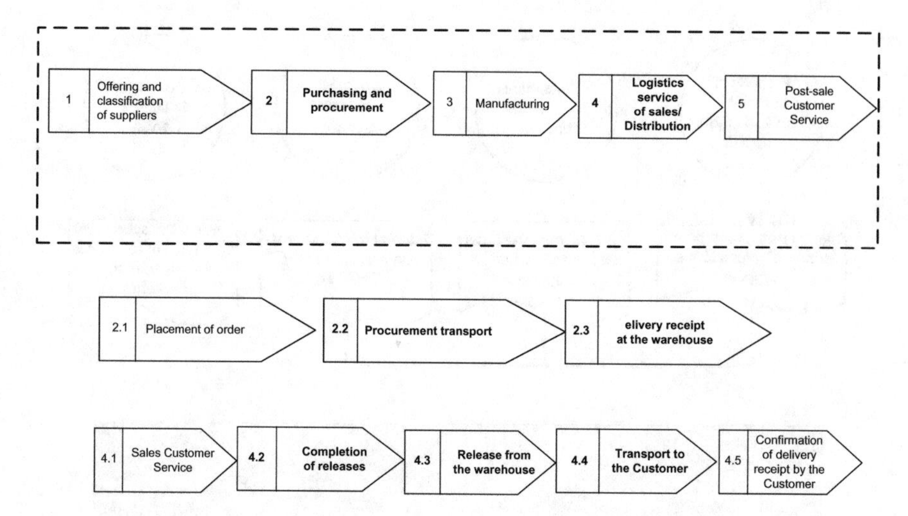

Fig. 6 General algorithm of information flow with the 4PL operator, own study

- (4) Logistical support of sales/distribution, in particular (4.2) completion of releases, (4.3) release from the warehouse, (4.4) transport to the Customer.

Figures 7, 8, 9, 10, 11, 12 and 13 present algorithms of information flow in identified subprocesses, which were considered as basic processes performed by the 4PL Operator. Figures showing the detailed implementation of a given process additionally highlight the potential places in the process where specific GS1 standards can be applied.

Algorithms of information flow developed in consultation with the companies highlighted basic events and messages implemented in logistic processes potentially implemented by the 4PL Operator. From the point of view of the analysis of individual processes, these algorithms should be called an attempt to standardize the processes. However, in terms of their implementation by the 4PL Operator, they should be considered from the point of view of their integration by means of information flow standards.

Table 7 presents a matrix of potential applications of GS1 standards in relations with the 4PL operator.

The proposal to apply GS1 standards in relations with the 4PL operator was developed on the basis of the outsourcing analysis in Poland, which resulted in the focusing of these studies on transport and storage processes, taking into account the integration needs of partners in the supply chain (Horzela et al. 2018). This list also includes the possibilities of applying GS1 standards taken from the good practice analysis, which means that this matrix also includes confirmations of individual implementations of GS1 standards in business practice.

The developed matrix contains the breakdown of events identified under consultations with companies into transport and storage processes. The scope of management of both transport and warehouse resources has also been separated. The developed

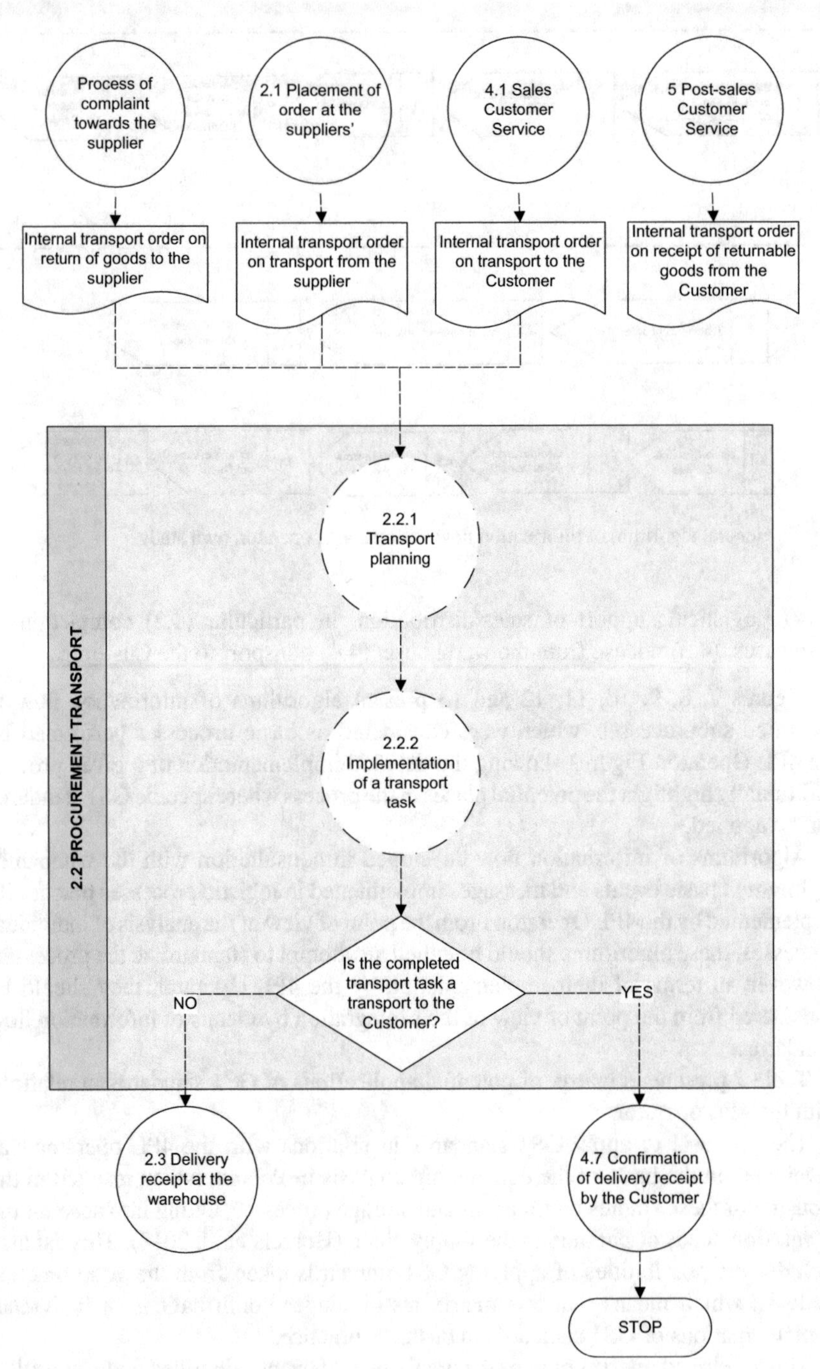

Fig. 7 Algorithm of information flow—procurement transport, own study

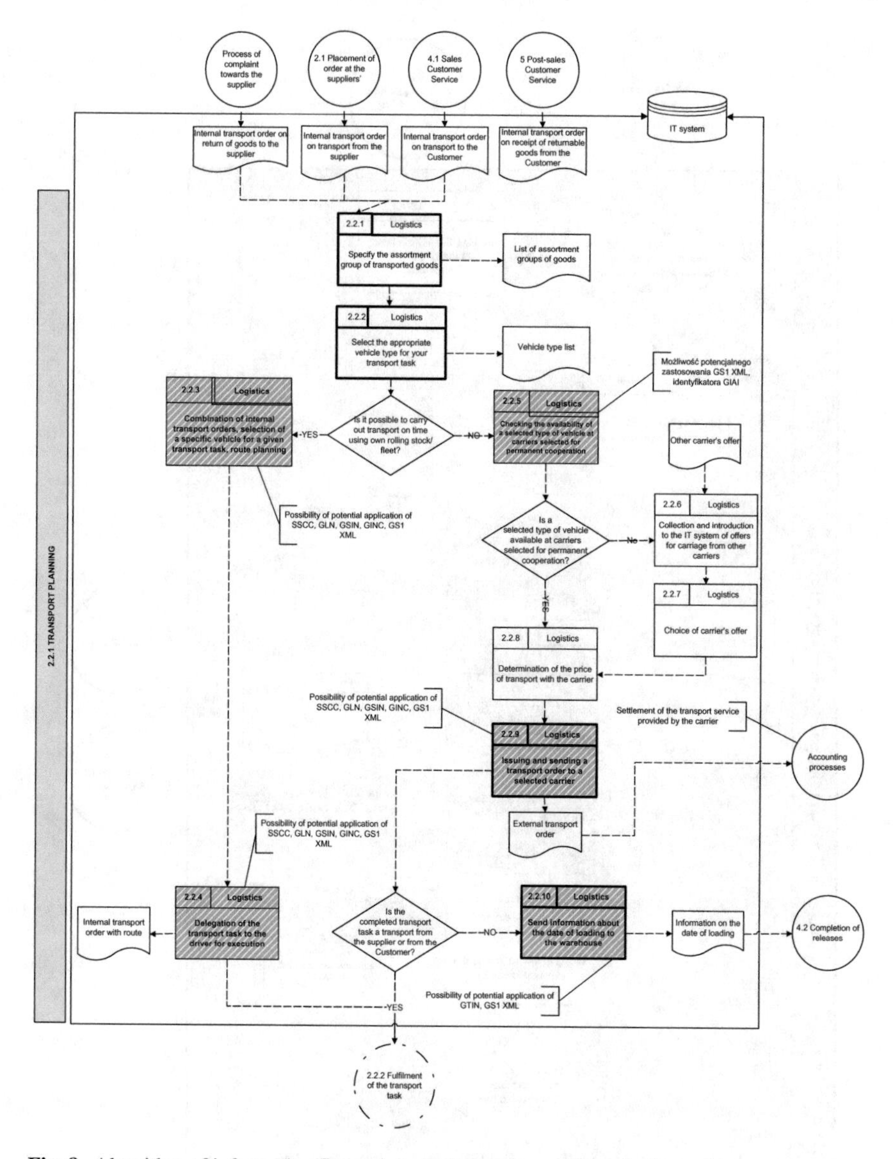

Fig. 8 Algorithm of information flow—transport planning, own study

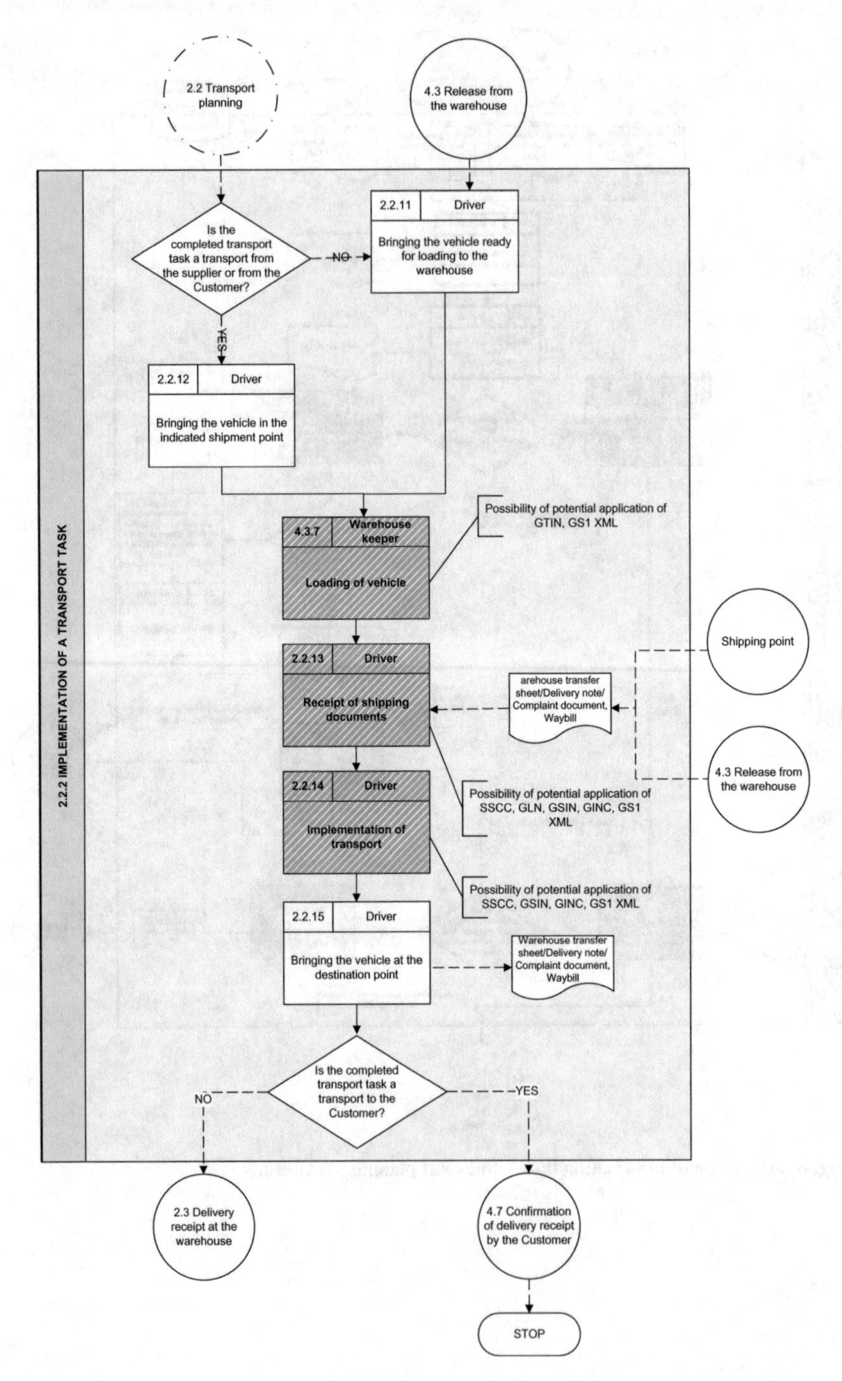

Fig. 9 Algorithm of information flow—implementation of a transport task, own study

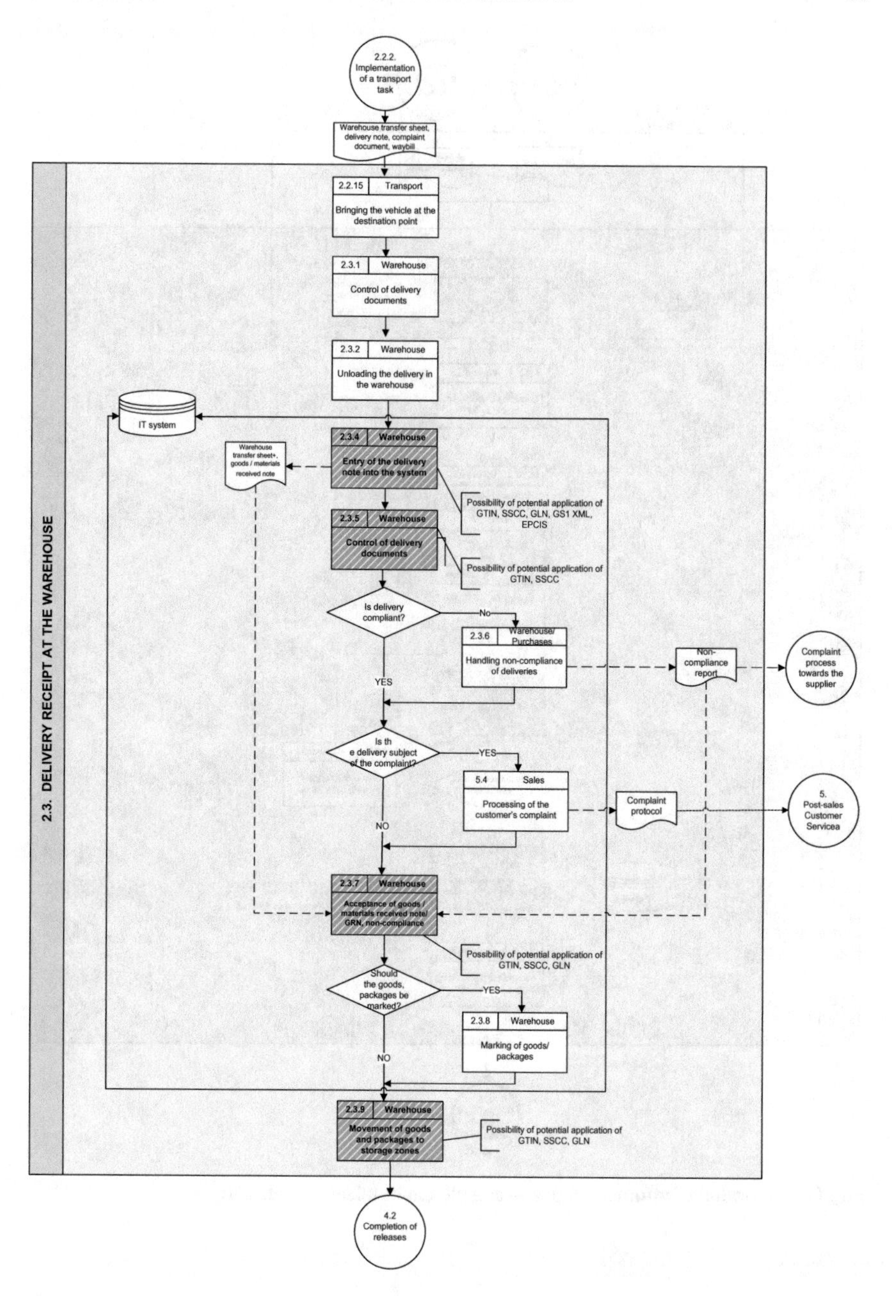

Fig. 10 Algorithm of information flow—acceptance of delivery to the warehouse, own study

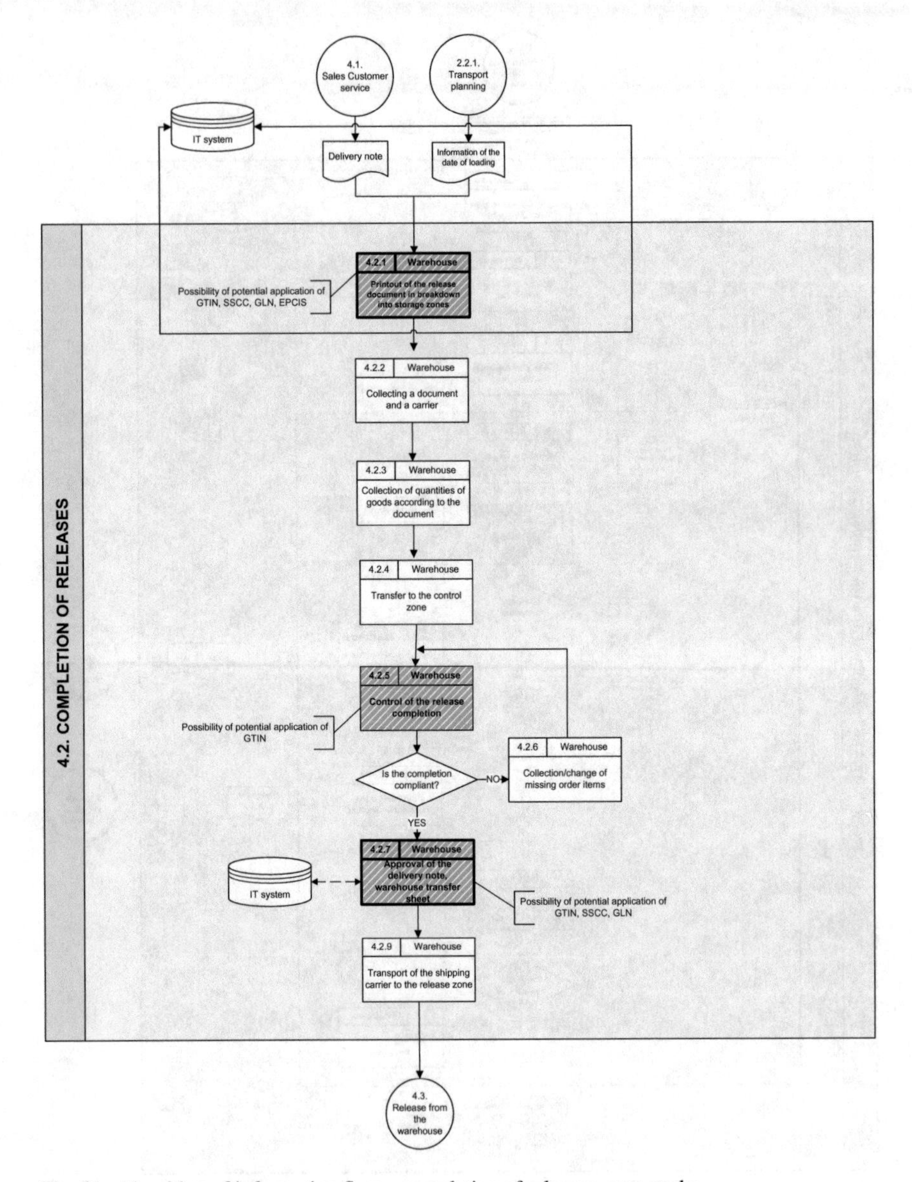

Fig. 11 Algorithm of information flow—completion of releases, own study

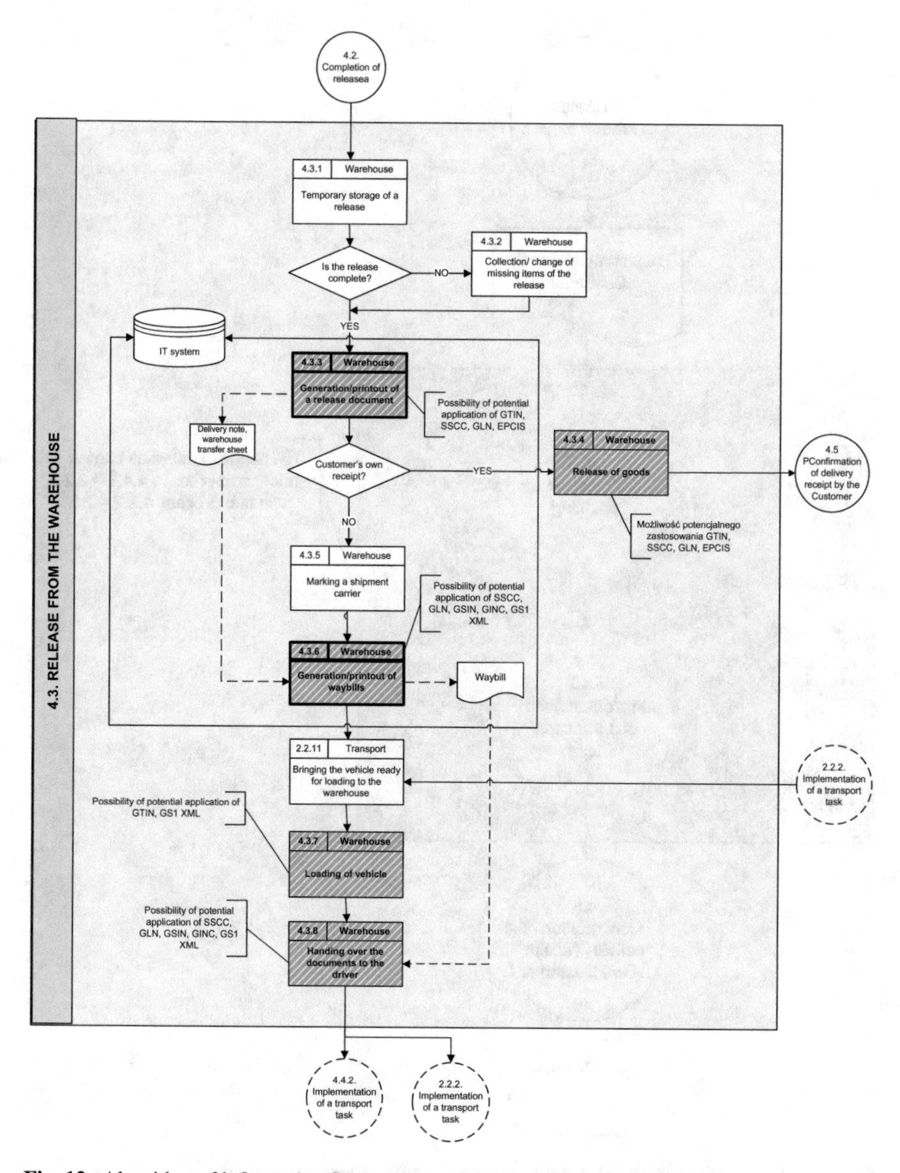

Fig. 12 Algorithm of information flow—release from the warehouse, own study

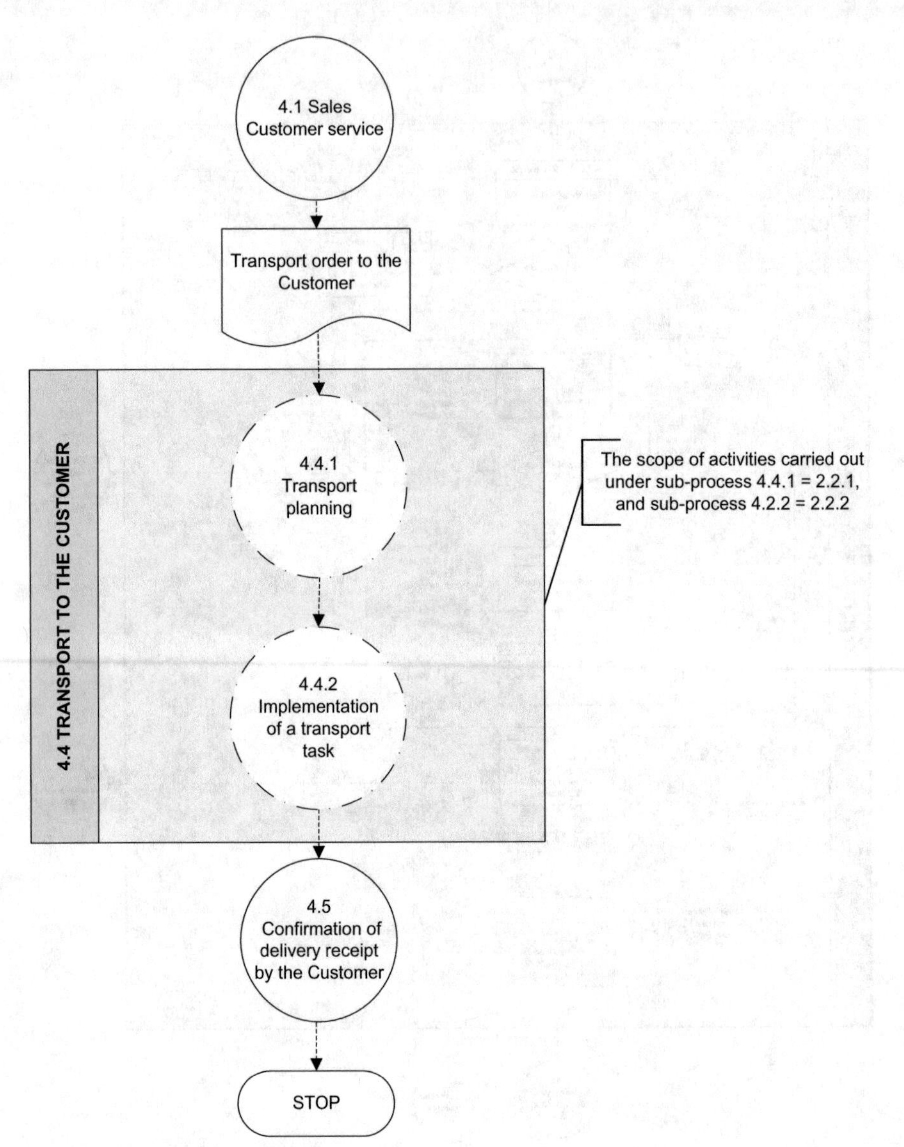

Fig. 13 Algorithm of information flow—transport to the customer, own study

Table 7 Matrix of potential applications of GS1 standards in relations with the 4PL operator

	Event	GTIN	SSCC	GLN	GRAI	GIAI	GSIN	GINC	EDI	EPCIS
Warehouse	Advice of delivery	X	X	X					X	
	Scheduling of receipts at the warehouse	X	X	X						
	Confirmation of delivery receipt at the warehouse	X	X	X					X	X
	Inventory analysis	X							X	
	Inventory movements	X	X	X					X	
	Quantitative control of receipts	X	X							
	Report on the quality control of receipts			X						
	Comparison of the picking list with the customer's order	X								
	Warehouse statuses								X	
	Location of cargo in the warehouse		X	X						
	Identification of storage places			X						
	Identification of the base unit in the formed cargo	X	X					X		
	Confirmation of release	X	X	X						X
	Schedule of releases from the warehouse	X	X	X						

(continued)

Table 7 (continued)

	Event	GTIN	SSCC	GLN	GRAI	GIAI	GSIN	GINC	EDI	EPCIS
Transport	Transport order		X	X			X	X	X	
	Visit of vehicle/vessel/train					X				X
	Loading list	X							X	
	Waybill	X	X	X			X		X	
	Delivery status		X				X	X	X	
	Transport statuses		X				X	X	X	
	Confirmation of receipt of delivery to the customer		X	X					X	
	Unloading list	X	X							
	A report on the completeness of deliveries	X	X	X						
	Current cargo tracing	X	X	X[a]			X	X		X
	Monitoring the conditions of carriage									X
	Customs handling of transported goods	X	X				X			
	Identification of the base unit in the formed cargo	X								

(continued)

Table 7 (continued)

	Event	GTIN	SSCC	GLN	GRAI	GIAI	GSIN	GINC	EDI	EPCIS
Management of resources	Identification of individual resources—semi-trailers, forklifts, containers, tractors, etc.					X				
	Identification of returnable resources—pallets, kegs, boxes, etc.				X					
	Ongoing tracing of containers/pallets/boxes				X					
	Ongoing tracing of container/railway wagon					X				
	Localization of the place of delivery (e.g. warehouse)			X						
	Ongoing monitoring of vehicles and trailers					X				

Source Own study
[a]If the location of the cargo is fixed

matrix contains a list of general events that can be treated as reference ones. It needs to be borne in mind, however, that within the frameworks of the detailed analyses of specific industries in business practice, this list of process events should be elaborated and adjusted to their specificity.

5 Conclusions and Further Research

The analyses allowed for identification of basic potential possibilities of application of GS1 standards in 4PL. It should be noted that the potential applications of GS1 standards are very broad, both from the point of view of the current possibilities as well as from the point of view of the identified new possibilities. The largest range of new possibilities of using GS1 standards is the implementation of logistics processes in procurement. These are processes analogous to the distribution processes, which are already applied in business practice, but due to the identified reluctance and lack of perception of potential benefits, especially from the point of view of manufacturers, this scope offers great opportunities for the development of GS1-based standardization, as well as the need to increase awareness of the mutual benefits of these solutions.

Works aimed at dissemination of the application of GS1 standards in 4PL should start by disseminating knowledge and demonstrating tangible benefits to manufacturing companies.

The reference model of information flow in relations with the 4PL Operator, developed as part of the conducted studies, may constitute not only a broader scope of analysis and assessment of the effectiveness of logistic processes in supply chains, but also a standardized method of monitoring information flows in the business practice of companies.

References

Awasthi A, Grzybowska K (2014) Barriers of the supply chain integration pro-cess. In: Golinska P (ed) Logistics operations, supply chain management and sustainability. Springer, Cham, pp 15–30

Chituc CM (2017) XML interoperability standards for seamless communica-tion: an analysis of industry-neutral and domain-specific initiatives. Comput Ind 92:118–136

Cyplik P, Hadas L, Adamczak M, Domanski R, Kupczyk M, Pruska Z (2014) Measuring the level of integration in a sustainable supply chain. IFAC Proc Vol 47(3):4465–4470

Domanski R, Adamczak M (2017) Analysis of the influence of the lot sizing on the efficiency of material flow in the supply chain. LogForum 13(3):339–351

Domanski R, Adamczak M, Cyplik P (2018) Physical internet (PI): a systematic literature review. LogForum 14(1):7–19

Dujak D, Sajter D (2019) Blockchain applications in supply chain. In: Kawa A, Maryniak A (eds) SMART supply network. Springer International Publishing AG

Elliott A, Woodward W (2007) Statistical analysis quick reference guidebook: with SPSS examples. Sage Publications Inc., Thousand Oaks

Gamst G, Meyers L, Guarino A (2008) Analysis of variance designs. A conceptual and computational approach with SPSS and SAS. Cambridge University Press, Cambridge

Gatignon H (2013) Statistical analysis of management data. Springer Science+Business Media, New York

Hadas L, Cyplik P, Adamczak M, Domanski R (2015) Dimensions for developing supply chain integration scenarios. Bus Logistics Mod Manage 15:225–239

Horzela A, Kolinski A, Domanski R, Osmolski W (2018) Analysis of use of communication standards on the implementation of distribution processes in fourth party logistics (4PL). Bus Logistics Mod Manage 18:299–315

Kawa A (2012) SMART logistics chain. In: Asian conference on intelligent information and database systems. Springer, Berlin, pp 432–438

Kawa A, Zdrenka W (2016) Conception of integrator in cross-border e-commerce. LogForum 12(1):63–73

Leuschner R, Rogers D, Charvet FF (2013) A meta-analysis of supply chain integration and firm performance. J Supply Chain Manage 49(2):34–57

Lopes-Martínez I, Paradela-Fournier L, Rodríguez-Acosta J, Castillo-Feu JL, Gómez-Acosta MI, Cruz-Ruiz A (2018) The use of GS1 standards to improve the drugs traceability system in a 3PL Logistic Service Provider. Dyna 85(206):39–48

Ozovaci E (2016) The new logistics methods. In: Proceedings of 3rd international conference on education and social sciences INTCESS, Istambul, p 411

Pavlić Skender H, Mirković PA, Prudky I (2017) The role of the 4PL model in a contemporary supply chain. Pomorstvo 31(2):96–101

Pedersen JT (2012) One common framework for information and communication systems in transport and logistics: facilitating interoperability. In: Golinska P, Hajdul M (eds) Sustainable transport. Springer, Berlin, pp 165–196

Power D, Gruner RL (2015) Exploring reduced global standards-based inter-organisational information technology adoption. Int J Oper Prod Manage 35(11):1488–1511

Prajogo D, Olhager J (2012) Supply chain integration and performance: the effects of long-term relationships, information technology and sharing, and logistics integration. Int J Prod Econ 135(1):514–522

Sliwczynski B, Hajdul M, Golinska P (2012) Standards for transport data exchange in the supply chain–pilot studies. In: KES international symposium on agent and multi-agent systems: technologies and applications. Springer, Berlin, pp 586–594

Speier C, Mollenkopf D, Stank TP (2008) The role of information integration in facilitating 21(st) century supply chains: a theory-based perspective. Transp J 47(2):21–38

Stajniak M, Guszczak B (2011) Analysis of logistics processes according to BPMN methodology. In: Golinska P, Fertsch M, Marx-Gomez J (eds) Information technologies in environmental engineering—new trends and challenges, ESE. Springer, Berlin, pp 537–549

Trojanowska J, Varela MLR, Machado J (2017) The tool supporting decision making process in area of job-shop scheduling. In: World conference on information systems and technologies. Springer, Cham, pp 490–498

Wong CW, Lai KH, Cheng TCE, Lun YV (2015) The role of IT-enabled collaborative decision making in inter-organizational information integration to improve customer service performance. Int J Prod Econ 159:56–65

Improving the Efficiency of Planning and Execution of Deliveries with the Use of a Location Register Based on GLN Identifiers

Marta Cudzilo and Roksolana Voronina

Abstract The efficiency of logistics processes largely depends on the effective and efficient flow of information in the supply chains. In the case of planning and implementation of distribution, the key information is the information about the operating conditions of the locations from/to which deliveries are made. In this chapter, the author presents the concept of functioning of the location register based on GLN identifiers. The register is a global, open and standardised database in which each location marked with the GLN identifier is described by a number of parameters and attributes characterising its physical features and the conditions for the implementation of logistic operations in a given location. The author describes the impact of using a location register based on GLN identifiers on the effectiveness of planning and execution of deliveries, presenting the results of the pilot implementation of the register in the form of a case study of a large logistics operator. The presented results of changes in the examined distribution process efficiency indicators recorded in the examined case confirm the benefits of using the location register. The last part of the chapter outlines further directions of development of the location register based on GLN identifiers and stresses the material importance of developing a functional and business model for the analysed solution. The model developed should take into account the diversity of roles and the links between actors in the supply chains affecting the conditions for the construction and use of the register.

Keywords GLN (Global Location Number) · Location register · GLN register · Distribution process · Process efficiency

M. Cudzilo (✉)
Institute of Logistics and Warehousing, Estkowskiego 6, 61-755 Poznan, Poland
e-mail: marta.cudzilo@ilim.poznan.pl

R. Voronina
National University "Lviv Polytechnic", Lviv, Ukraine
e-mail: roksolanavoronina@gmail.com

A. Kolinski et al. (eds.), *Integration of Information Flow for Greening Supply Chain Management*, Environmental Issues in Logistics and Manufacturing, https://doi.org/10.1007/978-3-030-24355-5_3

1 Introduction

Market observations indicate that the effectiveness of logistics activities is a very important issue for all types of companies co-creating supply chains. Effectiveness as such is a relatively difficult concept to unambiguously define, as confirmed by numerous publications (Li and O'Brien 1999; Beamon 1999; Mishra 2012; Lichocik and Sadowski 2013; Geunes et al. 2016; Brandenburg 2016; Govindan et al. 2017) covering various aspects of the efficiency of logistics processes in supply chains. The effectiveness of logistics processes depends not only on economic and organisational aspects of their implementation, but also on the effective flow of information, which is becoming an increasingly important element of competitive advantage on the market (Nakatani et al. 2006, p. 44; Śliwczyński et al. 2012, pp. 586–594). In the context of an efficient and productive flow of information, it is important to stress the enormous significance of availability of reliable information, understood in the same way by all cooperating entities (Bigaj and Koliński 2017, pp. 77–90). The guarantee of providing such defined conditions for effective information exchange is the global GS1 standards (Hałas 2012; Kisperska-Moroń and Krzyżaniak 2009; Horzela et al. 2018).

In the processes of planning and implementation of distribution, data and information on the locations from or to which deliveries are made are particularly important. Errors related to incorrect information about loading and unloading locations in the distribution planning and implementation processes generate a number of problems such as: late deliveries, incorrectly completed shipments resulting in the need for returns or re-deliveries, queues related to the unloading of goods in the customer's warehouse. Each of these situations is unfavourable from the point of view of the effectiveness of distribution processes and the elimination or attempt to minimize its effects is one of the key optimization activities observed in economic practice (Dujak et al. 2018).

The use of location information concerns both the information integration between contractors and its direct impact on the operational and economic activities of companies co-creating supply chains. Due to the cooperation between contractors in supply chains, data standardization is of key importance in the process of generating and flow of location information (and not only). In relation to this fact, the Institute of Logistics and Warehousing (ILiM) in cooperation with GS1 Polska undertook research and development activities aimed at building and implementing in practice a market location register based on GLN identifiers.

2 GLN Identifier (Global Location Number)

The GLN ID is one of the basic identification standards provided and promoted by GS1. GS1 is an international system of standards and business solutions. This system has existed and has been continuously developed since 1973. The GS1 standards

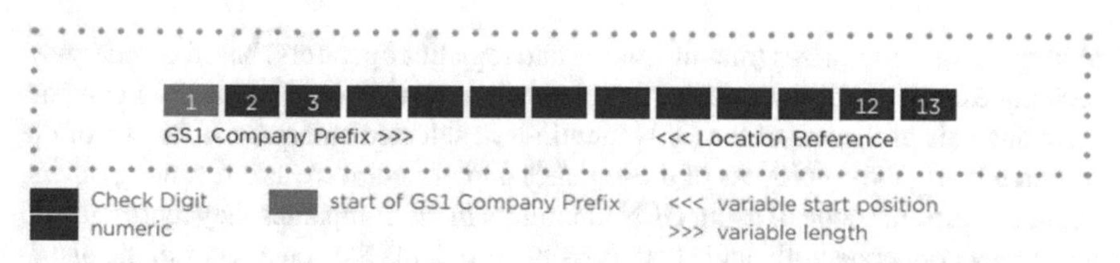

Fig. 1 Structure of the GLN identifier. *Source* http://www.gs1.se/en/our-standards/Identify/gln/

provide agreed principles and guidelines which are uniformly applied by all actors in order to improve supply chain operations in many industries. The GS1 system offers a number of tools and solutions for transport and logistics applications, including GLN. In addition to the global identification number, solutions such as global shipment identifiers, standard barcodes and radio tags, logistics labels and electronic messages for the exchange of information and data are also available.

The identifier is a response to the market needs for standard location marking and as a globally unique number it enables marking specific places/companies/objects in accordance with established principles of international communication (Dujak et al. 2017). GLN is used to recognize and describe any location, including physical, digital, functional or legal. Technically speaking, GLN is a 13-digit code consisting of a GS1 prefix, a reference to a specific location and a check digit. The structure of the GLN was presented in Fig. 1.

GLN assigned to legal persons and functions allows for unambiguous identification of these entities, while the GLN assigned to physical and digital locations allows to answer the question of where a given location is located and what are the conditions of its operation. The latter is particularly important from the perspective of increasing the efficiency of distribution processes and warehouse delivery service (Korzeniowski 2018; Kizyn 2011). Marking and unambiguous identification of locations in processes implemented in supply chains can be considered one of the key activities determining the effectiveness of these processes (Niemczyk 2016).

3 Current and Target Scope of Use of GLN Identifiers

As part of the research carried out for the purposes of this chapter, the author posed a question about the current scope of using GLN identifiers in economic practice. The source of information necessary to answer such a question were the results of the aforementioned research work carried out by the Institute of Logistics and Warehousing (ILiM), in which the author took part. The study was conducted under a framework cooperation agreement between ILiM and GS1 Poland. The subject of the agreement was scientific research and development work aimed at the development of the GS1 system in Poland. The detailed objective of the works was to popularise the use of GLN numbers for geolocation and to build a prototype of the location register. As part of the work carried out, in 2016 ILiM conducted surveys among 7 large enterprises (manufacturers and logistics operators) based on interview techniques. The studies were case studies and their results were the basis for comparative analysis of the use of the GLN identifier in selected enterprises. As part of the research work, the employees of the logistics department were asked, among others, about the current scope of using GLN identifiers in the companies they represent and those they cooperate with, and about possible problems they encounter in the distribution processes. One of the conclusions formulated within the conducted research was the statement that relatively few business entities on the Polish market use GLN identifiers, and the recorded cases of using these numbers are limited only to their use within the formal and legal register of entities. If the company has a GLN number or pools of GLN numbers, they are used in business transactions, such as placing orders or issuing invoices. In this context, the GLN identifier is described with such basic location information as the name of the company, the address of the company's registered office, the e-mail address of the company's office, etc. The surveyed companies also stressed the great impact of proper and up-to-date information about the locations of their customers (as collection and/or delivery points) on the efficiency of transport planning and execution. At the same time, emphasis was placed on the high additional costs incurred due to the lack of reliable information on the locations of customers to whom deliveries are made. As an example, the costs resulting from the long-term process of obtaining information about the location of customers and their operating conditions in the supply planning processes were given. Another example of costs in this respect was the costs incurred as a result of incorrect information on the functioning of the locations to which deliveries are made. These costs are often related to, among other things, the need to make a second delivery or to pay penalties for failure to deliver on time if the information about the working hours of a given delivery location was erroneous or there was no information about the possibility of handling certain types of cargo or vehicles (Cudziło and Niemczyk 2017, 49–53). All respondents stressed that the pool of parameters characterizing the location marked with the GLN number should be periodically updated, as well as they should describe the physical characteristics of the location. The inclusion of this type of information in a publicly available register of locations based on GLN identifiers will enable the improvement of efficiency indicators of planning and delivery processes.

In this place is worth to underline that the aim of the carried out research was to examine the level of use of GLN identifiers on the Polish market, the scope of use of this identifiers in other countries was not analysed. Nevertheless, the review study showed that other national organisations of GS1 are also conducting research and development works in the scope of increasing the use of GLN numbers in logistics. Moreover, the construction of data registers based on global identifiers is one of the elements of the GS1 Global strategy.

As part of the mentioned study, the author posed another question: What parameters should describe the locations identified by the GLN number? Numerous talks and interviews with representatives of companies from various industries conducted as part of ILiM and GS1 research work have shown that, in addition to the name and address of the location, it is important to provide reliable geographical coordinates and information describing the physical characteristics of the location, relevant for proper delivery planning and efficient loading/unloading. On the basis of the conducted research, a list of information and parameters was prepared, which should ultimately describe a given location, in order to enable efficient planning and execution of deliveries to it. Among the types of data mentioned above, the following parameters should be mentioned:

- Time periods of operation of the facility/location,
- Handled means of transport,
- Accepted logistics units in delivery,
- Recipient's requirements related to the implementation of loading/unloading operations, including, among others, the required car equipment or the required method of setting up logistic units on a semi-trailer,
- Conditions for notification,
- Documents necessary for the proper implementation of the operation,
- Guidelines for handling returns and replacement of packaging.

The list of parameters and information describing the physical characteristics of the location is illustrated in Fig. 2. All information listed should be identified by a GLN number, according to a specific data collection standard and using dictionary data to define specific types of information within each attribute.

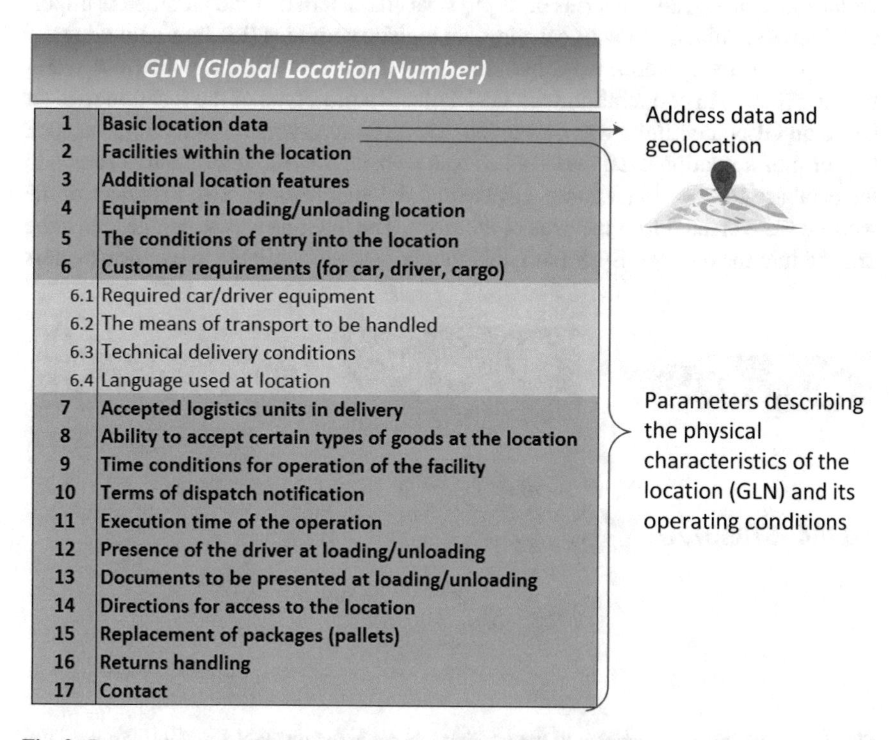

Fig. 2 Proposed groups of information on the location identified by a GLN number. *Source* Own study

As part of the aforementioned research work carried out by the Institute of Logistics and Warehousing (ILiM) in cooperation with GS1 Poland, a prototype of the location register based on GLN identifiers was developed, in which the mentioned parameters characterise each location. The GLN register is supposed to guarantee global access to the information exchanged, at the same time providing a stable, open and comprehensive database. The openness of the register, apart from issues related to access to it, should also mean the possibility of connecting additional software (API, website, mobile applications), which will guarantee easy and effective access to location data for different types of enterprises, depending on their role in the supply chain.

4 Location Register Based on GLN Identifiers

The location register based on GLN identifiers, as assumed, constitutes a stable, open and comprehensive database, which collects and makes available a wide pool of location data and information (Fig. 3). In accordance with the basic idea of the register's operation, the aforementioned data assigned to the GLN identifier comprehensively define the location, also in terms of its physical characteristics and parameters important from the point of view of carrying out logistic activities (loading/unloading).

As part of the research work carried out by the Institute of Logistics and Warehousing (ILiM) in cooperation with GS1 Poland, a prototype of the location register based on GLN identifiers was developed. The prototype version of the register took the form of a relational database linked to a website providing interactive access to the database for viewing, adding, editing and deleting records. Two instances of the website were created for the needs of the work. The first one was to enable data to be entered into the register. By default, this functionality is available to location owners

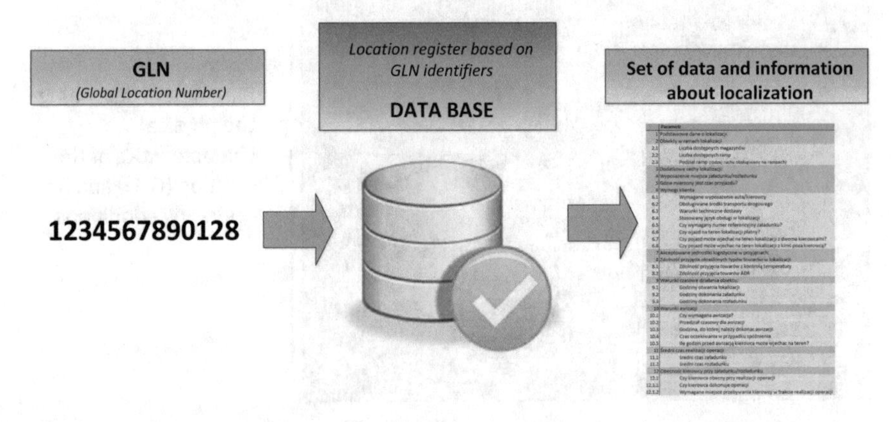

Fig. 3 The principle of operation of the location register based on GLN identifiers. *Source* Own study

(manufacturers, distributors), allowing them to define all the parameters describing the location signed with a specific GLN identifier. Of course, data entry is carried out each time in a strictly defined structure, in accordance with the developed data collection standard and with the use of prepared dictionary data to define specific types of information within individual attributes. After logging into a specific location (GLN identifier), the user can define and/or update parameters describing the location identified by a given GLN number. The second instance of the website allows users global access to the location database based on GLN identifiers, thus ensuring the ability to find and read a wide pool of information describing a given location. In the case of this instance, the user has the possibility to search for the required location according to 3 adopted criteria. The main users of this instance are the entities responsible for planning and execution of deliveries to specific location points, i.e. e.g. logistics operators, forwarders. The described main functionalities of both instances of the developed website linked to the database of the location register are illustrated visually in Fig. 4.

It is worth noting that as part of the continuation of research work on the localization register based on GLN identifiers, in cooperation between ILiM and GS1 Polska, the expansion and development of the localization register database is planned. Current research works carried out in the project include representatives of manufacturers, acting as owners of locations from/to which deliveries are made. This business support will be used to verify the compiled comprehensive list of parameters describing the locations and to fill the register database with the largest possible pool of data according to the developed format. In the next steps of the project it is planned to develop requirements specifications, as well as technical implementation of various types of software facilitating access to the GLN register (in cooperation with the business). Examples include mobile applications dedicated to different types of entities operating in supply chains, e.g. a mobile application for drivers or application programming interfaces enabling the integration of the location register database with external software, e.g. TMS *(Transport Management System)*. The idea of a comprehensive solution based on the location register using GLN identifiers is presented in Fig. 5.

5 Benefits Resulting from the Application of the GLN Register in Business Practice

The basis for defining the next steps for the development of the register of locations based on GLN identifiers was to conduct a pilot implementation of the register, confirming the possibility of increasing the efficiency of distribution as a result of its application in the process of planning and implementation of deliveries. As mentioned earlier, within the framework of research works of ILiM and GS1 Polska, a prototype of the location register was built as a relational base. The developed prototype was the subject of a pilot implementation, which was carried out with the

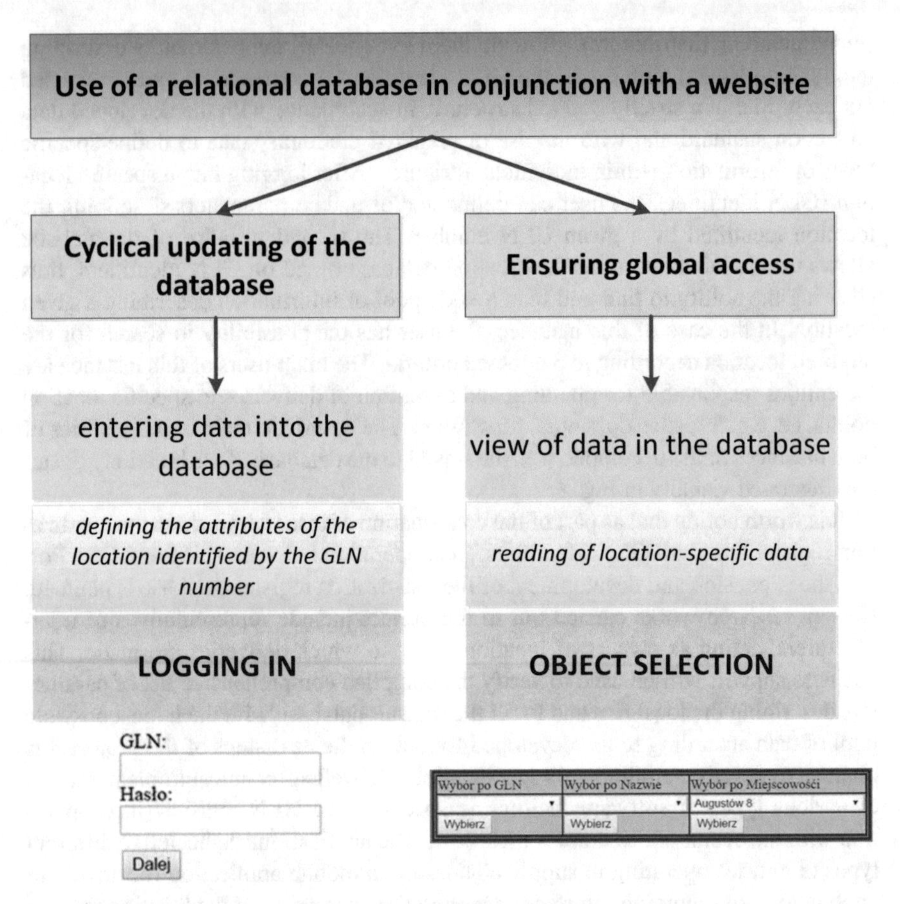

Fig. 4 Defined instances of a website linked to a database of locations identified by a GLN number. *Source* Own study

participation of a large logistics operator. The implementation carried out was a case study, which finally constituted a proof of concept. First of all, a set of indicators was developed to assess the effectiveness of distribution processes, taking into account operational and economic aspects, as well as information flow. The indicators used are presented in Table 1.

The system of indicators used to assess the impact of the implementation of the localization register based on GLN identifiers on the efficiency of transport processes was based on the basic assumptions of the Balanced Scorecard (BSC) and the classical dependence of economic efficiency (the ratio of effects to outlays, defined for the purposes of this analysis as the "a" to "b" ratio).

The pilot implementation was prepared precisely, based on the adopted methodology of the case study. The method was chosen because its application allows to discover the state of the studied phenomenon, which results of quantitative research can only suggest. The case study, compared to other research methods, offers a

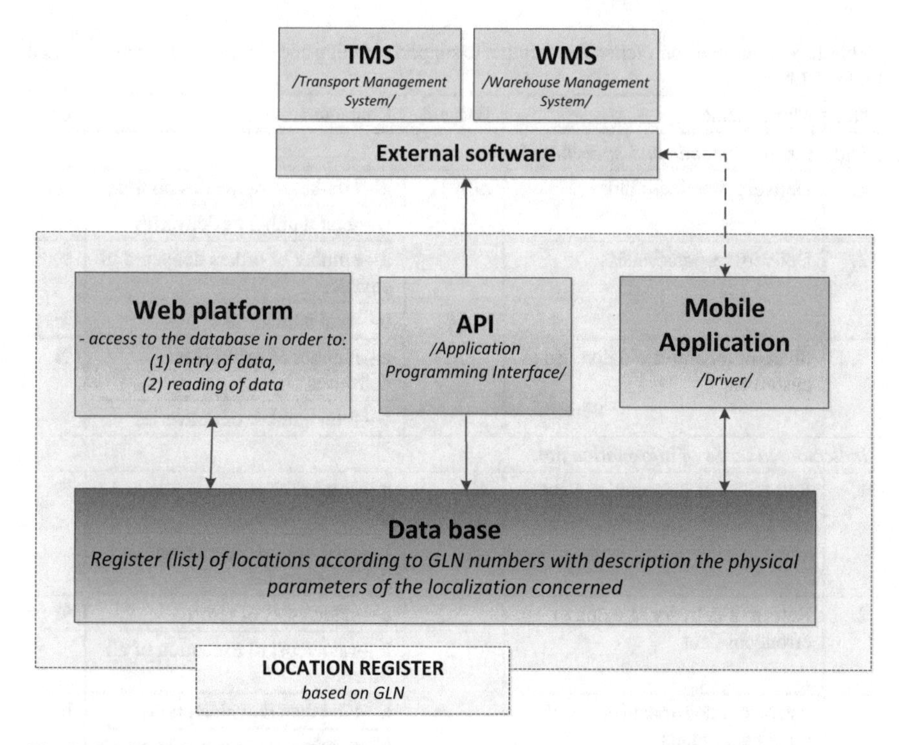

Fig. 5 Scheme of a comprehensive solution to support planning and execution of deliveries based on the GLN register. *Source* Own study

wider range of techniques and tools for data acquisition and analysis. Numerous possibilities of obtaining data mean that the case study method is not methodologically limited in terms of data analysis (Yin 2009). Research methods based on the case study are not subject to an assessment of the representativeness of the research sample (Siggelkow 2007, p. 21).

As part of the pilot study, multiple case studies were carried out to compare variants with contrasting features. In defining the variants, the extreme case method of selecting differentiated cases (from the extremes of scale) for the study sample, which allows comparisons and analysis of these variants, has been used (Yin 2009). Verification by means of multiple case studies is considered more reliable than a single case study (Eisenhardt and Graebner 2007, p. 27) as it enables differences and similarities between the analysed variants to be described in order to identify general trends (Yin 2009). The methodological assumptions adopted were the basis for the development of scenarios for the implementation of distribution processes in the examined business case, in order to monitor the use of the location register and to compare the status before and after implementation. The criteria for contrasting features of the case study were adopted as follows:

Table 1 Scheme of a comprehensive solution to support planning and execution of deliveries based on the GLN register

No.	Metric name	Formula	Characteristics	Unit
Indicators in economic and operational terms				
1.	Delivery timeliness index	*a/b*	a—number of deliveries on time	%
			b—total number of deliveries	
2.	Delivery responsiveness		a—number of orders delivered in advance	%
			b—total number of orders	
3.	Share of incomplete deliveries to customers		a—number of incomplete deliveries	%
			b—total number of deliveries	
Indicators in terms of information flow				
1.	Reliability of information flow	*a/b*	a—number of correctly planned deliveries/routes	%
			b—total number of planned deliveries/routes	
2.	Return of delivery rate due to erroneous data		a—value/costs of returns	%
			b—value/cost of execution of all orders	
3.	Average time for analysis of data on delivery plans		a—total data download time	h
			b—number of plans drawn up	

Source Own study

- Characteristics of the goods transported (in the case under consideration these were chemical and/or technical articles),
- Type of logistics units received (unit packaging, collective packaging, pallet packaging, containers),
- Delivery date (selected working day) and time (hourly ranges),
- Location of the recipient (kilometre ranges).

On the basis of the adopted criteria, a scheme of three pilot implementation scenarios has been developed. For each of the scenarios, 3 variants differentiating the adopted criteria of contrasting features were defined, in order to check the impact of the use of the register in various conditions of the planning and delivery process. The pilot study was carried out in 2017, in accordance with the developed scenarios. The data obtained as a result of the simulation were compared with the actual data in relation to the defined indicators of evaluation of the effectiveness of implementation. A comparison of the values of indicators before and after implementation showed the potential benefits of a location register based on GLN identifiers (Table 2).

As it is shown in Table 2, in the case of all the indicators examined, the change noted was positive. As a result of using a register of locations based on GLN identifiers, the timeliness of deliveries, which from the perspective of the surveyed company

Table 2 Results of the pilot implementation of the prototype location register

			Case study 1		
Indicators for assessing the impact of the implementation on the efficiency of transport processes in economic and operational terms		Type of change	BEFORE	AFTER	Change[a]
1	Delivery timeliness index (%)	Increase	94.92%	98.17%	3.25%
2	Delivery responsiveness (%)	Increase	0.80%	1.86%	1.06%
3	Share of incomplete deliveries to customers (%)	Decrease	2.82%	1.41%	−1.41%
Indicators for assessing the impact of the implementation on the efficiency of transport processes in terms of information flow					
4	Reliability of information flow (%)	Increase	66.67%	69.10%	2.43%
5	Return on delivery rate due to erroneous data (%)	Decrease	16.97%	12.46%	−4.51%
6	Average time for analysis of data on delivery plans (h)	Decrease	7.86	6.99	−0.87

Source Own study
[a]Change in the value of the indicator in its unit of measurement (AFTER minus BEFORE)

was of key importance, has significantly improved. The change of 3.25% recorded in this case translated into a significant improvement of the company's internal indicators related to the measurement of the Service Level. From the perspective of the companies surveyed, a significant benefit of the register was also a reduction in the return of delivery rate due to erroneous data. The 4.51% positive decrease in this case was mainly due to obtaining reliable information on the required picking conditions for deliveries, adjusted to the customer's unloading capacity.

The conducted research has shown that the GLN register can be an important support in the implementation of logistic processes. Therefore, the main beneficiary of this solution are companies from the TSL sector, but not only. The research has shown that the use of the GLN register results in the improvement of end-customer service indicators, therefore its use should also be of interest to companies having function of transport users, i.e. manufacturers and distributors. The impact of the types and roles of companies that they have in the supply chains on the implementation of the GLN register is discussed in the next section.

6 Supply Chain Logic—The Impact of Business Relations on the Construction and Operation of the Location Register

The very concept of functioning in business practice of the location register based on GLN identifiers, which is a global, open and credible solution, seems to be as profitable for business entities as it is easy to implement. However, attention should be paid to a very important element of the use of the register in cooperation between independent entities. Their independence, as well as their different roles and responsibilities in the use and construction of the register increase the level of practical application of this solution. In order to clarify this problem, actions have been taken to define the main beneficiaries of the use of the location register based on GLN identifiers (and the information it carries or could carry), as well as key entities from the point of view of defining the parameters of locations identified by the GLN tag. This task referred to the logic of supply chains and supply networks. The key objective was to identify location owners and "users" of locations that are subject to physical flows in supply chains.

Supply chain logistics covers everything from the sourcing of basic raw materials to the sale of the final product to the final buyer. The one-way flow of goods is accompanied by a two-way flow of information. A customer places an order for specific raw materials/materials or goods in a specific quantity and determines the date and location of their delivery (supply). The mere indication of a location, in addition to its geographical location, entails a range of information on how and when to deliver, as these factors are often conditioned by the physical characteristics of the location to which the delivery is to be made. In this context, the importance of information on the physical characteristics of the location is evident. It should be explored for which actors in the supply chains this information is crucial to achieve both individual and whole chain benefits and the efficiency of their flows (integrated supply chain management).

A very illustrative diagram showing types of entities in supply chains, physical flow of goods (raw materials, materials, finished products, goods), indicating the types of existing locations, is shown in Fig. 6.

As it follows from Fig. 6, at each stage of the flow of goods, there are locations which constitute the place of delivery of these goods. According to the idea of building and practical application of the localization register in order to increase the efficiency of logistic activities, these locations would be marked with GLN numbers, which would at the same time carry a lot of information about the physical characteristics of these locations. The use of location information would benefit certain types of actors in the supply chain. At the same time, it should be borne in mind that the proper functioning of the register of locations requires certain actions to be taken, which should involve different types of entities, owners of locations. The balance of benefits and actions to be taken (responsibilities of location owners) will be key to determining the key factors in the functioning of the location register.

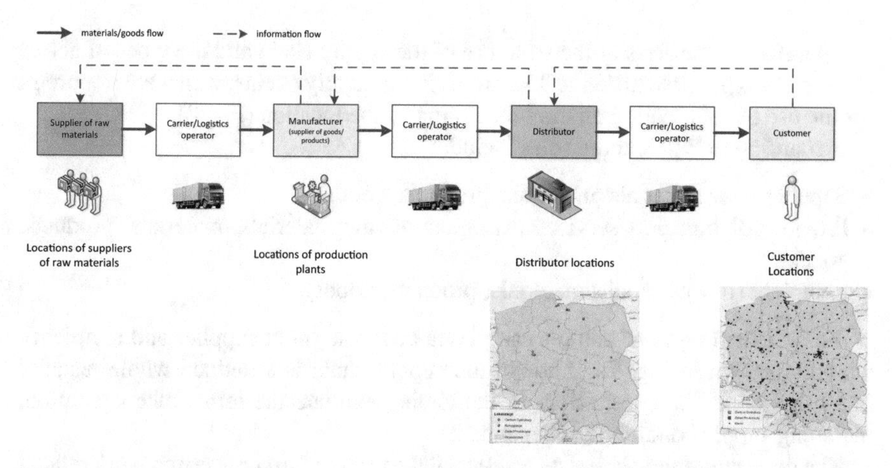

Fig. 6 Overview diagram of the supply chain with an indication of operator/location types. *Source* Own study

Graphics in Fig. 6 show the overall functioning of the supply chain. Within such a visualised chain, a wide variety of configurations of mutual cooperation between actors can be identified. One example is to show trilateral cooperation: manufacturer, distributor and carrier/logistic operator (Fig. 7). In the case displayed in Fig. 7, locations marked in green belong to one entity—the distributor. This figure shows that in the case of a distributor we can deal with an extensive network of locations, which perform various functions (distribution centre, regional warehouse, shop, etc.).

Regardless of the design of supply chains and their specificities, it is clear that economic operators involved in the supply process often have variable roles. For example, the wholesaler is the manufacturer's recipient and the retailer's supplier. Additionally, business entities dedicated to logistics operations are involved in handling the flow of goods within the supply chains. These are logistics operators providing services such as lease of warehouse space, handling of warehouse processes, transport service of deliveries. In the case of the latter service, carriers are also important for the proper functioning of supply chains.

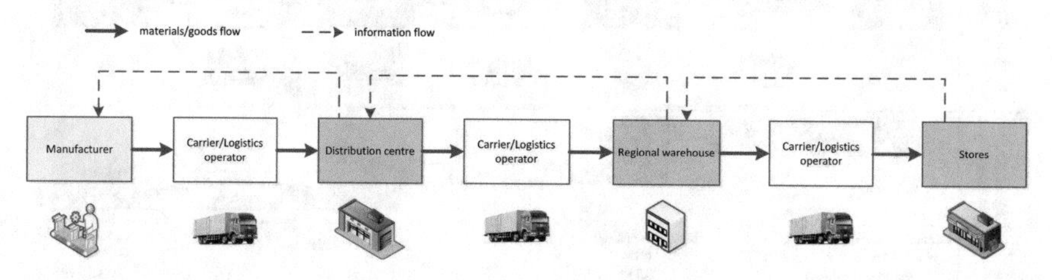

Fig. 7 Overview diagram of the supply chain with the selection of different location functions on the side of one type of entity (distributor). *Source* Own study

Therefore, regardless of the structure of the supply chain and the types of actors involved, it is possible to define 3 key roles in the supply chains, which are important for the use of information on the functioning of the location (Fig. 8).

According to Fig. 8, these roles include:

- Supplier (raw materials, materials, products, goods),
- Provider of transport services (transport of raw materials, materials, products, goods),
- Recipient (raw materials, materials, products, goods).

The selected types of entities can play a dual role (both supplier and recipient), which is shown in Fig. 9. Although the supply chain is a defined whole, usually in market conditions, the actors in the chain negotiate the terms of cooperation, including supply conditions, independently.

The described analysis and its results will be used in further research work related to the development of the GLN-based location register, aimed at defining a detailed

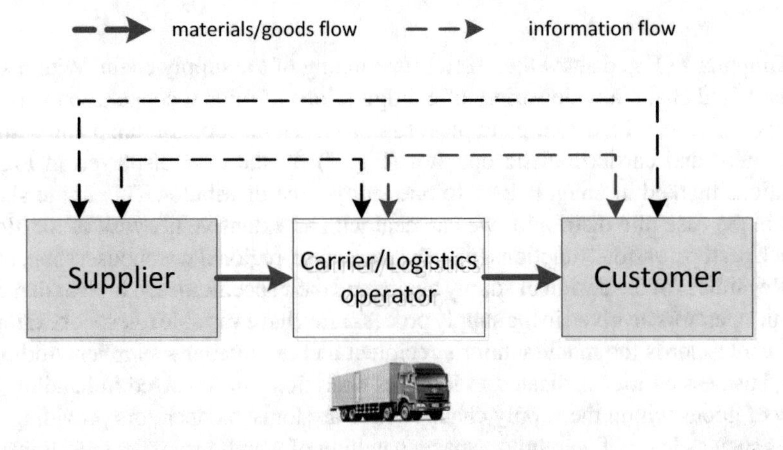

Fig. 8 Main roles and key relationship in supply chains. *Source* Own study

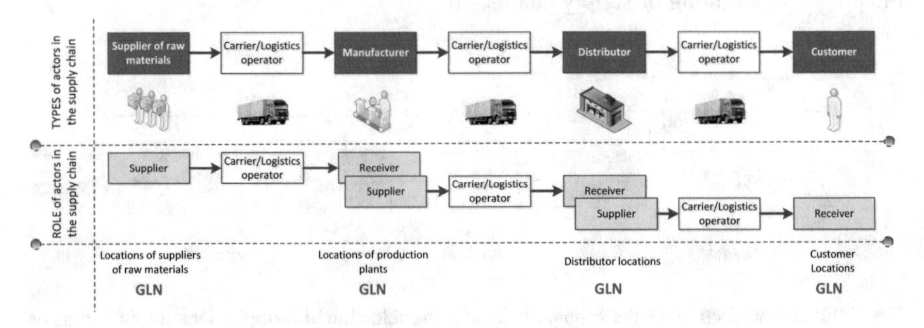

Fig. 9 An overview of the supply chain with an indication of the types of operators and their roles. *Source* Own study

functional and business model for the implementation of the register in business practice.

7 Summary

The results of the research described in this chapter confirm the potential for improving the efficiency of planning and delivery processes through the use of a location register based on GLN identifiers. The use of a global and standard GLN identifier describing the physical parameters of the location may, among other things, have a positive impact on the timeliness of deliveries and reduce the number of repeated and/or delayed deliveries. An important improvement that can be achieved through the use of a location register is also the shortening of delivery planning time and increasing the efficiency of this process. An important direction of the development of the register is the development and technical implementation of the concept of integration of technological solutions based on a standardised database, enabling easy and effective access to the register by individual entities performing various functions and cooperating in supply chains.

The research work carried out so far allows to define positive forecasts for the development of the GLN register in the supply chains activity. It should be borne in mind that, in order to increase its practical usefulness, the register should be linked to internal systems (ERP, TMS or WMS) used in enterprises.

The outlined directions of development of the location register based on GLN identifiers will be the subject of further research works conducted in cooperation between ILiM and GS1 Poland. The future development of the GLN register—construction of a integrated solution (advanced database based on GLN and related software) and its pilot implementation will allow for in-depth analysis of the conditions of GLN register use and the results of it.

References

Beamon BM (1999) Measuring supply chain performance. Int J Oper Prod Manage 19(3):275–292. https://doi.org/10.1108/01443579910249714

Bigaj Z, Koliński A (2017) The analysis of the cold supply chain efficiency with the use of mobile technology. LogForum 13(1):77–90. https://doi.org/10.17270/J.LOG.2017.1.7

Brandenburg M (2016) Supply chain efficiency, value creation and the economic crisis—an empirical assessment of the European automotive industry 2002–2010. Int J Prod Econ 171(3):321–335. https://doi.org/10.1016/j.ijpe.2015.07.039

Cudziło M, Niemczyk A (2017) Usprawnienie procesów dystrybucji i magazynowej obsługi dostaw przez wdrożenie standardów GS1, Wybrane problemy współczesnej logistyki w świetle badań naukowych i praktyki biznesowej, pod red. Cyplik P., Adamczak M., WSL, Poznań 2017, s. 49–63

Dujak D, Zdziarska M, Koliński A (2017) GLN standard as a facilitator of physical location identification within process of distribution. LogForum 13(3):247–261

Dujak D, Cudziło M, Voronina R, Koliński A (2018) Analysing the efficiency of logistic actions in complex supply chains—conceptual and methodological assumptions of research. LogForum 14(2):171–184

Eisenhardt K, Graebner M (2007) Theory building from cases: opportunities and challenges. Acad Manage J 50(1):25–32

Geunes J, Romeijn HE, van den Heuvel W (2016) Improving the efficiency of decentralized supply chains with fixed ordering costs. Eur J Oper Res 252(3):815–828. https://doi.org/10.1016/j.ejor.2016.02.004

Govindan K, Sarkis J, Jabbour CJC, Geng Y, Trandafir M (2017) Eco-efficiency based green supply chain management: current status and opportunities. Eur J Oper Res 10:57

Hałas E (red) (2012) Kody kreskowe i inne globalne standardy w biznesie, Instytut Logistyki i Magazynowania, Poznań

Horzela A, Kolinski A, Domanski R, Osmolski W (2018) Analysis of use of communication standards on the implementation of distribution processes in fourth party logistics (4PL). Bus Logistics Mod Manage 18:299–315

Kisperska-Moroń D, Krzyżaniak St (red) (2009) Logistyka, Instytut Logistyki i Magazynowania, Poznań

Kizyn M (2011) Poradnik przechowywania substancji niebezpiecznych zgodnie z wytycznymi unijnymi REACH i CLP. Instytut Logistyki i Magazynowania, Poznań

Korzeniowski A, Praiss A, Zmich J (2018) Comparative analysis of the quality of digitally printed barcodes. Bus Logistics Mod Manage 18:709–724

Li D, O'Brien C (1999) Integrated decision modelling of supply chain efficiency. Int J Prod Econ 59(1):147–157. https://doi.org/10.1016/S0925-5273(98)00097-8

Lichocik G, Sadowski A (2013) Efficiency of supply chain management Strategic and operational approach. LogForum 9(2):119–125

Mishra RK (2012) Measuring supply chain efficiency: a DEA approach. J Oper Supply Chain Manage 5(1):45–68. https://doi.org/10.12660/joscmv5n1p45-69

Nakatani K, Chuang TT, Zhou D (2006) Data synchronization technology: standards, business values and implications. Commun Assoc Inf Syst 17(1):962–994

Niemczyk A (2016) Warehouse processes in enterprises. In: Kolinski A (eds) Logistics management—modern development trends. Poznan School of Logistics Press, Poznan, pp 73–86

Siggelkow N (2007) Persuasions with case studies. Acad Manage J 50(1):20–24

Śliwczyński B, Hajdul M, Golińska P (2012) Standards for transport data exchange in the supply chain–pilot studies. In: KES international symposium on agent and multi-agent systems: technologies and applications. Springer, Berlin

Yin RK (2009) Case study research. Design and methods. SAGE Publications, Thousand Oaks

Supply Chain Digital Collaboration

Martyna Zdziarska and Nataliia Marhita

Abstract Digital integration of a supply chain is the result of changing the essence of existing relationships between enterprises from a transactional concept to a collaborative concept. The essence of it is based on mutual trust and implementation of a "win–win" strategy, which leads to the creation of a logistics partnership. Logistics partnerships between the participants of the integrated supply chain give an effective chance opposing competition and creating premises to meet the needs of customers. What is more, information sharing and coordination between buyers and vendors have been considered as useful strategies to improve supply chain performance (Huang 2003). The purpose of the article is to introduce the new collaboration paradigm called Physical Internet and its need of innovative IT solutions development that will support digital integration. Author presents the research results of European projects—MODULUSHCA and CLOUD that become a testing ground for the new tools in order to be evaluated and confronted with the real business environment and end user expectations.

Keywords Supply chain · Physical Internet · Distribution efficiency · Digital collaboration

1 Future of the Supply Network Cooperation

Nowadays, competition goes much further. Not only companies but whole supply chains compete between each other. The most successful firms are always the part of an efficient network. Willingness to adapt to the dynamics of the development of the modern economy requires enterprises to increase their competitiveness (Christopher 1998). Demanding customer expects to receive a high quality product

M. Zdziarska (✉)
Institute of Logistics and Warehousing, Estkowskiego 6, 61-755 Poznan, Poland
e-mail: martyna.zdziarska@ilim.poznan.pl

N. Marhita
Lviv Polytechnic National University, Lviv, Ukraine
e-mail: nataliia.o.marhita@lpnu.ua

A. Kolinski et al. (eds.), *Integration of Information Flow for Greening Supply Chain Management*, Environmental Issues in Logistics and Manufacturing, https://doi.org/10.1007/978-3-030-24355-5_4

at the lowest price possible in the shortest possible time. Moreover, universal access to information allows him to compare any number of listings and to choose the most convenient for him. Logistics is an integrated process of shaping and control of physical movement of products and their determinants in order to obtain information similar to the optimal relationship between the level of services provided and the level and structure of the costs (Abzari et al. 2009). However, even holistic approach to the logistics management is insufficient if it focuses only on the level of a single enterprise. Partnership in the supply chain is a key factor of success when it comes to the long-term mutual growth and business development of producers, suppliers and logistics service providers (Simchi-Levi et al. 2003). Only effective supply chain management allows companies to fully respond to the expectations of their customers while building a competitive advantage (Fawcett et al. 2008). Collaboration with external supply chain entities influences increased internal collaboration, which in turn improves service performance (Stank et al. 2001).

The incorporation of economic, social and environmental/climate dimensions is important in order to improve the characteristics of the current supply system, increase its robustness and resilience against man-made and natural disasters and support safety, security and quality of life. According to European Commission, research projects and activities should be now directed towards (Scoping Paper for Horizon 2020 work programme 2018–2020):

- more efficient and effective deployment and maintenance of assets, infrastructure and traffic management including leveraging the potential of hubs through innovative ICT enabling collaboration among stakeholders
- improving health and the environment with pollution and emission-free, low-noise transport and mobility solutions
- increasing the transport system's resilience and security.

In response to these challenges, in 2015, three years of work under the European Horizon 2020 program called Modulushca ended, in which the Institute of Logistics and Warehousing from Poznań took active part as a partner from Poland. Modulushca was a project that in its activities referred to the innovative vision of the implementation of logistic processes called Physical Internet (Ballot et al. 2014). Its main task was to develop a universal framework for cooperation in an open logistic environment, assuming a total and unwavering flow of information and cooperation that goes far beyond standard diagrams. The concept of Physical Internet (Montreuil 2011; Domanski et al. 2018; Osmolski et al. 2019) is based on full sharing of supply chains, information, resources and infrastructure.

Promotion of cooperation in the field of logistic services is dictated by inefficiency and instability of the current organization of the logistics system. Transport, due to the incomplete use of cargo space and limited information on the availability of resources in real time, generates significant costs, both economic, social and environmental.

According to that, Modulushca project, following new trends and meeting contemporary problems, influenced the change of approach to cooperation in the logistics system. Thanks to joint activities of the world of science and business, the consortium members developed innovative solutions based both on research work and practical experience of enterprises from all over Europe (Procter & Gamble, CHEP, ITENE, ILIM, EPFL, Jan De Rejik, Poste Italiane, ARMINES, Uni. Laval Canada, PTV, MEWERE, TU Graz, TU Berlin, Kirsen Global Security). One of the main objectives of the project was to develop a model and prototype of a new idea of full cooperation and effective logistics organization within the FMCG distribution network. The results of previous development works will be presented below.

Logistic networks intensely use means of transportation and storage facilities to deliver goods. However, these logistic networks are still poorly interconnected and this fragmentation is responsible for a lack of consolidation and thus efficiency. To cope with the seeming contradiction of just-in-time deliveries and challenging emissions targets, a major improvement in supply networks is needed. This new organization is based on the universal interconnection of logistics services, namely a Physical Internet where goods travel in modular containers for the sake of interconnection in open networks (Sarraj et al. 2014).

The definition of Physical Internet, proposed by Montreuil (2011) is as follows: "The Physical Internet is a global logistics system based on the interconnection of logistics networks by a standardized set of collaboration protocols, modular containers and smart interfaces for increased efficiency and sustainability".

The concept of Physical Internet aims to create a logistics system in which there is unwavering flow of information and cooperation that goes far beyond the standard schemas. Physical Internet is based on the full sharing of the supply network, resources and infrastructure, while leveraging standard, modular packaging. It is planned to replace the existing models. Its foundation is the cooperation of all entities involved in the distribution of goods and the full flow of information between them. Physical Internet aims at transforming handling, storage, distribution and implementation of the supply of goods, aimed at increasing the efficiency of global logistics and sustainable development (Zdziarska 2015).

This innovative concept is based on three main pillars. The combined infrastructure means that companies start to take action aimed at optimizing the operation of such resources like storage space, vehicles capacities and production systems through sharing. The current situation shows that most companies are not in a position to fully exploit its potential, thereby freezing their capital. The market of logistics services will strive to create a common infrastructure. Logistics centers, hubs and transit points located all over the world will be widely available to all operators, thus creating one global network. The ability to use a large amount of docs will increase the efficiency of transport.

The second area is the introduction of modular cargo units. Trying to be achieved with analogy of the Digital Internet data distribution in physical processes in the real world. Digital Internet does not provide the information but only transmits packets with embedded data. These packages are designed in such a way as to be easily recognizable by internet networks. Information in the package is closed and is not directly decoded by the network. The packet header contains all the information necessary for the identification and designation of transit routes to the destination. Digital Internet is based on protocols that structure the data packets regardless of the mode of transmission. In this way, they can be processed in different systems and networks such as modems, fiber optic cables, routers, local area networks, Intranet, Extranet and virtual private networks. Similarly to the Physical Internet (open logistics network) will not handle the goods directly (whether they are raw materials, components or finished products), but only manipulated specially designed modular containers that allow an encapsulation of these goods. Target solution involves a complete change of pallet system into modular loading units. This involves, of course, the adaptation of vehicles, handling equipment and warehouse space that will allow handling this type of packaging. However, simulations conducted for research projects clearly demonstrate that the investments made in the long term will help to significantly reduce logistics costs and losses related to the movement of goods. Containers thanks to the folding panels can create boxes of various sizes tailored to the individual needs of the sender. M-Boxes are easy for handling, storage, transport, loading and composition. They have a standard phrases recognizable throughout the system and are equipped with sensors and transmitters to maintaining full control during the transportation process. As a result, shipping safety is maintained throughout the journey, and all actors involved in the distribution have full overview of the status of the order. Moreover, the package is reusable and easy to recycle.

The last pillar is the exchange of data. This is the most crucial element of the whole concept. Physical flow of information in the Physical Internet will operate through an integration of infrastructure. In the PI you would be able to report and organize the individual orders from your own ERP system in a standardized format, which will be processed into 'the cloud' and decrypted by the other participants in the process. An important aspect in this data exchange is the access level. The architecture concept, developed so far, has designated four areas. Information on the container (its designation, dimensions, special conditions of carriage) will be available to all, then the data associated with the transport process (detailed route and delivery address), reserved only for the carrier. Another area is an information covering the delivery data such as sender and recipient, description of goods, value of the contract and the terms and time of delivery. For this type of data only logistics operators and customs will get an access. Most sensitive information will be used only by the sender and recipient, and will be associated with contracts, number of orders, invoicing or discrepancies in the delivery.

Logistics service providers, carriers and owners of the storage infrastructure will also share their detailed information. They will provide information on the availability of their resources, capacity and the status of implementation of orders. By combining all these data, the system will optimize the process and suggest the best possible solution for minimizing the cost of each of the participants in the process. Physical Internet is called the concept of win-win-win, because it allows the balanced growth of all actors in the supply chain (Zdziarska 2015).

Essential part of the Physical Internet trend is information flow and digital link between the cooperating parties.

2 Supply Chain Digital Connection

Competitiveness strongly relies on leading on technological developments, but also increasingly depends on suitable organizational and framework conditions for successful market rollout. Synergies on collaborative design and processes as well as manufacturing methods and supply chains in all transport modes should enable innovation breakthroughs that will keep the European producing industry, including SME, competitive in the decades to come. To create new markets and better respond to societal needs, research in new mobility concepts that integrate products and services should lead to the emergence of new innovative solutions and business models. Further knowledge is needed to support innovation-friendly standards, regulation and framework conditions that allow the emergence of successful new competitive operational models and business models. The provision and use of big data has a significant potential for the optimization of freight transport but further knowledge is needed in the areas of governance, security and data quality, availability and privacy.

The tendency of enterprises to further integration in supply chains

During Modulushca project, implementation team did the research, that was carried out in 2015 on a sample of 106 business entities. This test was chosen in a targeted manner. Participants of the study completed an online survey. The group of respondents was represented by producers (26%), distributors (34%) and Logistic service providers (40%). Fraction was dominated by enterprises whose employment exceeded 250 people. Employees in the SME sector constituted just over 40% of the population. The companies with domestic capital (46%) dominated the study. Foreign capital was declared by 39% of respondents, and mixed—14%. Due to geographical coverage, the largest segment were companies operating globally and internationally. In turn, the activity on the domestic market was indicated by 27% of the survey participants. The FMCG industry was the basic area of activity.

The research results confirm readiness to engage in digital corporate arrangements among the surveyed entities. More precisely, the participants were asked the question: “Do you agree with the opinion that digital integration/cooperation with other enterprises within the supply network will help you to gain a competitive advantage?”. In response to this question, as many as 60% of respondents indicated the answer “I definitely agree”, and in total 92% of the sample confirmed the view of the superiority of cooperation over its lack. The distribution of responses clearly shows a high level of support for the organization and implementation of joint activities. What is worth emphasizing, the study did not report opposing opinions, and only 8% of responses concerned the answer “neither agree nor agree”. Such a high level of support can be largely explained by positive experiences in the area of digital cooperation.

Against this background, the results of research on the readiness of enterprises for further digital integration with the participants of supply chains are interesting (Table 1). As you can see, the interest in strengthening inter-organizational ties dominates in all cases examined, i.e. in relationships with: recipients, suppliers and

Table 1 Readiness of respondents to digital integration in the supply chain

No.	In particular	$\overline{x}$	σ	1	2	3	4	5	Total
				Very low	Low	Average	High	Very high	
				%					
1.	As part of cooperation with suppliers	3.84	0.91	2.0	6.0	20.0	50.0	22.0	100.0
2.	As part of cooperation with recipients	4.18	0.85	2.0	2.0	10.0	48.0	38.0	100.0
3.	As part of cooperation with subcontractors	4.00	0.85	0.0	6.0	18.0	46.0	30.0	100.0

Statistical description: 1.1.–1.2. $Z = -2.456$, $p = 0.01$*; 1.1.–1.3. $Z = -1.095$ $p = 0.274$, 1.2.–1.3. $Z = -1.578$ $p = 0.115$

subcontractors. Relatively small standard deviations indicate a small dispersion of the analyzed results.

Generally speaking, in each case the average grade on a 5-point scale oscillated in the vicinity of the value of 4—which indicates high motivation of the respondents to establish a digital partnership with the participants of supply chains, the first as to the frequency of marking confirmation answers (high and very high level of readiness), respondents' readiness for further electronic integration with cargo recipients proved to be (86%), which is important, in comparison with the results regarding the development of relationships with suppliers (72%), the differences in assessments are statistically significant.

The research also shows that readiness for further digital integration is not diversified in subgroups distinguished by characteristics such as: area of activity, origin of capital or geographical coverage. To some extent, the type of cooperation and the role that the surveyed enterprises play in the supply chain determine it. It turns out that respondents pointing to a partnership model of cooperation express a higher willingness to develop electronic/digital relations with their partners than companies declaring a transactional approach.

The respondents were also asked to indicate the areas of logistics in which they would be willing to electronically cooperate in the future. In all cases examined, i.e. in relation to transport, warehousing, purchasing and inventory, packaging and reverse logistics, the dominance of high over low ratings was observed, which confirms the tendency on the market to extract many functions of logistics outside. Over 80% of respondents also declared high and very high readiness for further digital integration in the field of transport and storage services.

Although the readiness level is statistically high, a set of obstacles encumbers digital collaboration. One of the major barriers preventing stakeholders to enter in automated data exchange in B2B and B2Platforms interaction concerns data confidentiality, unauthorized intrusion and usage. According to European Commission EU projects should now focus on ensuring data exchange and access in a secure, controllable and trusted way but at the same time, providing easy connectivity facilitating generic data exchange. It also requires that supply chain data and information is visible and standardized on an international scale.

A trusted and seamless integrated system is a requirement for supply chain visibility, which in turn leads to safe and secure controlled logistics systems in the Physical Internet. The challenge is in itself to ensure the 'trusted' and 'seamless' access for supply chain stakeholders. Stakeholders need to be sure that data exchange can be under their control. The way stakeholders provide data (remain the owner, no owner or other models) and setting the conditions for use is key. This will be a prerequisite for stakeholders to 'trust' any IT solution developed for supply chains. It requires solid models for data ownership.

The other challenge is to organize the flow of information throughout supply chains in a seamless manner. This requires new ways of making use of existing data and provide tools to use data that is not per se uniform. Creating resilient supply chains based on integrated use of information (ex. RFID, sensor technology, open data, etc.) is required in a secure manner. The seamless and trusted exchange of data is to be governed and monitored to avoid misuse (and mistrust). National and international standards and legislation prohibit the development at this point in time as they are different and no common focus. It will be a challenge to provide a governance and monitoring framework for secure data exchange in international supply chain environments.

The huge data bases updated every day, the increasingly complex logistics processes, their international character as well as multi-channel sales make technology in the supply chain management a key element. The processing of a large number of data, effective planning, capturing trends and making proper business decisions based on them would not be possible without modern IT solutions. The dynamic development and constant evolution of integrated information systems make them more and more complex and functional. Initially, they focused on stock management, a bit later they were enriched with the possibility of planning material demand, and in the next years with the functionality of resource planning, distribution resource planning and finally customer relationship management system. IT systems that are associated with the logistics industry are now mainly dedicated to the single company—ERP, MRP and SCM systems. Already more than 70% of large logistics companies have a solution of Enterprise Resource Planning class. ERP systems are slowly becoming a standard in large logistics companies. For medium and small enterprises, this indicator is unfortunately still much lower and is at the level around 30 and 10%, respectively.

Despite that, international supply chains are becoming more and more complex. The pursuit of integrated services in the supply chain is a trend, which has evolved under the influence of customers' expectations. It is accompanied by the development of technical capabilities, in particular due to the progress that is being made in computer science.

3 Integrated ICT Solutions

In years 2016–2018, in reference to the previous issues CLOUD (Collaboration in Logistics Operations and Urban Distribution) project founded by the National Centers for Research and Development worked on the new integrated ICT solution. The idea behind the CLOUD project was to develop a virtual Logistic Single Window (LSW) as an ecosystem with services and applications for all kind of transport and supply chain stakeholders, which aims to provide the building blocks for supply chain management over door-to-door operations:

- sourcing, booking and planning of logistics services
- planning of trans-European logistics chains including first/last mile optimization
- in-transit execution management
- tracking and monitoring (visibility).

The project provided a solution that improves cooperation in logistics through more efficient B2B, M2M and M2B communication. Abovementioned solution was developed and demonstrated within the project. The Logistics Single Window supports a federation of regional/national community platforms such that the prospective user views it as a single logistics services platform even though it emerges through collaboration between individual initiatives—each of them with their own focus. The LSW brings together value added services from different angles: From end-to-end logistics chain management to operational systems focused on resource optimization for individual modes and/or hubs. Incorporation of end-to-end concepts address issues like synchro modality, network balancing and reliability of delivery while Physical Internet (PI) concepts address the use of local knowledge for resource optimization on the last mile. CLOUD aimed for the practical development of the Physical Internet (PI) concept. PI offers the potential for substantial improvements in supply process. As the adoption of the PI increases, more loads are available for sharing among transportation service providers and optimal use is made of local intelligence.

The approach taken towards communities, connectivity and interoperability lowers the costs and complexity of using value added services significantly. Realization of the project objectives allowed logistics and supply chains to develop towards a more efficient, but at the same time more customized and service-oriented sector, supported by full integration and synchronization of processes between involved parties and transport assets. Thanks to the international collaboration between Flanders, Poland, Sweden and Norway the critical market penetration and competitiveness of the solution was higher.

Project objectives directly corresponded with the main domains of the ERA-NET programme:

1. Cross-border freight transport corridors:
   - synchro-modality
   - optimal use of empty containers
2. Hub development
   - Connection of regional networks of inland waterway, rail and road services
3. Urban/last mile logistics
   - innovative solutions for first/last mile logistics in urban and suburban areas
   - development of coordinated and consolidated procurement and ordering strategies
4. Organizational innovations & new business models
   - business models for horizontal and vertical cooperation
   - collaborative chain control strategies
   - advanced supply chain risk management and resilience
5. Information infrastructure and services for logistics
   - rationalization of workflow/processes by early data acquisition
   - deviation management
   - inclusion of tracking and tracing as a service

Users' expectations regarding ICT tools

There is a strong interaction between requirements analysis and designing the solution. Since the CLOUD concept is innovative, the end users were not capable of expressing very well their requirements. Due to that project team had to come up with concepts and prototypes, sketching possibilities, in order to arrive at a solution that the end users recognize as useful. After the presentations and workshops with the potential stakeholders, the development team was able to identify the following most common expectations regarding logistics ICT integration tools:

1. **Unwavering data exchange**

The efficiency and reliability of logistics processes of sourcing, production, distribution and transport depend largely on the speed and efficiency of information processing, which in the present times is determined primarily by the efficiency of computer technology.

Effective data exchange between supply chain partners is now the basis of solutions provided by the developers of IT tools. At the same time, the large variety of programs supporting logistic activities enforces the introduction of effective communication between systems at all stages of the process. Therefore, it is important to use advanced IT technologies functioning within the framework of electronic data exchange standards.

2. **Integrity**

Integration is the next step in data exchange between partners as part of logistics processes. During the flow of goods in the supply chain, information is very often processed many times. Integration enables data exchange in real time, which increases the efficiency of the entire supply chain.

The integration of internal processes in enterprises is now becoming a standard, many companies use ERP or SCM class systems, which is the starting point for consolidation within the entire distribution network. Therefore, in line with current company trends, there is interest not only in the internal integration of processes, but integration between partners within the whole supply chain an it is becoming more and more important.

In a dynamic, complex logistics environment, EDI integration between contractors is still popular, but more often it also gives way to other solutions, such as communication platforms (Debicki and Kolinski 2018). This is due to the fact that many partners (customers, suppliers, LSPs) often work together within one network, and full integration with them would be time-consuming and costly. Therefore, a better option is to use an external EDI platform that provides unshakable data flow to all players through a single connection standard.

3. **Availability in the cloud**

Cloud computing is a service whose main advantage is the provision of IT services and solutions via the Internet. It is a very popular form of IT outsourcing and its main value is that there is no need to invest in your own infrastructure and software. Data processing in the cloud allows you to rent resources and make them available to many users.

The advantage of solutions in the cloud is its flexibility and security. A set of IT tools allows you to increase work efficiency while significantly reducing costs. Data processing in the cloud also means increasing data security, full availability, regardless of where the user is. All you need is access to the network and computer. The provider of this solution provides full support to all system users, quick implementation and flexibility in adapting to the growing number of users of the solution. The advantage of solutions in the cloud is access to all online applications using any web browser.

The offer of many companies such as SAP, Oracle, Inttra or Comarch, offering standard solutions implemented on the client's server, are also available in the Software as a Service (Cloud computing) model. Among the products available in the cloud are software supporting both logistic and financial-accounting processes.

4. **Complexity**

The contemporary market is becoming more and more demanding. The growing expectations of consumers and strong competition are the factors that encourage entrepreneurs to look for new technological solutions that offer fully comprehensive solutions.

Among IT systems supporting management of organizations, especially among integrated systems (e.g. ERP, SCM classes), standard software packages dominate. They usually have built-in adjustment mechanisms that allow the system to be adapted to the needs of a specific customer.

Increasingly, in addition to the standard system configuration for existing jobs, entrepreneurs require that specialized systems also support the company's development through easy integration with cooperators, enable work in the cloud and provide a mobile application. Considering this, IT solution providers develop their products in such a way that the customer, when buying a chosen tool, receives software that comprehensively supports multiple processes on as many devices that provide remote access.

5. **Ease of use**

"User-friendly" is a subjective term, which is why some common criteria used in user-friendly interfaces are presented below:

- simple to install—installation is the first point of contact for users, so it should be a friendly process. It does not matter if it is an operating system or a single client application, the installation should be simple and properly described in the manual.
- easy to update—as with installation, the application update process should be easy. Updates must be simple enough to ensure users continue to use the software without problems. When users do not update, and thus do not reveal problems, the software becomes less and less reliable and secure.

- intuitive—to be user-friendly, the interface must be understandable for the average user and should require the shortest user manual.
- efficient—not only software should work as expected, but should also be efficient. It should be optimized for the specific architecture, should be connected to all memory leaks and should work seamlessly with the underlying structures and subsystems. From the point of view of users, the software should be an effective means to perform its tasks.
- transparent—a good user interface is organized so that you can easily locate the various tools and options you need.
- reliable—ownership of the system that gives information about whether it works correctly (it meets all the functions and functions entrusted to it) for the required time and under specific operating conditions, which directly affects the user's satisfaction.

6. **Mobile access**

Due to the constantly growing mobile technology, which allows quick and easy access to information or tools, more and more IT tools providers also offer a mobile version of the software. Mobile applications support software regardless of its basic functionality, it can be an ERP system, as well as an SAP system or a commodity exchange, for example TimoCom or Trans.eu. Mobile devices are everywhere, thanks to which supercomputers become available at any time—often as services and applications in the cloud—and a high level of security is standard in this mobile environment. While this mobile trend offers great opportunities, it also increases the pressure on IT providers who improve their products and services in the clouds. Many companies are currently expanding a subset of existing applications for mobile devices. In many cases, these are micro applications that offer a subset of the functions available in the applications on the computer.

More and more ICT tool providers are focusing on offering complex solutions that enable efficient data exchange and process integration, not only within the enterprise, but also between the participants in the supply chain. Currently, many solutions are offered in the cloud or using electronic data exchange platforms. It is also important that the software must be tailored to the needs of the customer, and the use of it should be user-friendly. These trends within the framework of ICT technology supporting supply chain management are examples of trends whose selection and implementation is conditioned by the scope of business and the degree of integration between participants in the logistics process. This significantly influences the application capabilities of the chosen technology that increases the effectiveness of processes, broadly understood data analytics and facilitates contacts between business partners and clients. The development of new distribution channels along with the dynamization of ICT in the supply chain are becoming the driving force behind the creation of a new, integrated approach to logistics.

4 Conclusions

Being a complex network of suppliers, factories, warehouses, distribution centers and retailers, the success of any SCMS depends on how well these system components are managed. To make supply chain management successful, management must be committed to high standard of performance and trust including long-term collaborative relationships that can deliver results independent of industry and sector type (Simatupang et al. 2002). In recent times, information has become a key player in determining the productivity of a complex enterprise. The enterprise's ability to process information and make rapid but right decisions promises growth (Abzari et al. 2009). Information sharing and coordination between the agents of a supply chain are considered to be an effective strategy for improving its global performance (Montoya-Torres et al. 2014). Essential part of the Physical Internet is information flow and digital link between the cooperating parties. Current trends indicate that integration on the digital level within the supply network is inevitable. Although each day more and more companies implement popular IT solutions supporting their logistics operations, there is still an enormous need of development of new digital solutions aiming integration of all of the supply chain cells. EU project such as Modulushca and CLOUD tend to build up a research and testing grounds for the new tools to be evaluated and confronted with the real business environment.

Based on the projects' results further research is required to provide detailed view of the technological needs in terms of digital standardization. Respondents pointed out that logistics ICT integration tools supporting planning and organization of transport processes lack standards of electronic data exchange. There is a great potential for using for example GS1 standards to support the effectiveness of these platforms and thus the logistics processes of enterprises. The GS1 XML and a potential standardized Web API could in many cases become a solution to integration problems, improving the quality of cooperation between companies in supply chains, favouring their digitalization. This may also be possible thanks to the use of solutions, such as the e-Freight and e-Delivery communication standard, which, thanks to access points and connectors, allows secure data exchange between independent systems without the need to interfere in the software.

Improving the exchange of information between platforms of participants in the supply chain is a key aspect that will increase the efficiency of the transport process and empower development of the Physical Internet. Results of such research will indicate the direction of the ICT solutions development that would accelerate further integration.

References

Abzari M, Mohammadzadeh A, Shavazi A (2009) A research in relationship between ICT and SCM. Proc World Acad Sci Eng Technol 38:92–101

Ballot E, Montreuil B, Meller RD (2014) The Physical Internet. The network of logistics networks. La Documentation française

Christopher M (1998) Logistics and supply chain management, strategies for reducing costs and improving service, II edn. Financial Times, Pitman Publishing, London

Debicki T, Kolinski A (2018) Influence Of EDI approach for complexity of information flow in global supply chains. Bus Logistics Mod Manage 18:683–694

Domanski R, Adamczak M, Cyplik P (2018) Physical Internet (PI): a systematic literature review. LogForum 14(1):7–19

Fawcett SE, Magnan GM, McCarter MW (2008) Benefits, barriers, and bridges to effective supply chain management. Supply Chain Manage Int J 13(1):35–48

Huang G, Lau J, Mak K (2003) The impacts of sharing production information on supply chain dynamics: a review of the literature. Int J Prod Res 41(7):1483–1517

Montoya-Torres JR, Ortiz-Vargas DA (2014) Collaboration and information sharing in dyadic supply chains: a literature review over the period 2000–2012. Estudios Gerenciales 30(133):343–354

Montreuil B (2011) Toward a Physical Internet: meeting the global logistics sustainability grand challenge. Logistics Res 3(2–3):71–87

Osmolski W, Voronina R, Kolinski A (2019) Verification of the possibilities of applying the principles of the Physical Internet in economic practice. LogForum 15(1) (In press)

Sarraj R, Ballot E, Pan S, Hakimi D, Montreuil B (2014) Interconnected logistic networks and protocols: simulation based efficiency assessment. Int J Prod Res 52(11):3185–3208

Simatupang TM, Sridharan R (2002) The collaborative supply chain. Int J Logistics Manage 13(1):15 (ABI/INFORM Global)

Simchi-Levi D, Kaminsky P, Simchi-Levi E (2003) Designing and managing the supply chain: concepts, strategies, and case studies. McGraw-Hill, New York

Stank TP, Keller SB, Daugherty PJ (2001) Supply chain collaboration and logistical service performance. J Bus Logistics 22(1):29–48

Zdziarska M (2015) New logistics approach towards distribution. In: Stajniak M, Kolinski A (eds) Innovation in logistics contemporary and future development trends. Spatium Publishing House, Random, pp 170–178

Perspectives of Blockchain Technology for Sustainable Supply Chains

Helga Pavlić Skender and Petra Adelajda Zaninović

Abstract Academic and business interest in blockchain and its potential benefits for businesses has risen considerably since 2016. Blockchain is considered as a distributed ledger technology that records transactions between parties in a protected and permanent way and enables transferring data and assets easily without relying on a third-party intermediary. Current supply chains are mostly linear and heavily complex networks with various stakeholders which often causes low transparency, data silos and unstandardized processes. The aim of this paper is first, to analyze the overall perspectives of blockchain technology and second, to investigate the potentials of blockchain to enable sustainable supply chains which would eliminate many shortcomings in current supply chains. This paper analyses three cases which exploited the potentials of blockchain technology with the aim of promoting sustainable sourcing from origin to consumers. The analysis compares three different cases of global supply chains, with a focus on corporate sustainability in agriculture, and manufactured goods. The results of our analysis show that blockchain has the potential to boost the supply chain sustainability, however the blockchain can't guarantee that all supply chain stakeholders provide accurate, uniform and verifiable data. This paper provides a conceptual framework to better understand the benefits and challenges of the blockchain technology. Both academics and practitioners in companies might find this framework useful, as it outlines important lines of research in the field.

Keywords Blockchain · Supply chains · Transparency · Sustainability

H. Pavlić Skender · P. A. Zaninović (✉)
Faculty of Economics and Business in Rijeka, Department of International Economics, University of Rijeka, Rijeka, Croatia
e-mail: petra.adelajda.zaninovic@efri.hr
URL: https://www.efri.uniri.hr/hr/petra_adelajda_zaninovic/196/72

H. Pavlić Skender
e-mail: helga.pavlic.skender@efri.hr
URL: https://www.efri.uniri.hr/hr/helga_pavlic_skender/89/72

A. Kolinski et al. (eds.), *Integration of Information Flow for Greening Supply Chain Management*, Environmental Issues in Logistics and Manufacturing, https://doi.org/10.1007/978-3-030-24355-5_5

1 Introduction

Supply chains are nowadays facing challenges such as trust between parties, global competition, never-ending strive for efficiency, aim to reduce operating costs, and above all to be sustainable. Numerous products are being manufactured through supply chains which are extended globally. Supply chains consist of numerous actors such as suppliers, transporters, distributors, retailers who participate in designing, producing, delivering, and selling the products, however in most of the cases, these physical and information flows stay invisible to final consumers (Abeyratne and Monfared 2016). Furthermore, beside the low product transparency and traceability, the mayor bottlenecks in supply chain business are slow information flow, expensive and time consuming hard-copy paperwork, duplication of procedures, time-consuming and costly money transfers etc. (Heutger and Chung 2018). Although blockchain technology in supply chain business is still in his infant phase, it has the potential to boost the supply chain sustainability. It is suitable for use in supply chains because the technology has the potential to provide a substantial level of transparency. The social and environmental performance of the companies within the supply chains is of great importance and focal companies might be held responsible for it (Seuring and Müller 2008). In order to meet the future head on, the whole supply chain needs to operate in a more sustainable manner. The sustainable supply chain means that companies are "making their products and delivering their goods or services in a sustainable way that doesn't impact the environment, that doesn't contribute to social inequalities or injustice, and that in general, is done the right way" (RMA 2018). The aim of this paper it to answer following research questions: what is blockchain and how does it work? How can the blockchain be implemented in supply chain and enable its sustainability? What is the current practice of blockchain implementation in different supply chains.

Paper is structured in six sections. After the introduction, the second section presents the literature review on blockchain technology. Third section analyses the theoretical integration of the blockchain within the current supply chain. Fourth section describes used case study method while the fifth section discusses the results. Finally, sixth section concludes the paper.

2 The Main Features of the Blockchain Technology

The history of blockchain technology started when Satoshi Nakomoto in 2009 (Buterin 2014) created Bitcoin and introduced the concept of a blockchain in order to create a decentralized ledger maintained by anonymous unity (Niforos et al. 2017). Later, around 2013–2014, the world leading financial institutions started to invest in start-ups based on blockchain technologies, exploring its novel applications and opportunities. Recently, in 2016, the blockchain started to attract attention and be explored across various industries (Heutger and Chung 2018). Basically, all industries that serve as intermediaries for processing financial or other kind of transactions will have to implement the blockchain (Kalinin and Berloff 2018).

2.1 What Is the Blockchain?

Blockchain was created based on the old template of ledger, which was used to log transactions. However, in contrast to the old ledger which was owned by one entity and updated by one administrator, the blockchain is a shared, distributed ledger across a network of stakeholders that can only be updated with the approval of all the participants in the network and all changes to the distributed ledger are supervised (Heutger and Chung 2018). Bogart and Rice (2015) state that blockchain is a distributed ledger which works as a chronological chain of 'blocks' where each block consists of a record of valid network activity since the last block was added to the chain. "Blockchains are structured as a shared, decentralized database with immutable, encrypted copies of the information stored on every server or "node" in the network" (Golden and Price 2018).

Hackius and Peterson (2017) classify three most important features of the blockchain:

- **decentralization**—network participants completely run the network without central authority or centralized infrastructure to establish trust. Intermediaries who formerly acted as trusted third parties to verify, record and coordinate transactions, are no longer needed. Any transaction must be shared within the peer-to-peer communication network and all network participants keep their own local copy of the ledger. The blockchain is based on a consensus system, which represents a trust-worthy authority (Xu 2016).
- **verification**—the network participants sign the transactions using public-private-key cryptography before sharing them with the entire network. The private keys can be initiated only by the owner and the network participants can stay anonymous because the keys are not linked to real-world identities. A private key is like the key or password required to unlock any other thing and it's protected by the owner and cannot be shared with third parties.
- **immutability**—gathered transactions form a new block and all the network participants can verify the transactions in the block. The transaction is valid, and the block is added in the chain if there is consensus among all network participants, however, if there is no consensus on the validity of the new block, then the block is rejected. Furthermore, for each block is developed a cryptographic hash. Beside holding the transaction records, each block holds the hash of the previous block.

Similarly, Heutger and Chung (2018) classify four blockchain most important characteristics:

- **data transparency**—blockchain assures that stored records are authentic, tamper-evident, and from a valid source. Consequently, every stakeholder receives controlled access to a shared and secure dataset. Blockchain actually replace intermediaries and thus create a single source of truth. The network participants can rely and feel safe using this data this way.

- **security**—using blockchain the individual transactions and messages are cryptographically signed. This decrease the high risks of hacking and data manipulation. Furthermore, the absence of a central server makes more difficult for a network to be attacked (Xu 2016).
- **asset management**—blockchain enables to track the origin of the product, to track the asset or land ownership of titles or rights, such as copyrights etc.
- **smart contracts**—the concept was discussed by Szabo in 1997 who states that in the blockchain context a smart contract is a script that is stored on the blockchain. Smart contract can automatically implement the rules agreed by the stakeholder and are completely autonomous (Schillebeeckx et al. 2016).

According to Pilkington (2016) two types of blockchain-based system exist, one that is open to the public which means permission less, and the other which is private, hence, permissioned blockchain. In the second, private one, the rights to read and write are restrained by the central authority. In case of the public one, anyone can join the network, submit transactions and broadcast network data, while in case of private system, only safe listed participants can join the network and broadcast network data. Furthermore, in public system all transactions are broadcast publicly whilst the access privilege exists (Heutger and Chung 2018).

It is not surprising that the companies tend to embrace private, permissioned blockchains because it enables strict access controls and privacy protections. However, the choice which system should be use, public versus private blockchains should be made depending on the nature of work and individual needs of each blockchain implementation.

2.2 How Does the Blockchain Work?

A blockchain enables records to be stored and sorted into blocks (Government Office for Science 2016) which is the most practical applications of blockchain. Figure 1 shows simple example of how the process operates in practice.

According to Drescher (2017) two distributed databases are present. First one representing a transaction set called "backlog" which insert and assign incoming transactions, and second one, a blockchain which store ordered transactions. "The registered user (node) in this network system can update backlog, blockchain, and transactions between them by voting process based on a Consensus Algorithm" (Tian 2018). In simple words, blockchain is one kind of a database and takes several records and puts them in a block. Each block is then added (chained) to the next block, using a cryptographic signature (De Caria 2017). Finally, blockchain can be explained as "the technology that powers the Internet of Transactions" (Mainelli and Milne 2016).

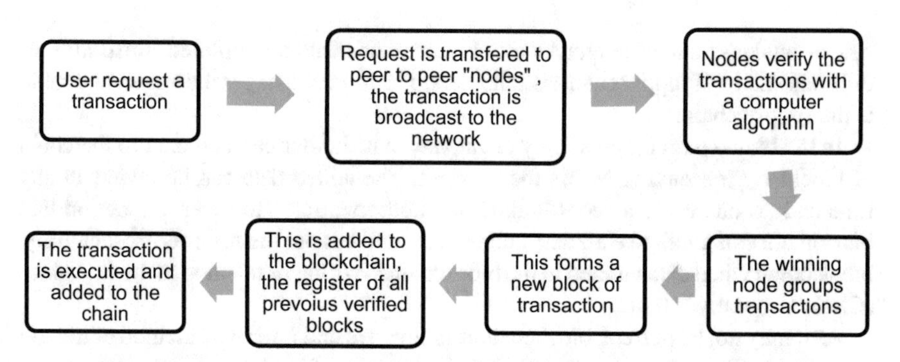

Fig. 1 Blockchain process. *Source* Chartered Accountants Australia and New Zealand (2017)

3 The Blockchain Implementation in Supply Chains

As previously mentioned, supply chains are facing many challenges considering its "complex structure emphasized by the number, position, the nature of relationships, activities, business objectives, capacity, information services and technology base of its participants" (Mesarić and Dujak 2013). Transparency is one of the most important and hardest improvement areas to achieve (Abeyratne and Monfared 2016), thus is clear that the blockchain applications may have one of the strongest impacts on the supply chain although it's still arguable innovation in the supply chain and many companies are sceptical regarding its implementation. However, as the supply chain issues take toll on company image and profit, the companies cannot afford to ignore it (Cottrill 2018).

There's no doubt the blockchain could boost efficiency and sustainability in the supply chain industry. For example, many producers do not receive fair prices for the value they create because of the supply chain opacity. Moreover, the costs imposed by intermediaries are sometimes enormous. Blockchain's decentralized ownership of data could help enhance the balance of power in the global supply chains by ensuring small producers and consumers sustainability and social responsibility. However, the blockchain may actually help to share information across all parties on and enhance visibility and traceability. On the other hand, most buyers and sellers do not know the true origins of the goods they have purchased which might cause huge problems, such as bad company image, profit loss and after all consumers heath issues. Blockchain can be a solution because the technology can track a product's life cycle and ownership transfer from origin to final destination (Heutger and Chung 2018). After all, just dealing with the paperwork and bureaucracy costs companies billions of dollars annually (Golden and Price 2018). By reducing bureaucracy and paperwork, blockchain could improve efficiency in international trade and facilitate business transactions.

Such a substantial number of stakeholders and partner's relationship involved in the supply chain often causes low transparency, loss of information, unstandardized processes and delays and non-payments in the delivery of products, loss of orders, risk of business database attacks etc. Figure 2 presents a simplified illustration of different stages of supply chain stakeholders. The more the stages the more vulnerable is the supply chain.

In the blockchain based supply chain, any stakeholder can add data to the chain of blocks by transacting across the network. The added data can be review at any time and no one can change it without the authorization. However, the notion that blockchain can guarantee all data authenticity and cut out the need for trust entirely, break down when data entered in the blockchain is wrong or falsified in the first place (Circle Economy 2018).

As it may not be perfect, blockchain is innovative and improved method of tracing product supplies. Moreover, the blockchain, because of its transparency, can provide consumers and other downstream partners higher standards and promotion of sustainable sourcing and production. Finally, the technology seems to be the right solution to solve core problems in supply chain transparency (Golden and Price 2018).

In addition, Fig. 3 illustrates the comparison between current linear supply chain directly or indirectly fulfils the supply needs and circular supply chain based on blockchain technology. Both linear and blockchain based circular supply chain consists of upstream and downstream activities, e.g. supply and demand side. Supply side includes upstream activities which provide inputs to a focal company while demand side includes downstream activities necessary for the product to reach final customer (Dujak and Sajter 2019). The main task of those activities is to improve

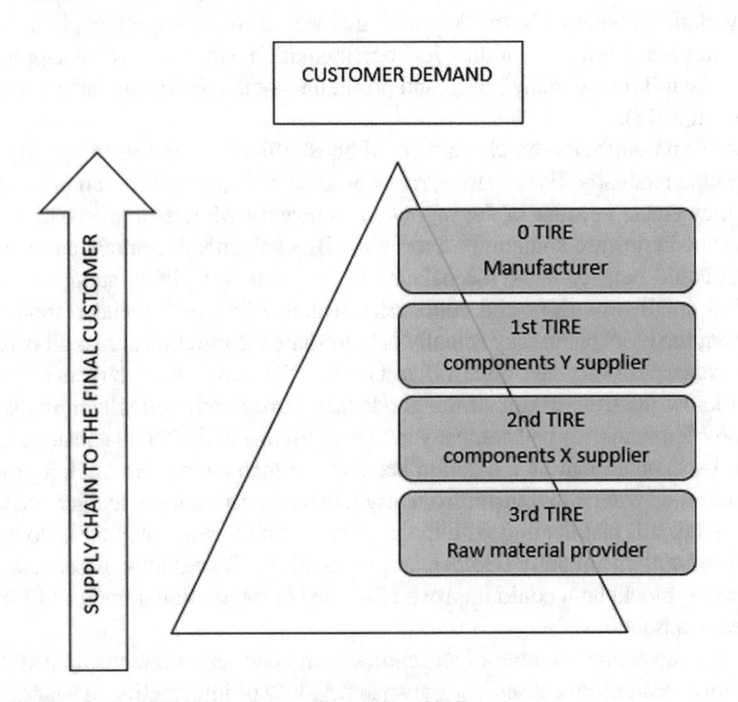

Fig. 2 Supply chain stakeholders. *Source* Consulting (2017)

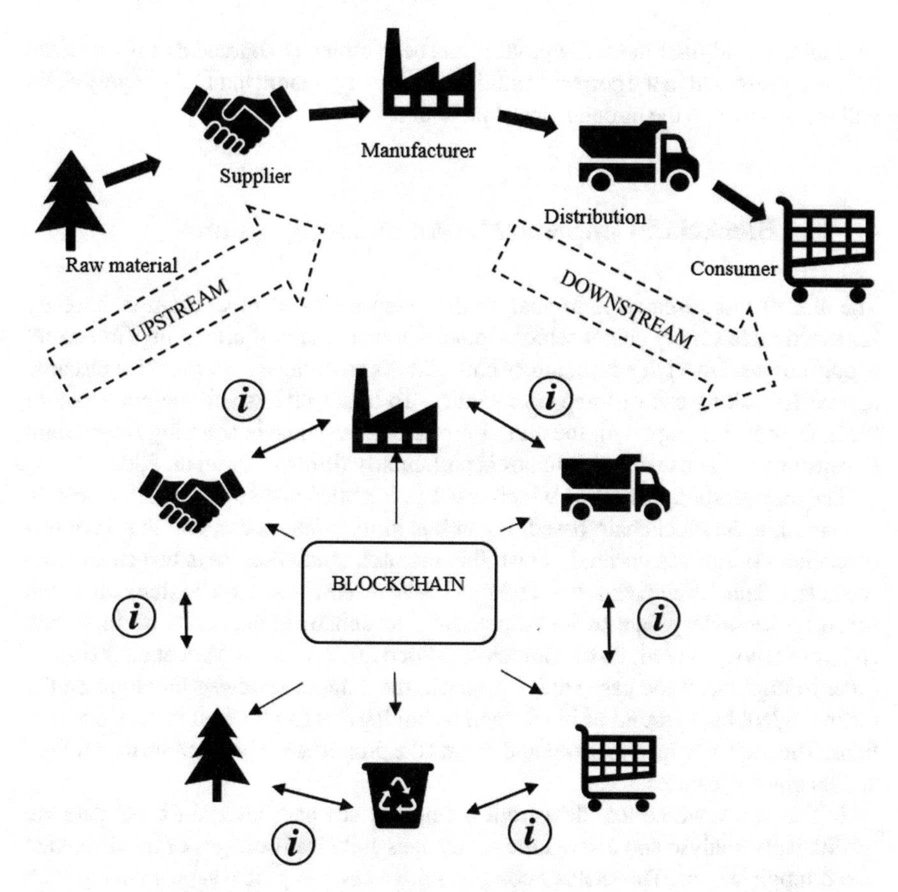

Fig. 3 Linear supply chain versus blockchain based circular supply chain. *Source* Revised according to Casado-Vara et al. (2018)

the flow of materials, information etc. within the supply chain. Blockchain based circular supply chain improves the flow of materials and information because the model enables all the parties included in the supply chain to have all the information at any time thus making supply chain more transparent and fairer. In the current supply chain situation, the stakeholders are unable to track the origin of materials, components, and products throughout the supply chain, so that anyone along the way can affirm their circularity, from the moment they were first extracted or created, all the way through their (many) life cycles (Circle Economy 2018).

In addition, blockchain based supply chain enables circular economy model which promote recycling. Namely, with the use of blockchain, all products could be tracked with blockchain and with traceability it is possible to give confidence to the final consumers about the origin of the products and whether they are recycled or first use, etc. (Casado-Vara et al. 2018). By using the blockchain, all the stakeholders in supply chain can find out more information about the products they are buying, if the product is an original item, if a product has been ethically sourced and if a product has been preserved in the correct conditions. This way is more likely the supply chain will transform into the circular, sustainable one.

4 The Blockchain Implementation in Supply Chains

The aim of this research is to analyse the perspectives of blockchain technology for sustainable supply chains which would eliminate many shortcomings in current supply chains. Given this exploratory aim, authors have chosen a qualitative research approach, making use of three case studies. To better understand the potentials of blockchain technology with the aim of promoting sustainable sourcing from origin to consumers, the use of case studies is particularly fitting (Voss et al. 2002).

The case study approach is widely used in logistics and supply chain research. Considering the blockchain based research is in its infant phase, and that very few blockchain initiatives currently exist, this research study considers use cases from secondary data. Moreover, several scholars have recently used a case study approach based on secondary data to investigate the blockchain implementation in supply chains (Verhoeven et al. 2018, Golden and Price 2018, Casado-Vara et al. 2018). In order to implement the case study approach, the authors reviewed literature on the technological background of blockchain technology as can be seen in previous sections. The review helped to better understand the properties of blockchain technology and its unique features.

In this study we collect data from online sources and based on these data we qualitatively analyse and discuss the advantages and disadvantages of blockchained based supply chains. The analysis compares three cases of global supply chains, with a focus on corporate sustainability in agriculture, and manufactured goods.

Authors searched online to find companies and cases that highlights the perspectives of blockchain technology for sustainable supply chains. The companies selected for this research paper are Bext Holdings Inc., IBM Food Trust and Provenance. Bext Holdings Inc. is a start-up focused on developing technologies to improve social sustainability in mostly food supply chains. The company developed a blockchain to track and trace coffee beans on the complete route from farmers to consumers. The IBM Food Trust uses blockchain technology to create unique visibility and accountability in the food supply and finally, the Provenance company helps brands and retailers to build customer trust through supply chain transparency.

Subsequently, three cases were chosen that portray most important features of blockchain technology and illustrate its potential to build-up sustainable supply chains. The cases illustrate how blockchain technology supports human rights and environmental sustainability.

5 Results and Discussion

In this section, the results from the case analysis are presented and afterwards discussed. First case shows how blockchain affects the producers and customers right. Second case demonstrates how blockchain can be used for food poisoning prevention and the third highlights how blockchain can help companies to operate in more ethical manner.

5.1 The Case of Bext360

Bext Holdings Inc. is a start-up based in Denver, Colorado. The company was established in 2016 and their main focus is on building technologies, bext360 that streamline critical supply chains in emerging economies (Bloomberg 2018). Bext360 focuses on supply chains such as coffee, seafood, timber, minerals, cotton and palm oil to provide a traceable fingerprint from producer to consumer. The aim of the Bext360 is to enhance traceability and transparency in the coffee supply chain and thus enable sustainability of supply chains in favour of both, producers and consumers.

5.1.1 Issues in the Coffee Supply Chains

Coffee is one of the most traded commodities in the world with the high economic impact and approximately 125 million people earn and live growing coffee (Nasdaq 2017). However, despite the importance of coffee industry, its supply chain remains out-of-date and opaque. Although the demand for coffee increases, small farmers in developing countries, who produce most of the world's coffee supply, are underpaid and often accept delayed payments. Most of those farmers live on less than $2 a day. Furthermore, coffee consumers also demand transparency for sourcing and origin which means there has been a shift in consumer preference and consumers are willing to pay for supply chain transparency in order to ensure the coffee they consume is sourced fairly and ethically (GlobeNewswire 2018) which means there is a demand for the coffee supply chain transparency from producers are consumers sides in order to enable sustainable coffee supply chains.

5.1.2 Blockchain Based Coffee Supply Chain

Bext360 blockchain technology was created to improve the supply chain flow of coffee. The technology allows the coffee to be traced from 'bean to brew' and ensures that the trade payments are fair. Data is collected at each level of the coffee supply chain and then blockchained for an immutable record while quantifying sustainable measures (Bext360 2018). Bext360 uses a distributed ledger technology which is is combined with artificial intelligence that intelligently sorts coffee beans and assesses a grade.

The coffee cherries and beans are evaluated by the bext360 technology and then divided into grades, based on the product. After the beans are sorted, the farmers can access the mobile application to view the number of each grade and accept payment offers. Furthermore, in order to enable payments, the Bext360 platform facilitate the collection of crucial data. The data are collected based on the quality and efficiency of farming practices. Afterwards, the supply chain flow is analysed in real time in order to create new business possibilities and enable supply chain efficiency (Scott 2017). Pay-outs are made in digital currencies using the Stellar protocol.

Bext360 permit the coffee supply chain stakeholders, e.g. farmers, roasters, and consumers, to access data through the whole supply chain flow. The technology provides a high level of transparency because having those data enables to identify the efficiencies in logistics processes. Furthermore, that information allows the opportunity for better compensation to farmers who produces higher quality products and for consumers who have the possibility to verify the origin of the coffee they bought is correct. However, it is unclear how data manipulation of further steps of supply chain is handled. It might happen that consumer don't have full insight into the whole supply chain because the data manipulation or lack of data can influence track quality (Verhoeven et al. 2018). Furthermore, there are some other issues with Bext360 technology. As previously mentioned immutability which ensures the transparency of data rely on a decentralized system of parties who store copies on a distributed ledger. However, the evidence on this case do not explain if the network is actually decentralized. Another issue is infrastructure. The evidence does not explain who will provide and ensure the constant internet connection to farmers mostly located in developing country. Although Bext360 is presented as a novel approach in solving everyday social and ethical problems in supply chain, the idea of tracking and tracing is expensive and not so simple. According to our knowledge, it's not clear yet who is responsible for the infrastructure, who will ensure constant internet connection, protection of robots from vandalism, robbery and other damaging exposures and who is responsible for network security and distributed system security.

5.2 The Case of IBM Food Trust

IBM Food Trust is the collaborative network which gathers growers, processors, wholesalers, distributors, manufacturers, retailers and others with the aim to enhance transparency and responsibility on each stage of the food supply chain. IBM Food Trust connects all the participants through a permissioned, permanent and shared record of food origin information, processing data, shipping information etc. (IBM 2018).

5.2.1 Issues in the Food Supply Chains

In 2017 was published that 20% of the eggs coming from the Netherlands were unsafe. In the same year, in Egyptian school lunches were contaminated. However, foodborne illness is not just a regional but rather the global issue. It's a problem including Netherlands, Egyptian schools, European and U.S. supermarkets etc. Due to contaminated food, every year one out of ten people in the word become ill, and approximately 400.000 people die. Every year around the world, one-in-10 people fall ill—and about 400,000 die (IBM 2017).

5.2.2 Blockchain Based Food Supply Chain

With the blockchain technology, companies included in the supply chain can easily track the food origin. In case it appears, it the food is not safe, the food can be eliminated from the offer. For example, with a blockchain can be traced each carton of eggs from the retailer back to the cage.

The project of IBM and Walmart in China confirmed that blockchain technology can be used to trace products from farm to retail shelf in seconds rather than days. All the participants across the food supply chain, which means from farmers and processors to retailers and regulators have access to a secure blockchain network. For example, a retailer in this food network is able to confirm in seconds if the food is coming from one of the affected farms and save their customers. The blockchain has the potential to massively simplify food traceability. The technology provides participants with a permission-based, shared view of food ecosystem information. The IBM Food Trust technology enables participants to enter and control access to their encrypted blockchain data. Transaction partners can only access the data they have permission to view. Furthermore, access controls secure that the organization which owns the data keep full control over who can access it on the network. A set of modules addressed to food safety, certification management etc. are incorporated by the IBM Food Trust. According to the available online sources, Blockchain Platform and Hyperledger Fabric ensures network and data security and does not require cryptocurrency or any other processor intensive computations. Hence, only the data can control who can see the data on a Hyperledger Fabric network. (IBM 2018).

Finally, a group of leading companies in the global food supply chain, including Dole, Driscoll's, Golden State Foods, Kroger, McCormick and Company, McLane Company, Nestlé, Tyson Foods, Unilever and Walmart, recognized the perspectives of IBM's blockchain and started the collaboration with IBM. The aim of this collaboration is to strengthen consumer confidence in the global food system and bring trust and transparency to food supply chains.

5.3 *The Case of Provenance*

Provenance is a platform which enables more transparent supply chains by tracing the origins and histories of products. Both people and planet are compromised because opaque supply chains present a high risk to businesses. The platform enables brands to bring the origin and stories of products to the point of sale (Blockchain for social impact 2018). Provenance platform have the opportunity to empower whole supply chain. Namely, Provenance helps to build trust between consumers and brands and gives consumers full supply chain transparency in order to build trust between consumers and brands (Provenance 2018a).

5.3.1 Issues in the Apparel Supply Chains

Apparel supply chains are particularly infamous for labour and human rights abuses. Apparel industry employs millions of workers worldwide and factory owners and managers often abuse workers, especially women. It's not a rare case where pregnant workers are fired or cannot take maternity leave. Furthermore, workers are forced to work overtime otherwise they risk losing their job etc. Because of incorrectly permitted factory in Dhaka, Bangladesh, around 1000 workers died, and more than 2000 workers were injured (NY Times 2013).

One of the key problems in apparel industry is lack of transparency. Workers and public do not know which factories are producing for which brands, what are the names, addresses, and other important information about factories manufacturing 'branded' products. Currently, retailers, buyers and consumers do not think much about the "Made in …" labels and who and under which condition cut, stitch, and glue the shoes, shirts, and pants (Human Rights Watch 2018). Global apparel companies that order products manufactured in factories should take measures to prevent human rights abuses throughout their supply chain.

5.3.2 Blockchain Based Apparel Supply Chain

The Provenance blockchain increase transparency in the apparel industry. Namely, in collaboration with Martine Jarlgaard, Provenance has introduced complete blockchain based apparel supply chain. The blockchain enables the brand to provide verified information about the materials, processes and people behind products (Provenance 2018b).

Provenance platform increase the transparency in the apparel supply chain with the aim to enable consumer the traceable and ethnical sourcing (Golden and Price 2018). The row materials can be registered on the Provenance application which enables to track the sustainable alpaca fleece from shearing in the farm, through to spinning, knitting, and finishing in Martine Jarlgaard studio in London. Customers who use Provenance application have a profile accessible with a private key which can be either public or private depending on use case and permissions. The system works on a Public-Private Key Infrastructure and allows for the trusted proof of ownership without revealing the identity of owners to the system. Customers could even use the system to sell a good on a secondary market, allowing the chain to continue post sale throughout the product lifecycle.

New solutions from initiatives like Provenance are setting up blockchain technology for sustainable and ethical sourcing, and consumers, regulators, and other stakeholders have more visibility into the inputs that comprise apparel supply chains.

In all three cases, the mail objective of technology is to enable tracking, tracing, and bring transparency between producers and consumers. The Bext360 technology aims to provide full transparency for procuders to be fairy paid for their work and products and for customers to be aware of the origin of the products they are consuming. Furthermore, millenials nowadays also seek for the products produced in a ethical way. Combination of automated data input is a good approach to create trust with customers (Verhoeven et al. 2018). Lack of transparency in food supply chain and inefficient processes lead to food contamination and customers illness. The IBM Food Trust is trying to solve the problem by using blockchain to track and identify its products. Creating a fully transparent food supply chain allows for quick reaction however, it's not clear who is responsible for supervising the data generation and who can guarantee that all supply chain stakeholders provide accurate, uniform and verifiable data. Provenance blockchain is targeting transparency in the apparel supply chain and it's main goal is to enable consumer the traceable and ethnical sourcing. The system works on a Public-Private Key Infrastructure and Provenance application enables to track the product from shearing in the farm, through to spinning, knitting, and finishing in studio.

Regardless the advantages of the blockchain technology, some questions remain unanswered. The blockchain technology seems as a promising platform to solve social and ethical issues of supply chains and based on the previous case study it supports the sustainability of the supply chains, however, much more needs to be done in the area of data and network security. It is not yet clear who is responsible that all supply chain stakeholders provide accurate, uniform and verifiable data and who will provide the infrastructure necessary for technology to operate.

6 Conclusion

Supply chain applications of blockchain technology should provide a significantly better way of doing business for both producers and consumers. By giving small producers, workers and consumers opportunity for sustainable and socially responsible and ethical environment, the technology could help improve the balance of power on the global market. The technology has the potential to help manufacturers, shippers and other stakeholders to deliver better results for people and the planet. Blockchain might enhance traceability and transparency in the coffee supply chain in favour of producers and consumers. In food supply chain, the participants from farmers and processers to retailers and regulators have access to a secure blockchain network which enables to track and trace the food across the whole supply chain and receive the information if the food is coming from one of the, for example, affected farms and prevent its customers from food poisoning. In case of the apparel industry, one of the major benefits of blockchain technology is prevention of labour and human rights abuses by enhancing the transparency. In conclusion, what is the value of blockchain to supply chains? Blockchain has the potential to enable accuracy and transparency. Furthermore, blockchain offers greater consumer security and trust by improving product visibility and provenance. Finally, through the shared data network, the technology offers real-time feedback and response. Each of these areas separately have the potential to deliver clear business benefits and customer satisfaction. As our findings show, blockchain technology is expected to have a considerable impact on the supply chain sustainability and should be considered value-add for all the stakeholders in the supply chains. This paper provides a conceptual framework to better understand the benefits and challenges of the blockchain technology and its perspectives for sustainable blockchain, however the paper does present the whole picture. Certainly, this study has many limitations, as our data collection was online only, and blockchain technology is at an early stage of adoption and there are many other issues regarding the implementation and function of blockchain which have not been published online. In order to determine the positive impact of blockchain on supply chain sustainability, future research should include more in-depth analysis meaning the online analysis should be broaden and include time span of several year experience of both, producers and consumers. However, both academics and practitioners in companies might find this framework useful, as it outlines important lines of research in the field.

References

Abeyratne SA, Monfared RP (2016) Blockchain ready manufacturing supply chain using distributed ledger. Int J Res Eng. 05(09):1–10. https://doi.org/10.15623/ijret.2016.0509001

Bext360 (2018) Retrived 10 Sept 2018 from: https://www.bext360.com/

Blockchain for social impact (2018) Provenance. Retrived 10 Sept 2018 from: https://www.blockchainforsocialimpact.com/provenance/

Bloomberg (2018). Retrived 11 Sept 2018 from: https://www.bloomberg.com/research/stocks/private/snapshot.asp?privcapId=531637748

Bogart S, Rice K (2015) The Blockchain report: welcome to the internet of value. Needham & Company, October, pp 1–57. https://doi.org/10.15623/ijret.2016.0509001

Buterin V (2014) A next-generation smart contract and decentralized application platform. Etherum, pp 1–36. https://doi.org/10.5663/aps.v1i1.10138

Casado-Vara R, Prieto J, De La Prieta F, Corchado JM (2018) How blockchain improves the supply chain: case study alimentary supply chain. Procedia Comput Sci 134:393–398. https://doi.org/10.1016/j.procs.2018.07.193

Chartered Accountants Australia and New Zealand (2017) The future of blockchain: applications and implications of distributed ledger technology

Circle Economy (2018) Blockchain and the circular economy: an exploration, "the best thing since the internet" or a solution looking for its problem? Could it be both? Retrived 15 Sept 2018 from: https://www.circle-economy.com/blockchain-and-the-circular-economy-an-exploration/#.W6IHHmgzaM8

Consulting DT (2017) Continuous interconnected supply chain: using blockchain and internet-of-things in supply chain traceability, pp 1–24

Cottrill K (2018) The benefits of blockchain: fact or wishful thinking? Supply Chain Manage Rev 22(1):20–25

De Caria R (2017) A digital revolution in international trade? The international legal framework for blockchain technologies, virtual currencies and smart contracts: challenges and opportunities. In: Modernizing international trade law to support innovation and sustainable development. UNCITRAL 50th anniversary congress, pp 1–18. Retrived 10 Sept 2018 from: http://www.uncitral.org/pdf/english/congress/Papers_for_Programme/5-DE_CARIA-A_Digital_Revolution_in_International_Trade.pdf

Drescher D (2017) Blockchain basics: a non-technical introduction in 25 steps. Blockchain basics: a non-technical introduction in 25 Steps. https://doi.org/10.1007/978-1-4842-2604-9

Dujak D, Sajter D (2019) Blockchain applications in supply chain. In: Kawa A, Maryniak A (eds) SMART supply network. ecoproduction (environmental issues in logistics and manufacturing), Springer, Cham. https://doi.org/10.1007/978-3-319-91668-2_2

GlobeNewswire (2018) Bext360 and coda coffee release the world's first blockchain-traced coffee from bean to cup. Retrived 18 Sept 2018 from: https://globenewswire.com/news-release/2018/04/16/1472230/0/en/bext360-and-Coda-Coffee-Release-The-World-s-First-Blockchain-traced-Coffee-from-Bean-to-Cup.html

Golden SLM, Price A (2018) Sustainable supply chains: better global outcomes with blockchain, (January). Retrieved 7 Sept 2018 from: https://newamerica.org/documents/2067/BTA_Supply_Chain_Report_r2.pdf

Government Office for Science (2016) Distributed ledger technology: beyond block chain. A report by the UK Government Chief Scientific Adviser. Retrived 1 Sept 2018 from: https://www.gov.uk/government/uploads/system/uploads/attachment_data/file/492972/gs-16-1-distributed-ledger-technology.pdf

Hackius N, Petersen M (2017) Blockchain in logistics and supply chain: Trick or treat? In: Proceedings of the hamburg international conference of logistics (HICL)-digitalization in supply chain management and logistics. Hamburg, epubli, pp 3–18

Heutger M, Chung G (2018) Blockchain in logistics, pp 1–28. Retrieved 7 Aug 2018 from https://www.logistics.dhl/content/dam/dhl/global/core/documents/pdf/glo-core-blockchain-trend-report.pdf

Human Rights Watch (2018) Soon there won't be much to hide, transparency in the apparel industry. Retrived 20 Sept 2018 from: https://www.hrw.org/world-report/2018/essay/transparency-in-apparel-industry

IBM (2018) IBM food trust: trust and transparency in our food. Retrived 20 Sept 2018 from: https://www.ibm.com/blockchain/solutions/food-trust

IBM (2017) IBM Blockchain: what can we solve together? Retrived 15 Sept 2018 from: https://www.ibm.com/blogs/think/2017/08/blockchain-ibm-food/

Kalinin KP, Berloff NG (2018) Blockchain platform with proof-of-work based on analog Hamiltonian optimisers, pp 1–10. https://doi.org/arXiv:1802.10091v1

Mainelli M, Milne A (2016) The impact and potential of block chain on the securities transaction lifecycle. Swift Institute Working Paper, 2015-007, p 81

Mesarić J, Dujak D (2013) Developing supply chain networks—status and trends. In: Proceedings of international scientific conference business logistics in modern management, faculty of economics in Osijek, Josip Juraj Strossmayer University of Osijek, Osijek, pp 59–71

Nasdaq (2017) Innovation percolates when coffee meets the blockchain. Retrived 15 Sept 2018 from: https://www.nasdaq.com/article/innovation-percolates-when-coffee-meets-the-blockchain-cm774790

Niforos M, Ramchandran V, Rehermann T (2017) Blockchain: opportunities for private enterprises in emerging markets. *Ifc*. Retrieved 25 Aug 2018 from: https://www.ifc.org/wps/wcm/connect/8a338a98-75cd-4771-b94c-5b6db01e2797/IFC-EMCompass-BlockchainReport_WebReady.pdf?MOD=AJPERES

NY Times (2013) Building collapse in Bangladesh leaves scores dead. Retrived 22 Sept 2018 from: https://www.nytimes.com/2013/04/25/world/asia/bangladesh-building-collapse.html

Pilkington M (2016) Blockchain technology: principles and applications. In: Olleros FX, Zhegu M (ed) Research handbook on digital transformations. Edward Elgar Publishing, pp 1–39

Provenance (2018a) Empowering the whole supply chain. Retrived 10 Sept 2018 from: https://www.provenance.org/

Provenance (2018b) Increasing transparency in fashion with blockchain. Retrived 10 Sept 2018 from: https://www.provenance.org/case-studies/martine-jarlgaard

RMA (2018) What is sustainable supply chain management? Retrived 26 Sept 2018 from: https://www.rmagreen.com/rma-blog/what-is-sustainable-supply-chain-management

Schillebeeckx et al (2016) Blockchain and smart contracts. Industry roundtable discussion paper. Retrieved 18 Sept 2018 from: https://www.smu.edu.sg/sites/business.smu.edu.sg/files/business/Strategy_Organisation/BlockChainReport_2016_02_highres.pdf

Scott DM (2017) Bext360 and the world of blockchain traceable coffee. Retrived 15 Sept 2018 from: https://medium.com/@theurbanejournalist/bext360-and-the-world-of-blockchain-traceable-coffee-4ee1d9bba560

Seuring S, Müller M (2008) From a literature review to a conceptual framework for sustainable supply chain management. J Clean Prod 16(15):1699–1710. https://doi.org/10.1016/j.jclepro.2008.04.020

Tian F (2018) An information system for food safety monitoring in supply chains based on HACCP, blockchain and internet of things. Doctoral thesis, WU Vienna University of Economics and Business, Vienna, Austria

Verhoeven P, Sinn F, Herden TT (2018) Examples from blockchain implementations in logistics and supply chain management: exploring the mindful use of a new technology. Logistics 20(2):1–19

Voss C, Tsikriktsis N, Frohlich M (2002) Case research in operations management. Int J Oper Prod Manage 22(2):195–219. https://doi.org/10.1108/01443570210414329

Xu JJ (2016) Are blockchains immune to all malicious attacks? Financ Innovation 25(2):1–9. https://doi.org/10.1186/s40854-016-0046-5

Challenges in Future Data Interchange in Transport and Logistics Sector

Tomasz Debicki and Carlos Guzman

Abstract Data integration in today's supply chain becomes more and more challenging for companies. In the age when technology is not so much obstacle for integration the few others appeared, related to: different systems in use, organization's aspects, standards in use, growing number of players in global supply chains and continues business changes. These article shows how data integration has been changing through years from one hand adopting new technologies and from the other adapting to business requirements. Thanks a wide spectrum of researches, surveys, consultation with IT managers responsible in their organization for data integration and EDI on-boarding projects. Authors were able to conclude on some of near future data integration aspects as well as to assess what solutions would have to be compromised and which bring true values for companies and trading partners. The professional experience of authors brings the true value to the article because it is based on real daily examples of data integration in global supply chains, also an experience in research projects for e-administration and idea of building blocks for partner integrations shows even more ways of where the future of data integration is heading to.

Keywords EDI · EDIFACT · XML · Data interchange · Web-service · Data integration · B2B · Supply chain information flow · Semantic for interoperability

1 Introduction

When we talk about electronic integration almost everyone links this EDI (Electronic Data Interchange) the 40 years old technology of exchanging standardized electronic messages of EDIFACT or X12 via the files from information system of one company

T. Debicki (✉)
Poznan University of Economics, Niepodleglosci 10, 61-875 Poznan, Poland
e-mail: tomasz.debicki@capgemini.com

C. Guzman
Capgemini Research Institute, 400 Broadacres Drive, Bloomfield, NJ 07003, USA
e-mail: carlos.guzman@capgemini.com

A. Kolinski et al. (eds.), *Integration of Information Flow for Greening Supply Chain Management*, Environmental Issues in Logistics and Manufacturing, https://doi.org/10.1007/978-3-030-24355-5_6

to another. Since, then a lot has changed and new technologies for data integration were developed. Conventionally architected ERP systems went as far as they could, resulting in a slow reaction if any changes are required. But let's start from the beginning. By American department of defence an Electronic Data Interchange is an information exchange between computers with use of commonly accepted standards (Hill and Ferguson 1989). The other definition says it is an Exchange of standard formatted messages between information systems (computers) of trade partners with minimal human intervention (Hill and Ferguson 1989). Both definition focuses on standardized messages formats. The standardized body like UN/CEFACT, American National Standards or GS1 from the beginning started to shape different types of messages related to business transaction between trading partners. Since then a lot's of addition sector and branch related standard have appeared like Odette for automotive, Hippa for health and insurance, SWIFT for bank sector, SMDG for Maritime and container flow sector, VICS a voluntary organisation for supporting x12 implementation in US and Canada. In worse situation are cross-industry and cross continental companies like for example logistics service providers (Horzela et al. 2018). They are in a situation where often have to support a lots of different branches, they offer their services across sectors and across borders. It leads to the situation of nature flexibility of logistics services providers, where from one side they adjust to the IT solutions of their customers and from other (mostly for small customers) they impost own solutions. Besides all of these standardized zone we have plenty of non-standardized messages or the standard is limited to very narrow group of companies (use the same software for example). Here the need of data integration has appeared but there was no awareness of existing global standards like EDIFACT, x12, GS1 or UBL or a software producers have implemented different messages into their solutions (Storz 2007). But sooner or later those companies have business relationships with bigger, where standards are met and they start use standardized solutions keeping old way of communications. In the following chapters Authors are going to help users to orient between different standards and solutions for data integration those from the past and those for the future.

2 The Problem of EDI Definition

Based on definitions, mentioned in the introduction EDI is the exchange of standardized information between information systems in automated way. Many people associate EDI with information exchange based on EDIFACT, X12 and their substandards which started in early 80. But that's wrong, based on definitions any data exchange in automated and standardized way is an EDI. Since the beginning of an EDI a lot more new standards organization started to develop their messages, a lot started to happen when an XML has been introduced. In case of XML there is another misunderstanding as it is very often called a message standard itself in relation to information exchange. Which is also not true. XML is tag-based language based on which we can describe surrounding us World. Objects, People and also business transactions or business documents. But in this case XML is just a file format with more complex structure than e.g. EDIFACT files and having much more possibilities as it can incorporate in nature way the common structures which could be the same for many messages types. But still there have to be strictly defined syntax and semantics for standards based on XML files. An examples of EDI message standards based on XML are GS1 XML, UN/CEFACT XML, UBL, e-Freight. Besides those standards provided by standardization bodies, we often have standards which are provided to the market together with IT systems providers like iDocs from SAP, XML files from Oracle Transport Management System and many, many others.

Another thing which has extended enormously since the beginning of first electronic data interchange are the communication channels. Since then and thanks to global access to the internet there has been a huge increase in communication capabilities between computer systems. Working for EDI consultant for big company we agree with business partners between more than 10 different communication protocols. The following chapters will show the most common used communication channels in B2B transactions (Leng and Parlar 2005).

What shouldn't be considered as an EDI? First the emails with the attachments like PDF, Docx and other unstructured files are not EDI. In case of such emails none of the EDI definitions rule is not supported neither is automatic neither is a standard of a business document. Even if emails are sent automatically from computer systems the attachment type makes impossible further automate processing of this file.[1] In EDI we often find an exchange of CSV, files or other types of txt files or any other structured files even XML, which structure is not existing anywhere else. In this case it works like an EDI but the second rule of EDI definition "Standard messages" is not supported, so again should not be treat as EDI. In the further chapters we will again take a closer look on the communication methods and data representation could no longer considered as files only.

In the next chapter the division of branch specific standardization organization can be found. When it comes to the transport and logistics sector and the big players have customers from different branches of industry the logistic service provider (LSP) have to deal with all different types of EDI branch specific standards, where we have different requirements based on the industry and the nature of the goods. What is more there are a number of specific messages related to moving and storage of the goods, so relation Buyer—Seller is less interesting for the LSP, but relations logistics service client (LSC) towards logistics service provider. In modern and global supply chains the LSP takes over a lot of other integration from LSC towards their suppliers and trading partners that's often a case for integrated logistics and 4 PL logistics scenarios.

The following pictures shows two scenarios with EDI implementation in transport and logistics (Fig. 1).

[1]There are some exception of this rule. When emails are sent automatically, attachments are automatically pass to information system, and attachment have defined structure where business information are stored. E.G. PDF forms files, Excel files. Those cases eventually could be treat as an EDI.

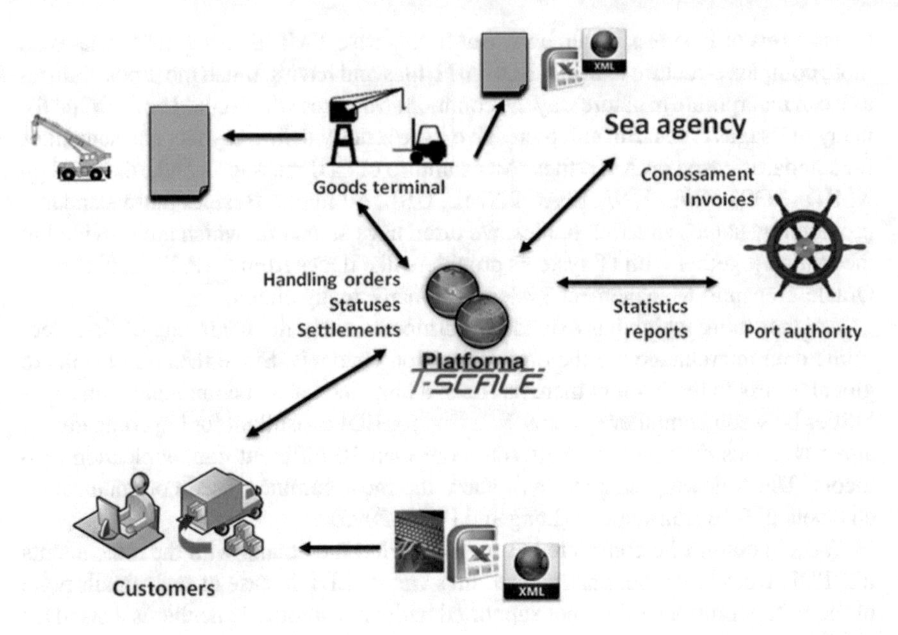

Fig. 1 EDI implementation in port's terminal with usage of integration platform t-scale, ACL (Amber Coast Logistics) project, ILIM 2014

On the above picture we can see implemented EDI messages between port's terminal and integration platform—t-scale. The integration platform ensures the one connectivity and one messages standard for terminal, independently how trading partners will place their orders and communicate with terminal. Terminal's trading partners are connected to integration platform, some of them may use web-EDI[2] if they are not able to handle electronic messages. Within this example handling (loading, unloading, re-loading of trucks, wagons, ships) orders for terminal are exchanged, the statuses of these operations and finally invoices (Fig. 2).

In the above scenario we can see a high level of information flow in 4Pl scenario. In this case LSP is exchanging data with its customer, which are production demands (purchase orders), master data alignment messages, and monitoring of the orders. LSP also exchange data with other partners in this supply chains which are other LSP (operators of cross-docks and warehouses, suppliers to agree details of picking the goods, custom agencies and different types of LSP's handling transports.

One thing is common for both examples without an EDI it would very difficult to maintain such big volume of data which needs to be exchange between business partners.

[2]WEB-EDI—The half automated EDI solutions when one side uses a manual process for providing and receiving the data. In most cases it uses a web page with forms for data provision.

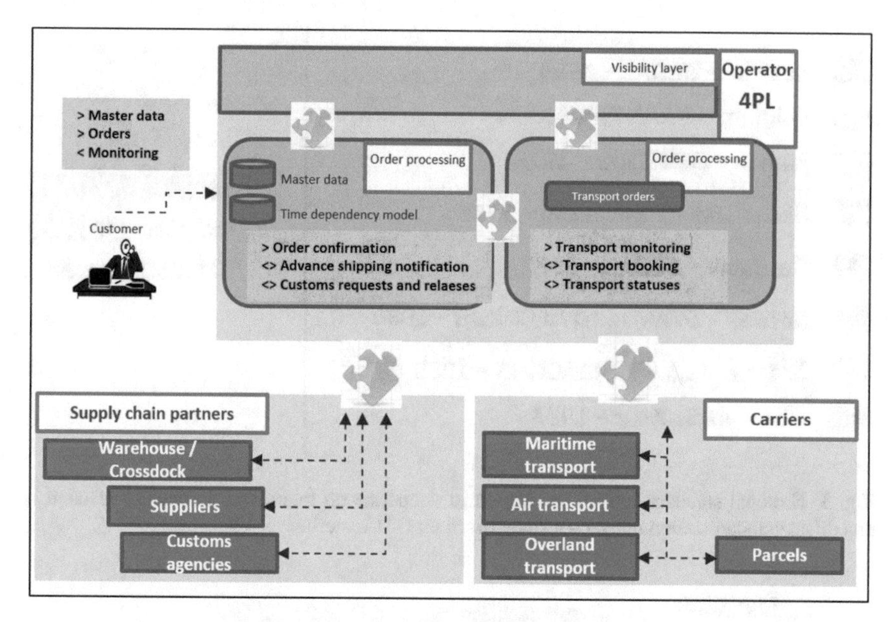

Fig. 2 Information flow in 4PL logistics scenario, own study based on business case of 4PL logistics service provider and its customer from automotive branch

3 Historical Development of EDI

In this chapter we will see a little bit of history and how EDI had been shaping up to what we have today. First when all of this started, it started when it was finally possible when computer networks started to appear so we have to go back to the 70-ties of XX century.

In the early 80s national standardization organisation started to work together on international standard (Fig. 3).

3.1 Standardized Way of Exchanging Messages

EDIFACT defined also the way messages will be exchanged those elements are especially important in VAN's network.[3] In data exchange the following elements are use envelope (exchange), messages (business documents), segments (records, fields, sentences), data elements (words, data, numbers), codes (abbreviation). On the below figure the EDIFACT terminology (Fig. 4).

[3]VAN's network—these are connectivity's and agreements about data interchange between VAN providers, which secures data interchange between a company having an agreement with VAN A and company having agreement with VAN B.

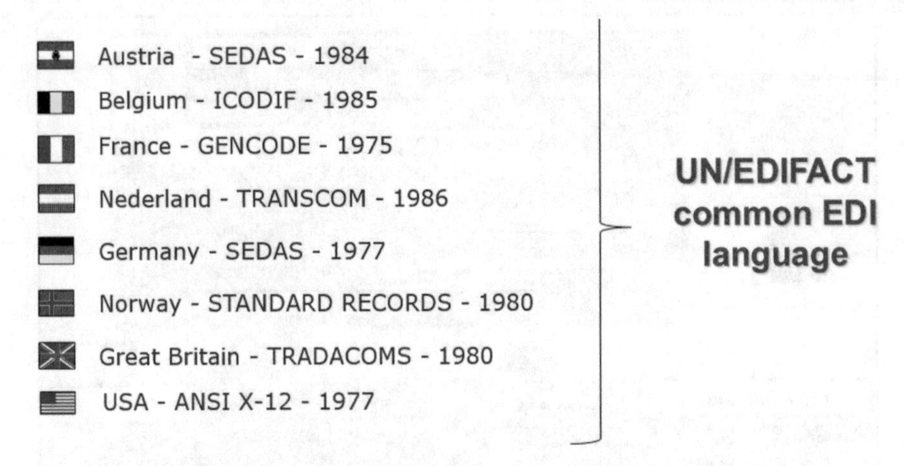

Fig. 3 National standard institutions started to cooperate on international standard resulted in un/edifact standardization body, own study based on GS1 materials

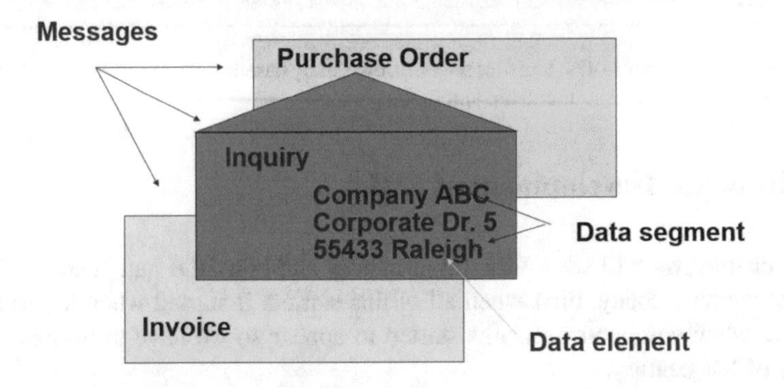

Fig. 4 EDIFACT terminology explanation, own study

VAN providers[4] played a huge role in the first day's of EDI, these providers only were able to connect company between each other in the days when internet was not so common or hardly available. Of course costs of messages exchange were enormous and only big companies could afford it. Since the many companies don't need VAN's providers anymore because of wideband internet access and communication software availability, and also competent IT departments. However VAN's still exists and through all the years they adapted their offer to the market requirements. Globalization and trade partners from around the World results that messages needs translation coming from one system to another it is especially seen in logistics where we often have cross-borders and cross industry technologies. This became one of the

[4]VAN—A value-added network (VAN) is a private, hosted service that provides companies with a secure way to send and share data with its counterparties. Value-added networks were a common way to facilitate electronic data interchange (EDI) between companies.

main services of VAN providers. They also offer WEB EDI for small and medium size companies, where one of the business partner's handled everything manually on website and the other receives and send messages like in full EDI. VAN also offers a full security of transferred data with encryption, originality and no changes during transfer. All of this could be achieved by digital signatures or maybe soon with blockchain (Dujak and Sajter 2019).[5] But anyway some companies prefer VAN's to do the service.

4 Researches Results on Nowadays EDI

In the following chapter there are results of one of the author research during the work at the Institute of Logistics and Warehousing in Poland. The survey's among companies from transport and logistics sectors, will show what is being used today for data interchange in logistics chains. One of them is about a survey carried out in 2014 among polish "T&L" companies. The second research has been made by Eye of Transport journal in 2016 called "Is EDI dead?" (EFT 2016) about what companies use nowadays and the future they think will be a future solution for data interchange.

4.1 Survey Results on Polish Market

The survey has been conducted between polish companies from transport and logistics sector which are participants of GS1. Two most important questions addressed in survey concerned of the type of used electronic messages and standard they use. The most use messages types are transport instructions used almost by 60% of respondents, despatch advice and transport statuses used by the half of respondents. Full coverage shows Fig. 5.

The most important message type in transport and logistics sector is transport instruction, which main function is to order a transport services. For the comparison in another survey for retail networks suppliers an order was most common used message type. Transport Instruction consists of moving goods order to the certain delivery place or places in given conditions and time frame. It concerns one or more goods type. Goods can be stored on logistics units which should be identified by (for example by SSCC[6]). Two other most important messages types are despatch advice and transport statuses. Statuses concerns of delivered goods, transport means or state of executed order. Status information can be exchanged depend on of the

[5]Blockchain—is a transaction database shared by all nodes participating in a system based on the Bitcoin protocol. A full copy of a currency's block chain contains every transaction ever executed in the currency. With this information, one can find out how much value belonged to each address at any point in history.

[6]SSCC—Single Shipping Container Code, Unique identifier for logistic unit In transportation and warehousing.

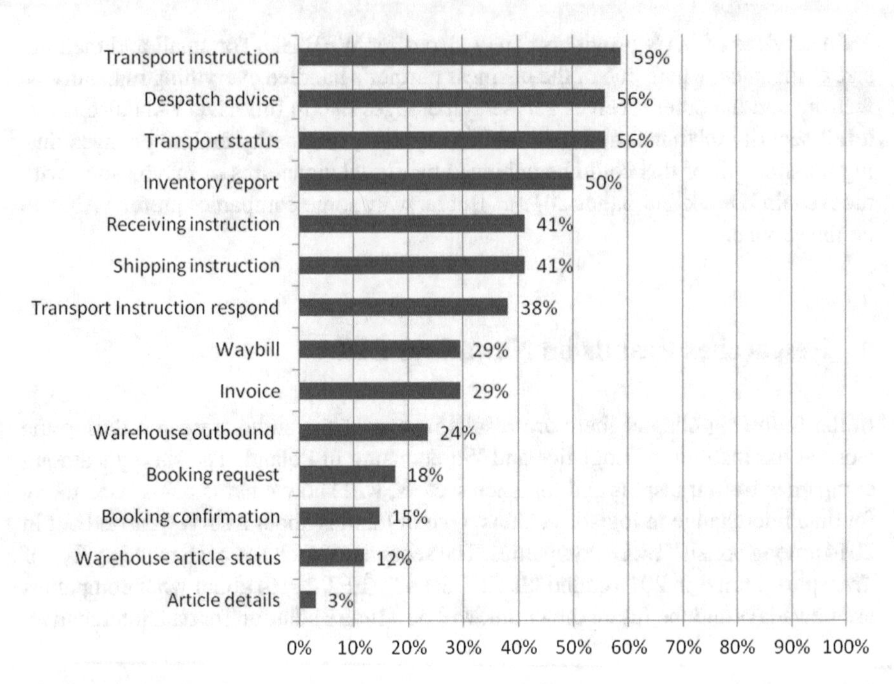

Fig. 5 Distribution of messages types in general survey, ILiM research (Debicki and Halas 2015)

requirements on each stage of transport chain. Despatch advice directly concerns goods is not a typical message for logistics service provider. Its high position is because nowadays LSP services are wider and often LSP uses despatch advice in the name of its customer.

The winner in category mostly used standard is EDIFACT,[7] which is used in 24 of 34 companies from the survey. The second position is standard Fortrass known from German market. It is due a lots of polish companies cooperate with German companies and industry and it has a big influence on polish transport and logistics sector. Figure 6 shows the most used messages standards.

Interesting group among the most often used messages standards is group other. In this group there are all other messages format, which have been declared by only one company. Often these are messages worked-out individually for the current needs of the companies or related with the software used by those companies. These companies should be the most interested of using standards. The following groups

[7]EDIFACT—Electronic Data Interchange For Administration, Commerce and Transport (UN/EDIFACT) international norm for data interchange created by United Nations.

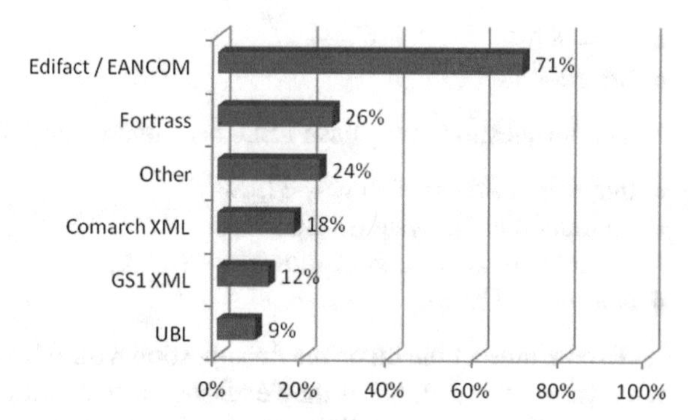

Fig. 6 Distribution of standards in general survey, ILiM research (Debicki and Halas 2015)

use XML[8] format in standards like GS1 XML,[9] UBL[10] or Comarch XML.[11] More often used XML formats are also related to the facts of using Webservices[12] and SOAP[13] protocol for exchanging files via http (Curbera et al. 2002).

4.2 Eye of Transport Survey Results

The Edi in conventional way is passing a messages between companies and nowadays it seems it's like a bad party game where noone wins. Data volumes grew 28 times in 2016. We are in era of machine to machine (M2M) messaging that will transform B2B relationships of last decades extremely developed EPR, MES, WMS or TMS. Enterprise systems will have to be redesigned for new forms of B2B and M2M connectivity. The Cloud technology could also be an enabler. The "Eye for Transport" presented a report "Is EDI dead?" (EFT 2016) shows results of the survey among logistics companies. On question "What alternatives to EDI are you considering?" they received the following answers:

- Web service APIs—52.5%,
- Manually checking websites—5.2%,
- Phone—0.6%,

[8]XML—Extensible Markup Language (XML) is a markup language that defines a set of rules for encoding documents in a format that is both human-readable and machine-readable.

[9]GS1 XML—GS1 set of electronic messages developed using XML, a language designed for information exchange over internet.

[10]UBL—Universal Business Language (UBL) is a library of standard electronic XML business documents such as purchase orders and invoices. UBL was developed by an OASIS.

[11]Comarch XML—set of XML messages standard maintained by Comarch company In Poland company, EDI provider.

[12]Web service—is a service offered by an electronic device to another electronic device, communicating with each other via the World Wide Web.

[13]SOAP—SOAP (Simple Object Access Protocol) is a protocol specification or exchanging structured information in the implementation of web services in computer networks.

- Fax—6.6%,
- Noneb—38.3%.

For the question "Why hasn't EDI been phased out? The answers were as follows:

- Industry is slow to change—31.6%,
- Engrained in the system—30.9%,
- Alternatives have been insufficient—26.5%,
- Not sure—11.0%.

To conclude a little bit on the first question we don't know exactly what they mean with webservice API but if they exchange a standard messages via webservice for me is still EDI it is just a different way of communication channel. Webservices are clear example how the EDI has changed it shape or can change. Companies around big supply chains starts to organize the information flow differently than pass the messages each to each, we can see that different platforms are in use like suppliers platforms where manufactures places orders which are transferred to supplied, confirmed, delivered and invoiced by suppliers. However we must say that platforms are not always best solutions for everyone in supply chain. For example when we are a global supplier and we have to integrate to several or more such platforms and often being charge for this is not an ideal situation.

5 The Future of EDI

One of the most interesting question was "Where do you think the future of data interchange lies?" The results are shown on Fig. 5. From the other questions, how ideal EDI should be? The answers were: reliable, fast, api, flexible, simple, secure, accurate, universal, standards, scalable, efficient. So why hasn't EDI been replaced, yet by other technologies? The answers lies a bit in the answers above but also if you look at the whole industry very big players who have systems up and running with thousands of transactions daily, they deal with hundreds to thousands of suppliers, vendors and each of them have the same EDI transactions operational, others gigantic manufactures who deals with warehouses and JIT logistics runs primarily EDI for their inventories pulled by robotics from automated manufactures. Then there are other 10 thousands regional bakeries, dairys, smaller manufactures linked with EDI to the supermarkets. It's just epic sized, monsterly complex standard. So just see how difficult is a replacement when a millions of EDI connection down if the new system is not reliable and up running from minute 1 (Fig. 7).

If we go back to the EDI definition from the first chapter where have stayed that EDI is an automatic data exchange between computer system with standardized messages, then we may say the shown new technologies with web services, api, mobile are still nothing else than EDI. EDI itself does not tell which standard to use it term itself should not only related with EDIFACT, GS1 or x12 standards. All those technologies are able to exchange data in automated way and they have to

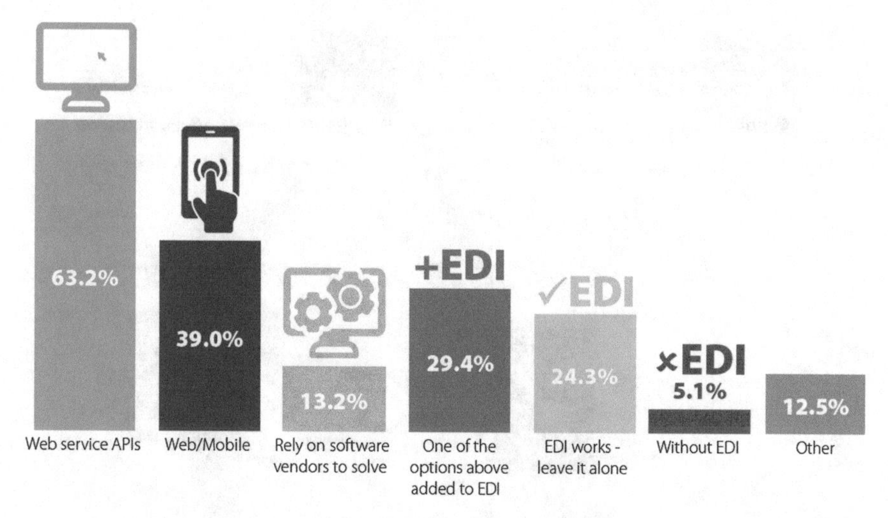

Fig. 7 The future of data interchange, own study based on: (EFT 2016)

use standards to be efficient. GS1 or UN/CEFACT or OASIS have worked on new technology XML for electronic messages for years, those standards are available for better adaptation new connectivity's. So for me personally the answer "One of the options above added to EDI" fits best. The EDI has changed its shape a little and it is using latest technologies.

5.1 Integration Platforms

As we mentioned already in previous chapter platforms are growing however it seems they are most beneficial to the strongest company in the supply chain, others have to adjust. But they are growing and for sure future will bring a lot more platforms. Some EU countries already established e-Invoicing national platforms where all invoices needs to be registered. The very good example of platform which would help with general data integration in supply chains is GS1 approach EPCIS.[14] Where data are shared between authorized supply chain partners (Fig. 8).

[14]EPCIS—Electronic Product Code Information Service—is a GS1 standard that enables trading partners to share information about the physical movement and status of products as they travel throughout the supply chain—from business to business and ultimately to consumers. It helps answer the "what, where, when and why" questions to meet consumer and regulatory demands for accurate and detailed product information.

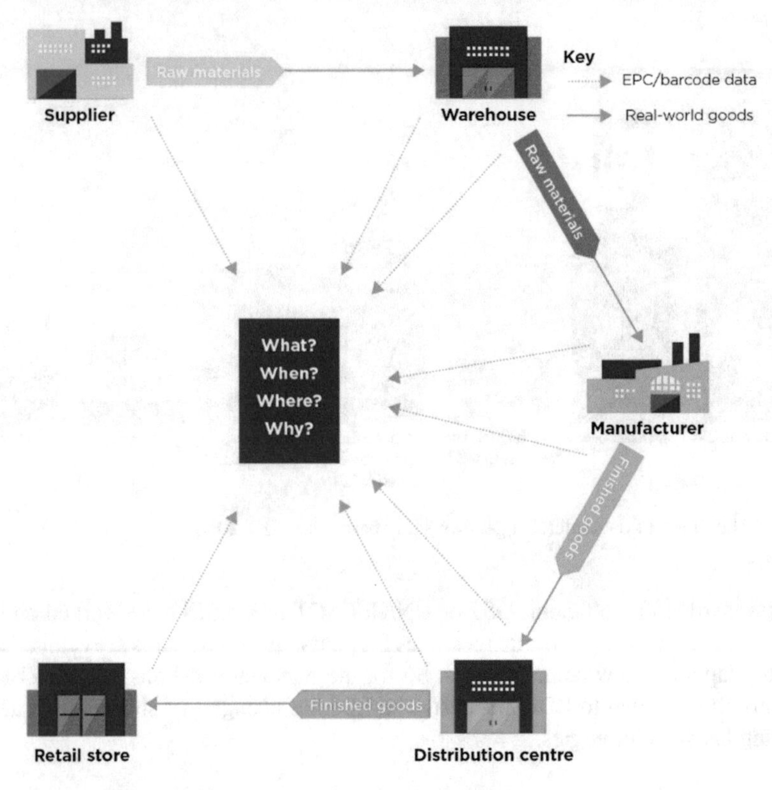

Fig. 8 EPCIS Model by GS1, own study based on GS1 materials

5.2 *Webservice API*

If web service API is just way of how electronic messages are being passed from one computer system to another it is nothing else than EDI. However there are solutions which allows to pass some piece of information which are being a part of data models. The data content then should be based on harmonized semantic models and business rules, attributes, identification keys, codes, etc. Technically those piece of information should be passed as kind of objects from one application to another always being a part of business process.

The growing API available in different systems which are also often part of supply chain management systems, part of LSP interfaces and integration platforms for transport and logistics are still nothing else that data exchange which is under the EDI definition. However although talking about API we have in mind two standards

SOAP[15] for web-services data exchange and JSON[16] for API rest technology. We are still facing before standardize the exchange data. SOAP may contain different payloads which could standardize EDI messages and as far as JSON is concerned this is just a notation which is not a business standard of any branch. If we would compare different API's interfaces of LSP's or integration platforms in transport and logistics sector we would come into conclusion these API's needs harmonization. Otherwise LSC have to build to different interfaces for two different LSP's. In other words JSON is just another data format like XML and needs or soon will need another standardization for business messages standards.

5.3 Future Trends of Integration

In big companies where there is large IT department, which is prepared for handling customers with different technologies and standards, they often intend to build an inhouse information model which could call canonic data model or semantic data model. In this case it means that all other formats received from outside are converted to the in-house format. The important thing is that external formats have comparable data elements which can be matched with internal one. In the logistics companies which are on global market, often services which element of main whole service are ordered at other logistics companies, carriers, rail, air or ocean carriers. From customer point of view for him there is still one service which he wants to monitor. But from the point of view of main logistic service provider there are many services of different companies which have to be integrated in standardized way. This is a common for lead logistics and 4PL providers where they establish such integration and services around so called control towers (Debicki and Halas 2015). Integration of information systems of logistics service providers and applying standardized solutions in electronic messages. Bring the logistics World to the idea of physical internet (Domanski et al. 2018), where the main assumption is optimization of transportation in the existing logistics networks and to offer best of the class services for customer (Osmolski et al. 2019).

[15]SOAP—Simple Object Access Protocol is a messaging protocol specification for exchanging structured information in the implementation of web services in computer networks. Its purpose is to induce extensibility, neutrality and independence.

[16]JSON—JavaScript Object Notation is an open-standard file format that uses human-readable text to transmit data objects consisting of attribute–value pairs and array data types (or any other value). JSON is a language-independent data format. It was derived from JavaScript, but as of 2017 many programming languages include code to generate and parse JSON-format data. The official Internet media type for JSON is application/json. JSON filenames use the extension .json.

6 Further EDI Developments

Today's EDI implementation projects may bring many people from companies to headache, because of different things. Unfortunately average time of such EDI implementation between partners is 3 months, which in today's technology is a lot too long. It seems that EDI in present form of syntaxes will exist for long time still but we will be able to seen much more API oriented integration which are based on semantic data model (Debicki and Kolinski 2018). The ontology of business documents will contain all necessary object which will be exchanged. Messages will be built from different piece of information often origin from different sources. Some of good examples already existing solutions going into this direction are isa2 projects and solutions like core vocabularies, Asset description metadata scheme, core public services vocabulary application profiles. Which can found under European commission websites. Figure below shows a layered approach for semantic interoperability in which the layers reflects the various company layers (Fig. 9).

Different organisation requires different information in their business process. Some organisations can be interested in concerns related to the transport of dangerous cargo, while customs organisations may want to look into details of import/export declarations. Within the logistics domain, several global standardisation bodies for information exchange in logistics supply chains exist (Kawa 2012), including UN/Cefact, the World Customs Organisation (WCO) and GS1. Individual services of different business actors requires a specific information or delivers specific information which are described in business community specific layer where also are the rules for logistics messages standards. Ontology's are also helpful with documents validation specially in customs or everywhere where legal rules are defining some formalities on documents ontology's with automated reasoning may help to validate documents before further processing it.

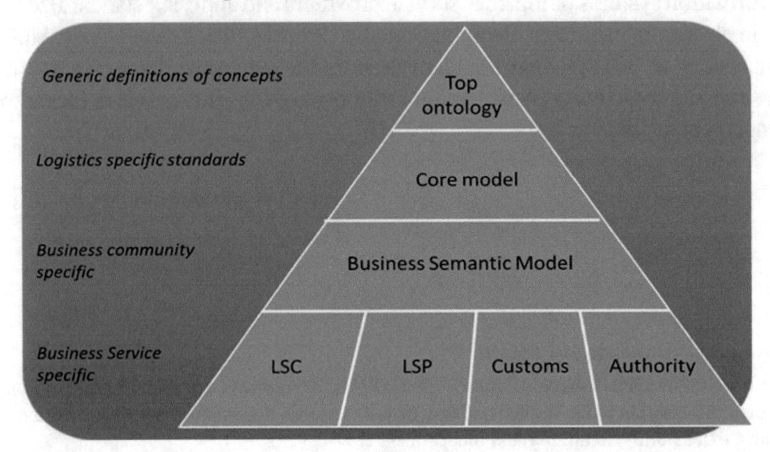

Fig. 9 Layered approach for semantic interoperability, own study based on i-Cargo project approach for logistics interoperability (i-Cargo—Intelligent Cargo in Efficient and Sustainable Global Logistics Operations European funded Project)

References

Curbera F, Duftler M, Khalaf R, Nagy W, Mukhi N, Weerawarana S (2002) Unraveling the Web services web: an introduction to SOAP, WSDL, and UDDI. IEEE Internet Comput 6(2):86–93

Debicki T, Halas E (2015) Electronic messages standards in transport and logistics sector in Poland. Mater Manage Logistics 2:22–26 (in Polish)

Debicki T, Kolinski A (2018) Influence of EDI approach for complexity of information flow in global supply chains. Bus Logistics Mod Manage 18:683–694

Domanski R, Adamczak M, Cyplik P (2018) Physical internet (PI): a systemat-ic literature review. LogForum 14(1):7–19

Dujak D, Sajter D (2019) Blockchain applications in supply chain. In: Kawa A, Maryniak A (eds) SMART supply network, Springer International Publishing AG

EFT (2016) Eye of transport. Is Edi dead?, https://www.eft.com/edi-dead-we-asked-200-supply-chain-executives-their-thoughts

Hill NC, Ferguson DM (1989) Electronic data interchange: a definition and perspective. EDI Forum: J Electronic Data Interchange 1(1):5–12

Horzela A, Kolinski A, Domanski R, Osmolski W (2018) Analysis of use of communication standards on the implementation of distribution processes in fourth party logistics (4pl). Bus Logistics Mod Manage 18:299–315

Kawa A (2012) SMART logistics chain. Asian conference on intelligent information and database systems. Springer, Berlin-Heidelberg, pp 432–438

Leng M, Parlar M (2005) Free shipping and purchasing decisions in B2B transactions: a game-theoretic analysis. IIE Trans 37(12):1119–1128

Osmolski W, Voronina R, Kolinski A (2019) Verification of the possibilities of applying the principles of the physical Internet in economic practice. LogForum 15(1):7–17

Storz C (2007) Compliance with international standards: the EDIFACT and ISO 9000 standards in Japan. Soc Sci Jpn J 10(2):217–241

Information Systems and Technological Solutions Integrating Information Flows in the Supply Chain

Logistics and Supply Chain Intelligence

Alexander Haas

Abstract Seizing digital challenges of contemporary logistics and supply chain management (LSCM) like globalization, ubiquitous data usage and availability or sustainability, a holistic perspective on the integration of Big Data and Business Intelligence (BI) techniques is still at the beginning. With the help of different model types like procedure and reference models, a holistic view using a meta model is created to present a new management approach in LSCM blending data and information aspects with traditional logistics and SCM approaches. Results of an international study executed in Bulgaria, Slovenia and Germany fortify the new management approach and deliver useful insights on practical application. Contributions to literature and research are made through a newly constructed management approach in LSCM and the opening of a new research direction.

Keywords Supply chain management · Logistics · Big data · Business intelligence

1 Introduction

Logistics and supply chain management (LSCM) are widely integrated nowadays and supported by IT systems. Nevertheless, research tackles mostly the increase of efficiency when it comes to view IT and LSCM together (Trkman et al. 2010). Business intelligence as a particular component of IT striving to transform data into decision support is addressed in this paper to get integrated into LSCM. Therefore, a management approach is suggested to merge BI methods and LSCM processes. By doing so, Intelligence systems as an overarching management approach is constructed and converted to LSCM in order to achieve a domain-oriented approach by using structures like the logistical chain and the House of SCM. Prior to that, foundations in LSCM and IT/BI are set followed by the constructional path into Intelligence Systems including results from a trinational study. After finalizing Logistics and Supply Chain Intelligence (LSCI) and its model family, two examples within the

---

A. Haas (✉)
SAP Deutschland SE & Co. KG, Dresden, Germany
e-mail: Alexander.Haas@hhl.de

A. Kolinski et al. (eds.), *Integration of Information Flow for Greening Supply Chain Management*, Environmental Issues in Logistics and Manufacturing, https://doi.org/10.1007/978-3-030-24355-5_7

domain concretization are given. The paper closes with a brief summary and an outlook into further research directions.

2 Theoretical Foundations

Logistics and supply chain management (LSCM) can be seen as two major concepts which are growing together from a content perspective. Orienting on strategy and flows, logistics is defined as "the process of strategically managing the procurement, movement and storage of materials, parts and finished inventory (and the related information flows) through the organization and its marketing channels in such a way that current and future profitability are maximized through the cost-effective fulfillment of orders" (Christopher 2005). Apart from operational aspects consisting of classical transport, transshipment and warehousing, logistics is merging with a management point of view. Therefore the Council of Supply Chain Management Professionals (CSCMP) defines logistics management as "that part of supply chain management [SCM] that plans, implements, and controls the efficient, effective forward and reverses flow and storage of goods, services and related information between the point of origin and the point of consumption in order to meet customers' requirements" (Council of Supply Chain Management Professionals 2018). Logistics can be viewed purely operational which is reflected in the depiction of a logistics chain (see Fig. 3).

In contrast to the operational background of logistics, the establishment of the term Supply Chain Management results from the integration of business processes, business departments and interlinked companies in value and supply chains (Stevens 1989). Although even during the increasing popularity of SCM in 1990s and 2000s, definitions of this holistic concept were hardly consolidated (Mentzer et al. 2001; Burgess et al. 2006). Taking the originally integrating perspective into account, CSCMP (2018) states that SCM integrates "supply and demand management within and across companies" (Council of Supply Chain Management Professionals 2018). Furthermore, this organization delivers a widely accepted definition of SCM which "encompasses the planning and management of all activities involved in sourcing and procurement, conversion, and all logistics management activities" (Council of Supply Chain Management Professionals 2018).

Concluding from the existence of both concepts which act complementary and contrary to certain extends, major effort was put into distinguishing them. Larson and Halldorsson (2004) identified four perspectives how logistics and SCM might be put into relationship. These perspectives are depicted in Fig. 1.

It is now widely adopted in research and practice that the Intersectionist and Unionist perspectives are the leading ones (Larson et al. 2007; Niine and Lend 2013). Following these perspectives, it is evident that logistics as well as SCM need to be taken into account if they should serve as a domain for applying for Intelligence Systems.

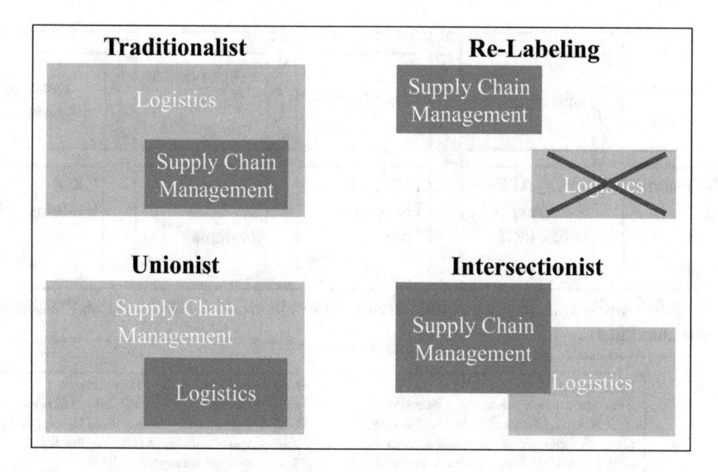

Fig. 1 Relationship between logistcs and SCM in four perspectives, Larson and Halldorsson (2004)

By reviewing this, different approaches come into place to deliver a broken down structure of each concept. Concerning logistics, a logistical process chain is used which divides logistics into a combination of procurement logistics, production logistics, distribution logistics, after sales logistics and reverse logistics (Fig. 3). When exposing SCM with such an approach, the broad and holistic manner of this concept becomes obvious. Therefore, the House of SCM integrates the four perspectives competitiveness, customer service, integration and coordination to display non-operational and supply chain-wide subjects (Stadtler 2010). The House of SCM is shown in Fig. 2.

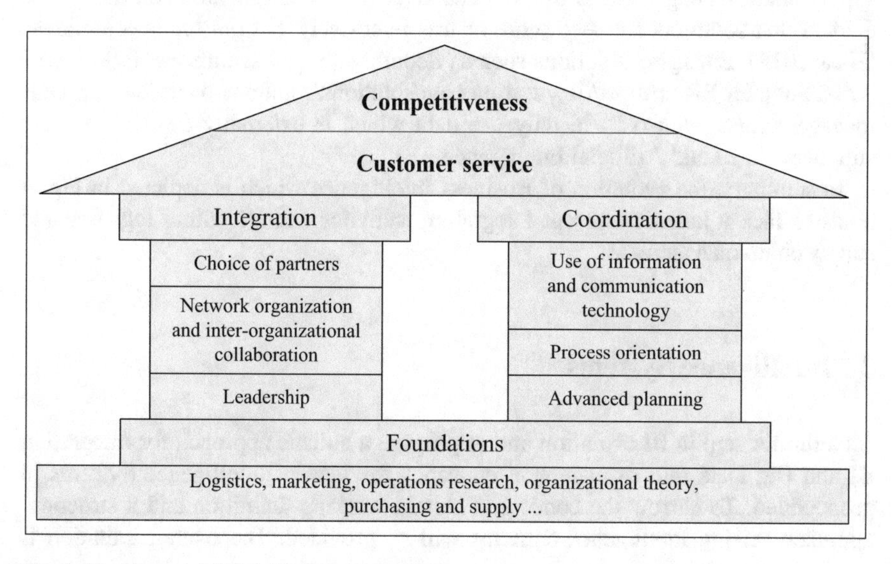

Fig. 2 The House of SCM, (Stadtler 2010)

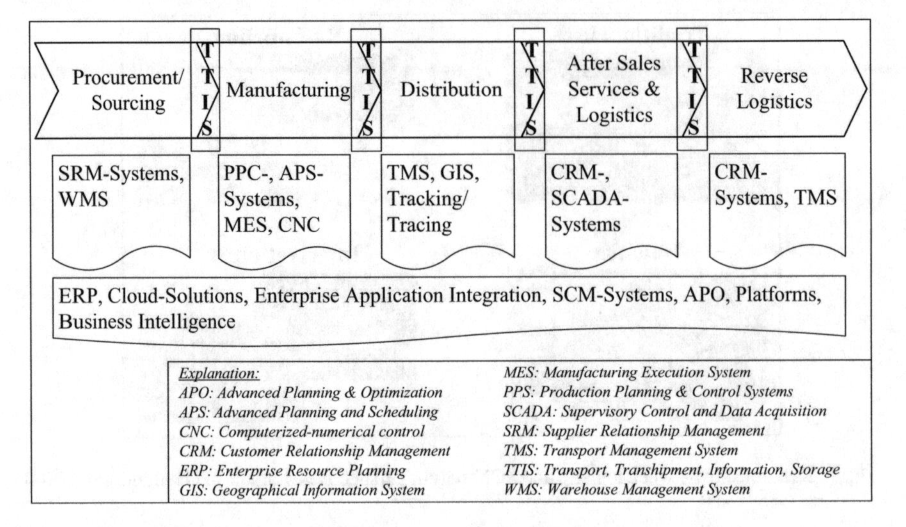

Fig. 3 IT systems in logistics and SCM, own study (Haas 2018)

As Introna (1991) puts in more than 25 years ago, a "logistic system converts material into products that are of value to a customer, and similarly the information system converts data into information that is meaningful to the user" (Introna 1991). The impact and contribution of IT systems is nowadays unquestionable and empirically proven (Koçoğlu et al. 2011; Trkman et al. 2010). An overview of different IT systems supporting logistics and supply chain processes is given in Fig. 3.

The usage of Business Intelligence and analytical instruments in logistics and supply chain management is by now characterized by fragmented solutions. It is evident that solutions are used more or less intensively, but only in specific areas (Haas 2018). Emerging solutions such as Scoutbee (https://scoutbee.com/) or SAP Ariba Supplier Risk (https://www.ariba.com/solutions/solutions-overview/supplier-management/supplier-risk) heavily use data which is externally available to rate suppliers via BI and Artificial Intelligence.

In summary, the evolution of Business Intelligence which is depicted in Fig. 4 tends to lack a holistic concept integrating analytics and BI within logistics and supply chain management.

3 Intelligence Systems

As a further step in BI evolution and to propose a holistic approach for integrating BI and Big Data into management concepts, the idea of Intelligence Systems in propounded. To narrow the concept, a certain working definition and a structural approach seizing Intelligence Systems will be provided. The overall intention is

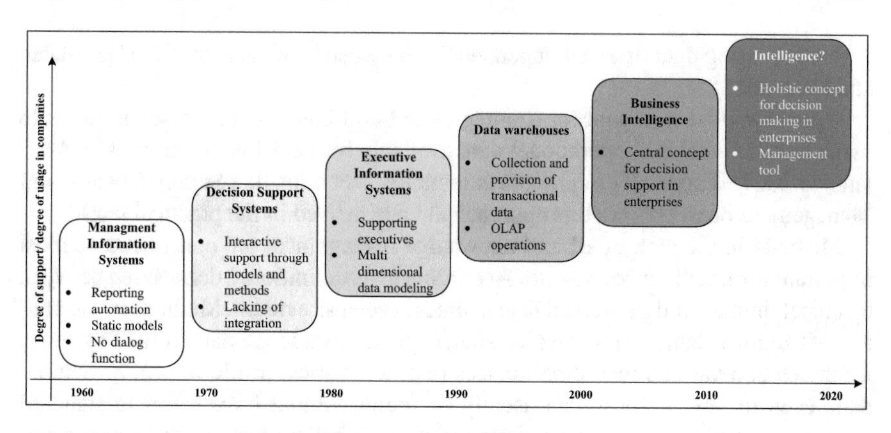

Fig. 4 Evolution path of BI and beyond, own study (Haas 2018)

to establish a Management by Information as a new perspective in management approaches.

Intelligence Systems describe a holistic management approach based on the usage of all given technical, human and peripheral data sources, optimally in real time, to execute process adjustments, process recommendations and contributions to strategic development. Intelligence Systems offer therefore a Management by Information (Haas 2018). To create a systematic concept, a structural approach is given in Fig. 5.

Business process notations are a substantial element to standardize business processes. Business process model & notation (BPMN) depicts the most common notation used in international environments. Hausladen and Haas (2015) developed a model to enhance this view by adding different perspectives such as the Supply Chain Operations Reference (SCOR) model to depict different hierarchical layers, Sankey flows to describe data volume intensities and entity relationship (ER) dia-

Definition	• Holistic management approach (by information) • Usage of all data sources to execute process adjustments, process recommendations and contributions to strategic development				
Subject	• Data/ Information carrier	• Furthermore data/ information sources			
Objective	• Operational: process adjustments	• Tactical: process recommendations	• Strategic: strategic development		
Guidelines	• Process orientation	• Real-time processing	• Usage of all data sources	• Dynamics	• Model orientation
Methods	• Business process notations	• Operational BI • CEP/ SCEM • Real-time BI	• BI solutions • Knowledge management	• Cybernetics • Lifecycle • Maturity • Principal agent theorie	• Model theory • Meta models • Process simulation

Fig. 5 A structural approach for Intelligence Systems, own study (Haas 2018)

grams providing data oriented dependencies between process activities (Hausladen and Haas 2015).

Operational BI is a concept to integrate BI and business processes in order to make transactional and operational data available for decisions in processes (Marjanovic 2007). Especially in process automation scenarios like Supply Chain Event Management or workflows this approach already arrived in the practical world.

Methods in the area of BI and knowledge management are considered as most important for Intelligence Systems. According to its definition, a delineation between technical, human and peripheral data sources is seen as evident. Methods in the context of technical data and information sources primarily address data from IT systems, machines and various recording devices (e.g. telematics, traffic flows). Structured data, even in tabular form, can usually be found without large effort in standard software systems for classic processes such as warehouse or order management. Machine data, which are also largely subject to a structure, can be called up either via machine-specific or standardized interfaces. As automation progresses, these processes are increasingly confronted with big data, meaning that machines or monitoring systems (e.g. driving behavior or temperature) generate large volumes of data.

Recording, processing and interpreting human data and information sources require completely different instruments than the technical extraction of data from tables and machines. Knowledge management is here the upper category of methods. However, differentiation must be made between online and offline data and methods.

Peripheral data and information sources as the last category can be considered to be very heterogeneous, in particular because here the classic corporate boundaries are broken and external data and information needs to be integrated.

Summing this up, all relevant methods are gathered and classified according to the management framework of Labrinidis and Jagadish (2012) in Table 1, whereas an x indicates the application of the particular method and an [x] indicates a possible but less prioritized application.

Maturity models can be viewed either from a lifecycle perspective to achieve a final evolution stage by evolving over the time or from a potential performance perspective, meaning that the maturity stage is assessed and can be enhanced by doing certain improvements (Wendler 2012; Odwazny et al. 2019). The evolutional stages are also the crucial element when delimitating maturity models from lifecycle approaches. A lifecycle shows different conditions during its steps as well as positive and/or negative alterations by shifting from phase to phase, whereas stages in maturity models evolve in a positive way (Hausladen and Haas 2014).

As maturity models and lifecycle models are viewed as eminent methods for Intelligence Systems, a generic and integrated lifecycle and maturity approach is suggested for a further application when it comes to domain concretization. The idea of merging the lifecycle and maturity perspective is based on overcoming the clear delineation of these two approaches. While the lifecycle model accompanies the object to be evaluated over the entire lifecycle, maturity approaches are only dedicated to the deployment phase and strive for optimal functionality of the object. Thus, it is obvious that the maturity model is integrated in the actual operating phase

Table 1 Summary and classification of Intelligence Systems methods

Methods	Framework				
	Data management			Analytics	
	Data acquisition	Data extraction	Data integration	Modeling/ Analysis	Interpretation
Data warehouse		[x]	x		
Data marts			x		
OLAP			x	x	
ETL		x	x		
Operational BI	x	x	x	x	x
Free data research	x				
Process mining	x	[x]			
Special software (e.g. Splunk)	[x]	x	x	x	
Predictive analytics				x	
Informal events	x				
Experience workshops	x	x			
Communities of practice	x	x			
Project briefings	x	x	x		
Expert interviews	x	x			
Best practice cases	x	x	x		
Knowledge broker	x	x	x		
Experience reports	x	x			
Databases	x	x	x		
Services	x		x		
Questionnaires	x				

(continued)

Table 1 (continued)

Methods	Framework				
	Data management			Analytics	
	Data acquisition	Data extraction	Data integration	Modeling/ Analysis	Interpretation
Knowledge mapping			x		
Knowledge scripting and profiling	[x]	[x]	x		
Task environment analysis	x	x	x		
Knowledge use and requirements analysis	x	x	x		
Critical knowledge function analysis			x	x	
Knowledge flow analysis				x	
Wikis	x	x	x		
Social media and social software	x	[x]	x		
Text mining	x	x		x	x
Audio analytics	x	x		[x]	
Video analytics	x	x		[x]	
Sentiment analyses				x	x
Structure analyses			x	x	x

Source Own study (Haas 2018)

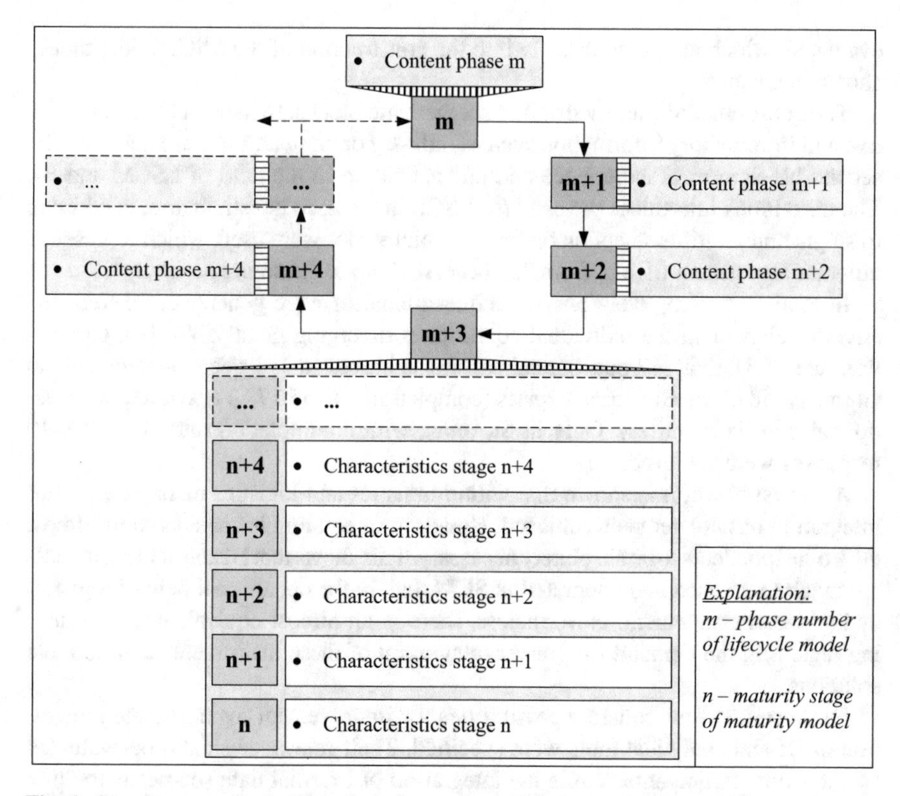

Fig. 6 Generic approach of a lifecycle and maturity model, own study (Haas 2018)

of the lifecycle. This procedure is presented as a meta model in Fig. 6. The upper part of the Figure shows the generic lifecycle model, which remains freely selectable in terms of its configuration (number of phases). In a selectable phase, usually the operating phase (in Fig. 6 Phase m + 3), the maturity model is then activated as an incremental improvement method. The maturity model also has a freely selectable stage configuration in accordance with the design specification.

4 Empirical Analysis

Through the execution of a trinational study in Germany, Croatia and Bulgaria valuable input is generated for constructing the meta model of Logistics and Supply Chain Intelligence. A detailed overview of the conducted study including all descriptive data can be found in (Haas 2018).

In the course of the survey, two goals are to be pursued. On the one hand, the need for a data- or information-driven LSCM in general and according to the LSCI in particular will be examined. On the other hand, ideas and hints are collected and evaluated, which are going directly into the construction of the LSCI model family shown in Chap. 5.

The questionnaire was divided into six thematic blocks. In the first block, descriptive and introductory information such as industry or company size are queried. The second block aims at experience and understanding in the field of LSCM and BI. The third block introduces the need for LSCI. In the used questionnaire, in addition to single and multiple selection options, 5-point scales were used, which represent a unipolar assessment of, e.g. from "no necessity" up to "absolute necessity".

In total, 105 completely answered questionnaires were generated, whereas the answer behavior in the individual countries is diverging greatly. While a completion rate of 31.13% (47 questionnaires) was achieved in 151 questionnaire calls in Germany, 48 of the Bulgarian experts (completion rate 55.17%) answered the questionnaire in their entirety. Only in Slovenia, with a completion rate of 16.4%, 10 responses were achieved.

As a first result, it is shown that within logistics and logistics management, full integration has not yet been achieved. However, integration is considered by almost 60% of respondents to be absolutely necessary. This shows that BI and analytical tools are much more used and integrated in SCM than in the operational field of logistics or logistics management. Nevertheless, there is an almost equivalent understanding regarding the demand for greater integration of these instruments and possible solutions.

Furthermore, four concrete possibilities for improvement regarding the integration of BI and analytical tools were specified. Their relevance should be evaluated by the study participants. While the integration of external data sources is the first suggestion to address the application of Big Data, the further improvement options directly target process changes, recommended actions and strategy recommendations to the operational, tactical and strategic objectives of the Intelligence Systems. Concerning the integration of external data sources, unexpectedly less than half of the respondents in SCM consider this as useful and necessary. In logistics/logistics management, however, almost 60% of respondents agree with this need. The immediate process change is the most popular one in both areas. Strategy recommendations are most demanded in SCM and thus confirm the strategic perspective of supply chain management. Likewise, three survey participants called for a "real-time" mapping of all existing data, and this requirement is already addressed by the real-time or near real-time demands of the Intelligence Systems.

Finally, the relevance of the different models, which play a significant role in the meta model of LSCI, should be assessed. For this purpose, the use of a lifecycle and maturity model, a procedure model and a reference model as well as a fundamental integration of modularization was examined. It has been shown that all four aspects have been perceived as very positive across all national borders.

In order to achieve a successful application, potential success factors of LSCI were compiled. Specifically, the models of LSCI are intended to guarantee economic efficiency in the sense of an effective cost/benefit ratio as well as a concrete target orientation. In addition, the models should meet the quality criteria of universality, clarity and simplicity. The models themselves should be comparable and reusable in

Table 2 Allocation of 4 LSCI success components to 12 LSCI success factors

Success factors	Success components			
Modularity	Process component			
Process orientation				
Nomenclature				
Use of modeling notations				
Multiperspectivity				
Comparability		Objectivity component		
Simplicity				
Clarity				
Economic efficiency			Business component	
Target orientation				
Universality				Usability component
Reusability				

Source Own study (Haas 2018)

order to maintain the claim of standardization. This is supported by the demand for a system with regard to the model design and representation using multiple perspectives as well as a modular nature of the models. Furthermore, strict process thinking in logistics and supply chain management is to be maintained through the focal demand for process orientation and the use of process modeling notations using standard nomenclatures such as BPMN. Table 2 addresses these twelve success factors and their correlation. According to an executed factor analysis (see Appendix 3), four success components are derived.

5 Meta Model of Logistics and Supply Chain Intelligence

After the generic description of Intelligence Systems, a specific application to logistics and SCM will be accomplished. Therefore, a meta model respectively a meta model family is constructed which is furthermore detailed through a domain concretization to logistics and SCM domains.

To reduce complexity, the meta model approach is applied to the three elements of the model family—lifecycle and maturity model, procedure model and reference model. Because meta models are frame-works for designing models themselves, they basically describe the structure, elements, and relationships of the elements, providing a blueprint for the three elements of the model family, which is then applied in the context of domain modularization.

In accordance with the generic approach presented in Sect. 3, the maturity model is switched on in the operational phase 4 of the lifecycle in order to frame the iterative improvement. Based on best practice, five stages of maturity are integrated to progress within the maturity model. The meta model of the lifecycle and maturity model is illustrated in Figure 7. The structure of these maturity stages should also be prescribed in the sense of meta-sizing. Each maturity stage is described qualitatively. This description complies with the claim of a multiperspectivity which is intended to allow different views of a construct. However, the maturity model remains in the generic stages and structures. The three perspectives to be concretized by the LSCI in the meta model of the lifecycle and maturity model are functionality, data and integration. The functional point of view directly addresses LSCM, while from the data perspective relevant aspects of data, information, knowledge and technical aspects such as software are summarized. The integrative view represents the status of the link between the two previous views.

The meta model of the procedure model describes the structure of the to be concretized procedure models of the LSCI. By focusing on development and implementation processes, the procedure model is suitable for the specification of the first three phases of the lifecycle and maturity model and covers together with the reference model and maturity model the lifecycle model completely. The core of the meta model of the procedure model is the description of the three phases of design, modeling and configuration, whereas these phases referring to the requirement of multiperspectivity should incorporate the perspectives functionality, data and integration. The design phase includes targets, objects and strategies. Furthermore, questions concerning the determination of the relevant data or information are

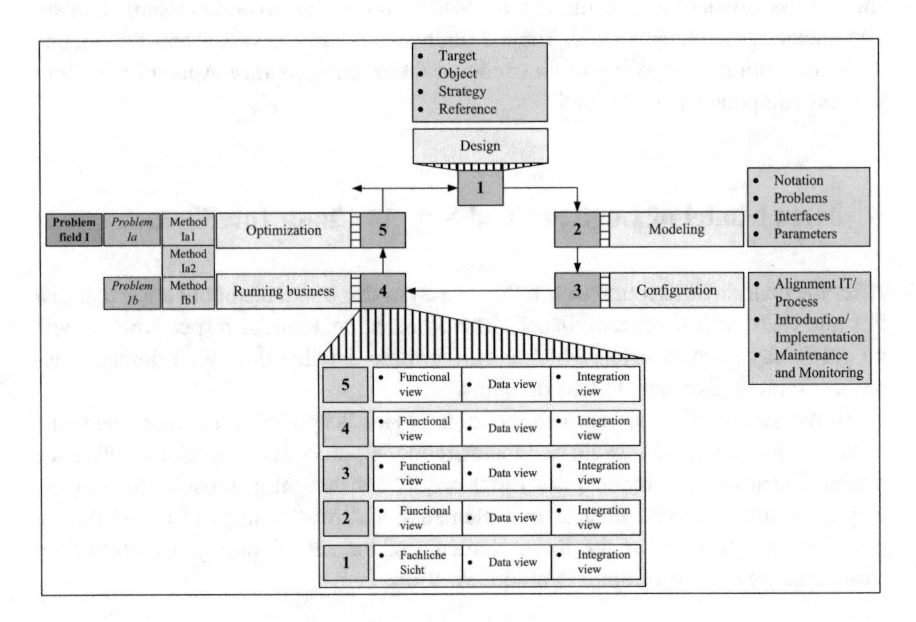

Fig. 7 Meta model family, own study (Haas 2018)

clarified. The modeling phase clarifies all the theoretical fundamentals prior to the introduction and provides a blueprint for the actual use. It should be noted that this is not a business process modeling in the narrow sense, but also content-related aspects are addressed. Essential points of this phase are appropriate modeling as well as the clarification of the notations to be used, fields of action (problem fields) as well as the identification of interfaces to adjacent processes. The definition of the fields of action provides a crucial preparatory work for the downstream reference model, which takes up these and connects them with possible solutions.

The meta model of the reference model describes the structure and composition of the reference model of the LSCI to be concretized in the same way as the procedure model and also follows the already identified requirement criteria of multiperspectivity, variant management and reusability. Similar to the meta model of the lifecycle and maturity model, the views of functionality, data and integration are favored in the sense of multiperspectivity. The requirement of variant management should also be supported, namely in the sense that different fields of action (problem fields) are assigned to the respective reference model in the state of domain modularization, which summarize individual problems and then propose corresponding methods as possible solutions including alternatives as variants. Due to the generic nature of the reference model as well as the different method and solution variants, the claim of reusability should also be met. Due to the large number of different reference process models in the field of LSCM, the specific concretized reference model primarily targets problems within the respective application domain (e.g. procurement in logistics or coordination in SCM) and overcomes these problems by means of corresponding data- and information-driven reference methods.

The idea of domain concretization is utilized to transform the meta models described in the previous Sect. 5 into concrete and detailed models within LSCM. In order to adequately address the application domain of LSCM, both logistics-related and supply-chain-related ele-ments are included. To systematize and standardize the structure as well, a modularization of the domains is pursued by breaking into the areas Logistics Intelligence (LI) and Supply Chain Intelligence (SCI). The functional conceptualization of LI is provided by the logistical process chain illustrated in Fig. 3, which contains main process functions as well as interface functions of logistics management. Although the SCOR model appears to be an obvious conceptualization for the SCI, the processes (source, make, deliver, return) presented in its first level have a strong similarity to the logistical process chain. Therefore, the House of SCM, which emphasizes the strategic component of the SCM with its customer and competition orientation as well as with integration and coordination aspects, proves to be useful. The meta model family is transferred from the meta level to the concrete level by conceptualizing LI and SCI, resulting in many modular constructs that follow a set structure and have concrete content. The final step in the transformation will be the conversation to the company-specific sector. Similar to the SCOR model, the concrete models can be used for the respective application domain. The transformation process from the meta-model family to domain individualization is illustrated in Fig. 8.

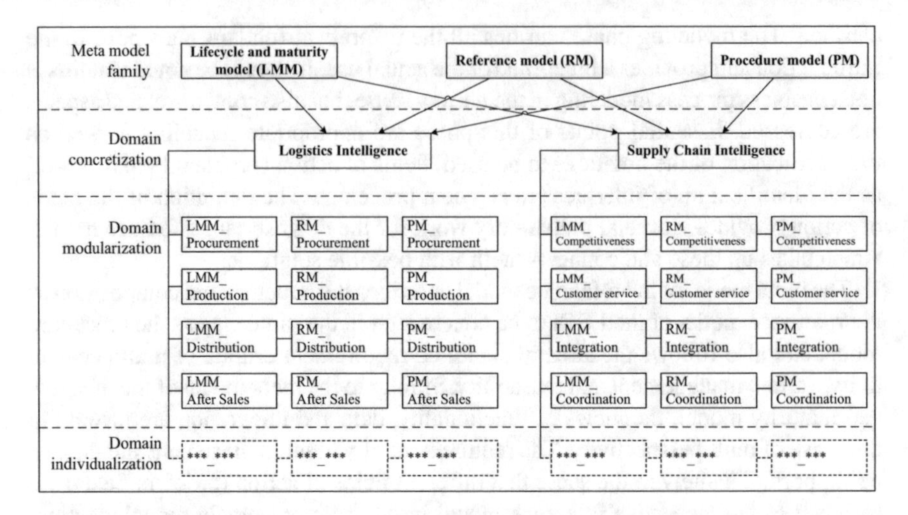

Fig. 8 Transformation from meta model family to modularized LSCI models, own study (Haas 2018)

The result of this process is a set of different models within the domain modularization. The model family for reference and procedure models for the specific domain of customer service within SCI is presented in the following as an example. An excerpt of the procedure model is shown in Table 3. Here, the phase is derived from the lifecycle model and the element incorporating classes and subclasses are objects within the meta model of the procedure model. In the last column the description for domain-specific is integrated, in the example case for customer service.

Furthermore, an excerpt of the reference model for customer service is shown in Table 4. Different problems subsumed within a problem field are encountered with methods and recommendations for aspiring more efficient processes in LSCM.

Several aspects can be given for a contribution of LSCI to green supply chains which are listed below:

- Information technology solutions contribute to sustainability in LCSM. This has been analyzed in general by providing a procedure and an assessment model (Hausladen and Haas 2016) and in particular for emerging IT solutions like Cloud Computing which depict now standard/best practice solutions (Hausladen et al. 2013).
- Kayikci (2018) shows in a case study the huge impact of digitization in sustainability in logistics (Kayikci 2018). LSCI can work here as a guiding principle to incorporate IT in LSCM for sustainability goals.
- Klumpp and Zijm (2019) discuss the influence of Artificial Intelligence for the social aspect of sustainable in LSCM (Klumpp and Zijm 2019). LSCI can support this view by delivering a framework approach from a management perspective on how to incorporate future social aspects in LSCM.

Table 3 SCI-Procedure model of customer service

Phase	Element	Class	Subclass	Customer service
Design	Target	Process changes		Integration of data and information into all operational processes of the supply chain, which have a direct influence on the customer service
		Process recommendations		Integration of data and information in tactical decision support and templates in the context of customer orientation
		Contribution to strategy		Integration of data and information in the strategic planning of the supply chain under aspects of customer service
	Object	Information carrier	Transactional data	Sales data, complaint data, click-through rates, webshop data (length of stay, cancellation rates)
			Analytical data	Payment behavior (defaults, payment terms, financings, credit ratings), buying behavior (competing products, (life) cycles, jumps, potential complaints)
			Unstructured data	Ratings, trends as video and audio materials
			Linked data	Customer and Sales Knowledge, Point-of-Contact Knowledge (Last Mile)

Source Own study

Table 4 SCI-reference model of customer service

Problem field	Problem	Method/Recommendation
Customer focus	Organizational customer focus	• Detection and monitoring of historical, current and planned sales order quantities • Deriving measures for deliveries and locations
	Informational customer focus	• Detection of efficient communication channels and information access channels for specific customer and/or customer groups • Implementation of communication channels and integration into the affected processes (data integration from customer communication)
	Customer communication	• Analysis of customer groups/segments for communication and interaction behavior (proactive vs. reactive behavior) • Derivation and flexible adaptation of the communication and interaction structures
Customer flexibility	Awareness	• Complete integration of all customer information via direct (e.g. complaint e-mails) or indirect (e.g. rating portals or Facebook comments) into CRM and SCM systems
	Integration	• Integration of flexibility initiatives • Anticipation of customer requests and direct transfer into the processes

Source Own study

6 Conclusions and Further Research

To merge the perspective of logistics and supply chain management and IT/BI, Intelligence Systems as a newly constructed management approach with its domain concretization was presented. The meta model of Intelligence systems consists of three different models—lifecycle and maturity model, procedure model and reference model—which are bound together within one construct. As a specific application towards LSCM, the generic meta model family was then specified into a set of models to cover four areas for logistics known as the logistical chain (Procurement, Production, Distribution, After Sales) and SCM (competitiveness, customer service, integration, coordination).

Further research areas emerge mostly in the direction of the application of the model in order to solve practical issues and problems. Practical application is the foundation to measure success and contribution to efficiency increase of LSCI. Potential metrics could be standard ones like process cycle times, costs and quality. Furthermore, more and more methods and recommendations can be empirically gathered and evaluated to provide a higher value to the reference model part of LSCI. Hence, a new study can be executed including more countries.

Appendix 1: KMO Results

KMO and Bartlett's test		
Kaiser-Meyer-Olkin measure of sampling adequacy		0.707
Bartlett's test of sphericity	Approx. Chi-Square	341.207
	df	66
	Sig.	0.000

Appendix 2: Scree Test

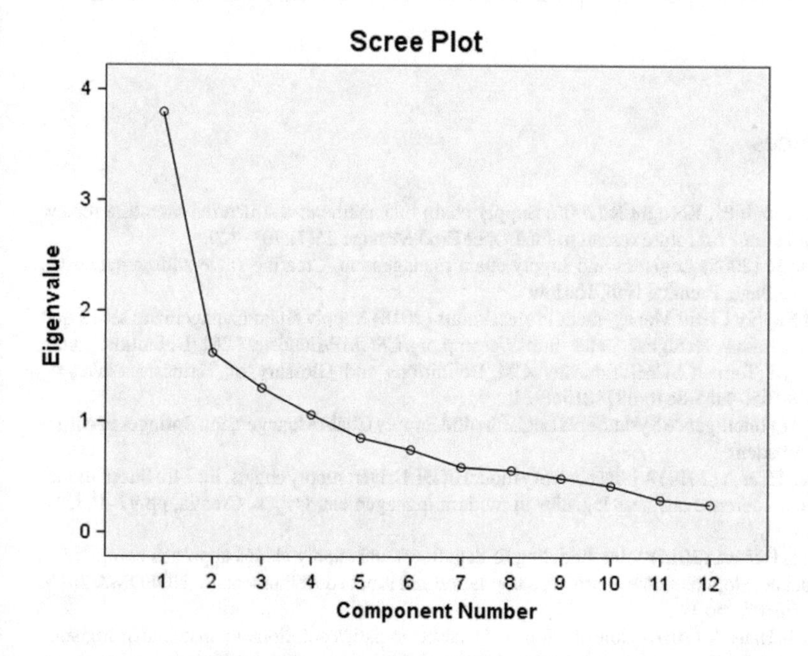

Appendix 3: Loading Results of Factor Analysis

Success factors	Success components			
	1	2	3	4
Modularity	0.752	0.155	0.012	0.170
Process orientation	0.733	−0.149	0.414	−0.156
Nomenclature	0.723	0.195	0.095	0.107
Use of modeling notations	0.666	0.102	0.094	0.016
Multiperspectivity	0.600	−0.128	0.063	0.474
Comparability	0.493	0.413	−0.156	0.254
Simplicity	0.006	0.871	0.165	0.058
Clarity	0.315	0.653	0.301	−0.003
Economic efficiency	0.011	0.250	0.803	−0.008
Target orientation	0.197	0.087	0.781	0.205
Universality	−0.086	0.333	0.026	0.822
Reusability	0.360	−0.170	0.180	0.674

References

Burgess K, Singh PJ, Koroglu R (2006) Supply chain management: a structured literature review and implications for future research. Int J Oper Prod Manage 26(7):703–729

Christopher M (2005) Logistics and supply chain management. Creating value-adding networks, 3th edn. Auflage, Prentice Hall, Harlow

Council of Supply Chain Management Professionals (2018) Supply chain management: terms and glossary. Online verfügbar unter https://cscmp.org/CSCMP/Educate/SCM_Definitions_and_Glossary_of_Terms/CSCMP/Educate/SCM_Definitions_and_Glossary_of_Terms.aspx?hkey=60879588-f65f-4ab5-8c4b-6878815ef921

Haas A (2018) Intelligence Systeme im Logistik- und Supply Chain Management. Springer Fachmedien Wiesbaden

Hausladen I, Haas A (2014) A joint maturity model of BI-Driven supply chains. In: 14th International scientific conference business logistics in modern management, Osijek, Croatia, pp 97–108, 16 Oct 2014

Hausladen I, Haas A (2015) Joint modeling of data flows and supply chains approach compilation and model development, line approach compilation and model development. In: NOFOMA 2015, Molde, Norway, pp 1–17

Hausladen I, Haas A (2016) Contribution of IT-based logistics solutions to sustainable logistics management. In: Kramberger T, Potocan V, Ipavec VM (eds) Sustainable logistics and strategic transportation planning, Bd. 74. Business Science Reference, An Imprint of IGI Global (Advances in logistics, operations, and management science (ALOMS) book series), pp 128–147

Hausladen I, Haas A, Lichtenberg A (2013) Contribution of emerging IT solutions to sustainable logistics and supply chain management–a theoretical framework analysis. In: 10th International conference on logistics & sustainable transport, pp 164–174

Introna LD (1991) the impact of information technology on logistics. Int J Phys Distrib Logistics Manage 21(5):32–37

Kayikci Y (2018) Sustainability impact of digitization in logistics. Procedia Manuf 21:782–789

Klumpp M, Zijm H (2019) logistics innovation and social sustainability: how to prevent an artificial divide in human-computer interaction. J Bus Logistics 29(7):733

Koçoğlu İ, İmamoğlu SZ, İnce H, Keskin H (2011) The effect of supply chain integration on information sharing: enhancing the supply chain performance. Procedia Soc Behav Sci 24:1630–1649

Labrinidis A, Jagadish HV (2012) Challenges and opportunities with big data. Proc VLDB Endowment 5(12):2032–2033

Larson P, Halldorsson A (2004) Logistics versus supply chain management: an international survey. Int J Logistics Res Appl 7(1):17–31

Larson P, Poist RF, Halldórsson Á (2007) Perspectives on logistics vs. SCM: a survey of SCM professionals. J Bus Logistics 28(1):1–24

Marjanovic O (2007) The next stage of operational business intelligence: creating new challenges for business process management. In: 40th Annual Hawaii international conference on system sciences, 2007

Mentzer JT, DeWitt W, Keebler JS, Min S, Nix NW, Smith CD, Zacharia ZG (2001) Defining supply chain management. J Bus Logistics 22(2):1–25

Niine T, Lend E (2013) Logistics management versus supply chain management—the crystallization of debate for academic and practical clarity. In: 10th International conference on logistics & sustainable transport, Celje, Slovenia, pp 209–230

Odwazny F, Wojtkowiak D, Cyplik P, Adamczak M (2019) Concept for measuring organizational maturity supporting sustainable development goals. LogForum 15(2):237–247

Stadtler H (2010) Supply chain management—an overview. In: Stadtler H, Kilger Ch (eds) Supply chain management and advanced planning. Concepts, models, software, and case studies, 4th edn. Springer, Berlin, pp 9–36

Stevens GC (1989) Integrating the supply chain. Int J Phys Distrib Logistics Manage 19(8):3–8

Trkman P, McCormack K, de Oliveira MPV, Ladeira MB (2010) The impact of business analytics on supply chain performance. Decis Support Syst 49(3):318–327

Wendler R (2012) The maturity of maturity model research: a systematic mapping study. Inf Softw Technol 54(12):1317–1339

The Impact of Communication Platforms and Information Exchange Technologies on the Integration of the Intermodal Supply Chain

Johannes Betz, Ewa Jaskolska, Marcin Foltynski and Tomasz Debicki

Abstract Information and communication technologies (ICT) are a very important element of modern supply chains, they facilitate access to information, allow for more efficient processing of information and, consequently, for making more accurate decisions. This chapter is focuses on the available ICT solutions which support intermodal transport operations, such as communication platforms and information exchange technologies. The authors presented current applied technologies, connection types, message standards, as well as future trends and their impact on intermodal supply chains, based on the results of international research work carried out in cooperation between HAFEN HAMBURG and the Institute of Logistics and Warehousing in the NSB Core—North Sea Baltic Connector of Regions project.

Keywords Intermodal supply chain · Information flows · EDI · EDIFACT · API · Data integration · Data exchange · Clouds solution · Mobile application · Logistic platforms · ICT

1 Introduction

Nowadays, it often happens that companies are not interested in big EDI project connecting two partners bilaterally. If companies have hundreds or thousands cooperative partners, customers, suppliers, logistics operators and others. They are look-

J. Betz
HAFEN HAMBURG Marketing e.V, Pickhuben 6, 20457 Hamburg, Germany
e-mail: betz@hafen-hamburg.de

E. Jaskolska · M. Foltynski (✉)
Institute of Logistics and Warehousing, Estkowskiego 6, 61-755 Poznan, Poland
e-mail: marcin.foltynski@ilim.poznan.pl

E. Jaskolska
e-mail: ewa.jaskolska@ilim.poznan.pl

T. Debicki
Independent Consultant GS1 Poland, Estkowskiego 6, 61-755 Poznan, Poland
e-mail: tomasz.debicki@gs1pl.org

A. Kolinski et al. (eds.), *Integration of Information Flow for Greening Supply Chain Management*, Environmental Issues in Logistics and Manufacturing, https://doi.org/10.1007/978-3-030-24355-5_8

ing for alternatives in Electronic Data Interchange. They do not want bother with their EDI, they are looking for one single connections to communicate from their IT landscapes to their cooperatives companies. Such approach can be possible with use of an external EDI platform which ensures one standard connection to all players, that means platform needs to take care about different connectivity's and different messages standards and much more like issuing paper documents or images, manual information inputs and master data synchronization. Good commercial example of such platforms are Transporeon and AXIT operating in Europe mostly in logistics. The methodology of the conducted research takes into account both the literature review presented in this chapter, as well as the study of economic practice carried out in 2017–2018. The conducted research is presented in the form of best practices. A detailed description of best practices can be found in the NSB Core project report (NSB Core 2018). This chapter focuses on the impact of ICT solutions on the efficiency of information flow.

On Fig. 1 present a logical integration architecture for Intermodal Platform being implemented in Polish seaports. All the integration to the Intermodal Platform will be done via access point. The assumption is that all integration between Intermodal Platform are 100% done with e-Freight messages. For some of the message with other actors the connectors with messages translations will be built. Setting up architecture this way, Polish users could be easily connected to the e-Delivery infrastructure where more users are connected via access points. It means that they will able to share their electronic service and connect to other electronic services of other users without additional development on their side as this functionality brings the e-delivery infrastructure.

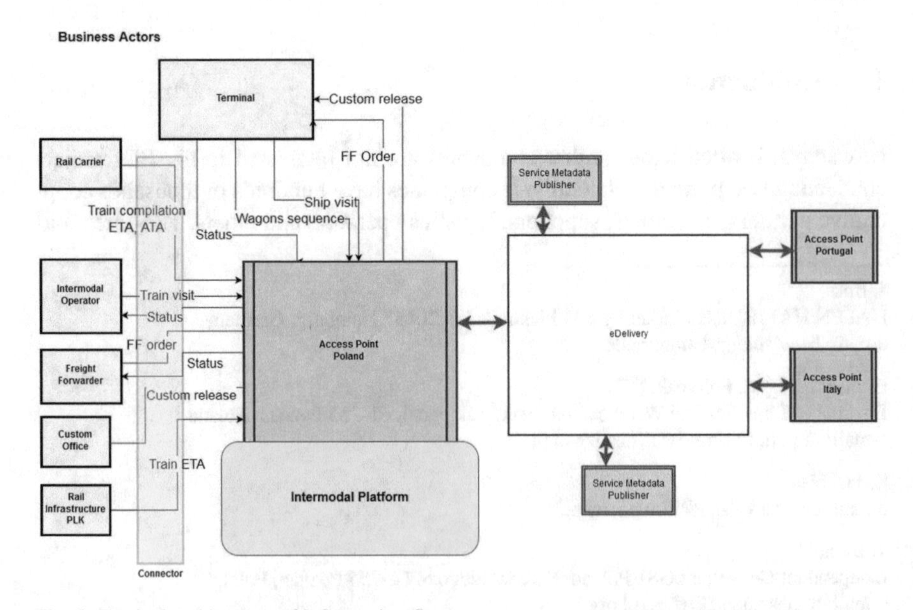

Fig. 1 Logical architecture of information flows

The companies are also implementing more and more API—which calls directly objects in companies databases or to companies semantics data models. With today's technologies for integration more important are well described semantics data models and methods to get or insert data than messages syntax themselves and there are already companies which are changing their approach to integration. World changes and conception of Industry 4.0 requires to exchange more and more information, this trend is specially seen in supply chains. Necessity of data exchange between business partners impact with a lots organizational activities inside companies. This all influence on costs of EDI implementation and maintenance. One of the direction which allows companies with reducing EDI costs are creation of automatic and semi-automatic integration solutions, this approach will let companies to:

- savings because of long term integration projects,
- reduction of human resources involved in testing and go-live process,
- reduction number of many EDI projects executed simultaneously,
- reduction of manual configuration and maintenance many configurations,

The report of EFT (Eye For Transport) organizations—'Is EDI dead—The future of Data Interchange',[1] may prove this direction. It shows that companies are investing money in integration via API and WebServices, mobile solutions, outsource everything to EDI specialized companies or EDI platforms, Only 25% of respondents are happy with current situation. The example of such approach in supply chain is Kuehne + Nagel Integrated Logistics where in global supply chain, 4PL service provider integrates with ten's, hundred's subcontractors, suppliers and consignees.[2] Application of this approach could adapted not only to logistics but all other branches where integration requirements are growing.

1.1 Data Interchange—Technology

To assess the interoperability possibilities of ICT tools two factors are important. First is a connectivity the established communication protocol like a pipe connecting two sides of the transactions and second a message (document) standard a syntax of the message within a content (interoperable data) are transferred. So more less like an order of the people in wagons which are transported on above mentioned pipe. There is one more thing in this context a wagon type which is an envelope where data are wrapped and send as a whole. Thanks to the this envelope which contains a metadata and basic information about sender, receiver and document type this message can be properly routed to via intermediate systems and information hubs to the receiver. Two most common used envelopes are SBDH—Single Business Document Header(UN/CEFACT) and eBXML SOAP Envelope Extension(OASIS).

[1] http://events.eft.com/cio/pdf/InfographicAlt.pdf.

[2] Dębicki T., Logistyka zintegrowana—moda czy konieczność?, Gospodarka Magazynowa i Logistyka 5/2017.

1.2 Connectivity Types

FTP, SFTP—Technologies based on File Transfer Protocol (He et al. 2015) which is a standard network protocol used for the transfer of computer files between a client and server on a computer network. SFTP is a secured with data encryption with use of encryption technology SSL/TLS (FTPS) or SSH (SFTP).

OFTP—Odette File Transfer Protocol is an FTP protocol used by standardization body of EDIFACT messages in automotive industry (Engel et al. 2012). As it all started in the beginning of EDI and EDIFACT, the protocol was designed to point or in-directly via VAN (Value Added Networks) to secure file transfer of business documents over the internet, ISDN and X.25 networks. Till today two specifications of OFTP exists OFTP 1 and OFTP 2. OFTP 2 can work in a push or pull mode, as opposed to AS2, which can only work in a push mode. OFTP 2 can encrypt and digitally sign message data, request signed receipts and also offers high levels of data compression. All of these services are available when using OFTP 2 over TCP/IP, X.25/ISDN or native X.25. When used over a TCP/IP network such as the Internet, additional session level security is available by using OFTP 2 over TLS (Transport Layer Security).

HTTP, HTTPS—these are web browsers protocols to transfer hyper-text (HTML) files over internet (Kohout et al. 2018). It can also be used for exchanging business documents which can be exchange via get, put and post methods. The encrypted version is HTPS which use TSL (Transport Layer Security) or SSL(Secure Socket Layer) security standards. Mostly used in latest system for data interchange via web-services or API via http.

X400—is a message handling service used widely for emails. But it was applied as an EDI protocol specially in aviation and military (Sherif 2016).

SOAP—Simple Object Acces Protocol, for exchange structured information about objects used for web-services implementation (Curbera et al. 2002). It uses xml information set for message format and uses application protocol http, SMPT for message negotiation and transmission.

RFC—Remote Function Call, is the standard SAP interface for communication between SAP systems. The RFC calls a function to be executed in a remote system. Remote function calls may be associated with SAP software and ABAP programming and provide a way for an external program (written in languages such as PHP, ASP, Java, or C, C++) to use data returned from the server. Data transactions are not limited to getting data from the server, but can insert data into server records as well. SAP can act as the Client or Server in an RFC call.

AS2—Is a specification about how to transport data securely and reliably over the Internet. Security is achieved by using digital certificates and encryption. Very popular and preferable always uses http and https.

AS4—Majority of the AS4 are based upon the functional requirements of the AS2 specification. By scaling back ebMS 3.0 by using AS2 as a blueprint. It is a Conformance Profile of the OASISebMS 3.0 specification, and represents an open standard for the secure and payload-agnostic exchange of Business-to-business documents using Web services (Hodge 2013). Secure document exchange is governed by aspects of WS-Security, including XML Encryption and XML Digital Signatures. Payload agnosticism refers to the document type (e.g. purchase order, invoice, etc.) not being tied to any defined SOAP action or operation.

Web-Services—A web service is a software system designed to support interoperable machine-to-machine interaction over a network. In a Web service, Web technology such as HTTP, originally designed for human-to-machine communication, is utilized for machine-to-machine communication, more specifically for transferring machine readable file formats such as XML and JSON. In practice, the web service typically provides an object-oriented web-based interface to a database server, utilized for example by another web server, or by a mobile application, that provides a user interface to the end user.

1.3 Messages Standards

EDIFACT—(Electronic Data Interchange for Administration, Commerce and Transport) by UN. Is the international standard for EDI and one of the oldest that's is widely used all over the world. It has lots of sub-standards created by other standardizing bodies (Storz 2007), national and branch organizations. It often happens that companies interpret EDIFACT by themselves and share its specification all over their customers and business partners. Example of EDIFACT subs standards: GS1/EANCOM standard of GS1 organisation, SMDG—maritime and containerized transportation, ODETTE—automotive and many others.

GS1/XML—a set of modern and well prepared XML standards by GS1 organizations. a GS1 set of electronic messages developed using XML (Herfurth and Weiß 2010), a language designed for information exchange over internet. GS1 XML is based on UN/CEFACT Core Component Technical Specification (CCTS) and UN/CEFACT Modelling Methodology (UMM).

UBL—Universal Business Language, open library of standard electronic XML business documents for procurement and transportation such as purchase orders, invoices, transport logistics and waybills. UBL was developed by an OASIS Technical Committee with participation from a variety of industry data standards organizations (. UBL is designed to plug directly into existing business, legal, auditing, and records management practices. It is designed to eliminate the re-keying of data in existing fax- and paper-based business correspondence and provide an entry point into electronic commerce for small and medium-sized businesses.

e-Freight—a set of standards for transport and logistics based on UBL specifications and approach. e-Freight framework offers a level of standardisation that is different from the type of standardisation currently offered such that it will be possible for small, medium and large sized enterprises to implement it and connect to and/or be part of efficient multimodal logistics networks (Osmolski et al. 2018). The e-Freight framework has been adopted as part of the ISO/IEC DIS 19845 Standard

Ansi.X12—Probably as mature as EDIFACT widely used in North America. American National Standards Institute (ANSI) in 1979, it develops and maintains the X12 Electronic data interchange (EDI) and Context Inspired Component Architecture (CICA) standards along with XML schemas which drive business processes globally (Zabinski 2017). The membership of ASC X12 includes technologists and business process experts, encompassing health care, insurance, transportation, finance, government, supply chain and other industries. It has a different branch version for Finance sector, Transportation, Supply chain, Insurance etc.

iDOC SAP—Intermediate Document, is a SAP document format for business transaction data transfers. Non SAP-systems can use IDocs as the standard interface (computing) for data transfer (Saha et al. 2016). IDoc is similar to XML in purpose, but differs in syntax. Both serve the purpose of data exchange and automation in computer systems, but the iDoc-Technology takes a different approach.

VDA—For automotive industry in Germany. Initially used for CAD systems data exchange format for the transfer of surface models from one CAD system to another. Its name is an abbreviation of "Verband der Automobilindustrie—Flächenschnittstelle", which translates to the "automotive industry association—surface data interface". Standard was specified by the German organization VDA.

CSV—a simple file where data elements are separated with an agreed separator mostly comma or semicolon.

2 Current Trends in the Development of ICT Solutions

Identified participants in the supply chain perform specific roles depending on which they have a different range of activities, and also their expectations towards software. Each of them will use the IT system best suited to the tasks performed by the company. Therefore, IT systems of various classes are used in the supply chain (Kolinski and Jaskolska 2018): related to customer service, container reloading, transport ordering, tracking of deliveries, etc. Therefore, a wide range of ICT tools is available, adapted to the functionality required by the user related to the position in the logistics process. However, when analysing the current trends related to the development of ICT tools supporting intermodal transport and the results of surveys, 6 significant software identifiers common to all participants in the supply chain were identified, they are intensively developed by solution providers in this area. Details are provided in the following sections.

2.1 Electronic Data Exchange

The efficiency and reliability of logistics processes of supply, production and distribution to a large stage depend on the speed and efficiency of information processing, which nowadays is determined primarily by the possibilities of modern computer technology. Every day, business partners in the supply chain exchange innumerable information, which until recently were sent in paper form, currently progressing digitalization significantly influenced the way of data exchange, i.e. orders, orders, invoices, demand data. Efficient exchange of information between supply chain partners is currently in line with the expectations of the participants of the supply chain, the base of solutions provided by the developers of IT tools. At the same time, a large variety of programs to handle processes enforces the introduction of effective communication and data exchange at all stages of the logistics process. That is why it is important to use advanced IT technologies operating in a standard electronic data exchange environment.

2.2 *Integration*

Integration is the next step in the exchange of data between logistic partners. During the flow of goods in the supply chain information is very often modified. For this reason, the IT systems of individual business partners supervising the flow of information and goods must guarantee the integrity of all of its links. Integration enables data exchange between partners in real time, which affects the optimization of the entire supply chain. Integration of internal processes in enterprises is now becoming a standard, many companies use ERP or SCM systems, which is the starting point for consolidation within the whole supply chain. Therefore, in line with current company trends, there is interest not only in the internal integration of processes, but integration between partners throughout the supply chain is becoming more and more important. Thanks to the integration of logistic processes, it is possible to achieve synergy effect, for example by mutual adjustment of processes related to the container movement between participants of the intermodal supply chain, achieving a higher level of customer service, higher and even use of resources, while optimizing the time needed to perform activities. As a result of joint planning of processes and scheduling the use of resources in the supply chain, it is possible to reduce the time of logistic processes and the response time needed in the situation of variable demand. Currently, many ICT tools are standardized for integration between business partner systems. Way of communication, changes very dynamically, for example integration by EDI between two partners (Debicki and Kolinski 2018), is still popular, but it gives way to other solutions, like communication platforms. Because if companies have a lot of cooperative partners (customers, suppliers, LSP) integration with each will be time-consuming and expensive, so better solution is to use an external EDI platform, which provide one connection standard to all players.

2.3 Clouds Solutions

Cloud solutions include alike disks in the cloud for storing files, but more and more often solutions based on cloud computing. Cloud Computing is a business model whose main advantage is the provision of IT services and solutions via the Internet. It is a very popular form of IT outsourcing and its main advantage is that there is no need to purchase own infrastructure and software. Data processing in the cloud allows to rent all these resources and make them available to end users. Thanks to the cloud, every user can use both e-mail and display documents, settle accounts with contractors, make backup copies and sell their products/services from any place and at any time. The advantage of solutions in the cloud is its flexibility and security. It is a set of IT tools that allows user to increase work efficiency while significantly reducing costs. Data processing in the cloud also means increasing data security, full availability, regardless of where the user is. All user need is access to the network and computer. The provider of this solution ensures full support to all system users, quick implementation and flexibility in adapting to the growing number of users of the solution. The advantage of solutions in the cloud is access to all online applications using any web browser. The most common applications of cloud solutions should be mentioned: website hosting, data security services—backup (backup), CRM, e-commerce services, bookkeeping, settlements with contractors and many others. The offer of many companies such as SAP, Oracle, Inttra or Comarch, offering standard solutions implemented on the client's server, are also available in the Software as a Service model (cloud computing). The products available in the cloud include software supporting logistics processes (ERP, load calculators, electronic transaction platforms) as well as financial and accounting services.

2.4 Complexity

The contemporary market is becoming more and more demanding. The growing expectations of consumers and strong competition are the factors that encourage entrepreneurs to look for new technological solutions that offer fully comprehensive solutions. Between IT systems supporting management of organizations, especially among integrated systems (e.g. ERP and SCM classes), are dominating standard software packages. They usually have built-in adjustment mechanisms that allow the system to be adapted to the special needs of a specific customer. However, offered solutions always require personalization, in each company, activities performed during the implementation of tasks are customizing to the individual requirements of the company and process users. Recent, Increasingly, in addition to the standard system configuration for existing processes, entrepreneurs require that specialized systems support the development of the company through easy integration with co-operators, enable work in the cloud, provide the application on the phone. Therefore, IT solution providers develop their products in such a way that the customer, when he's buying a chosen solution, he receives software that fully supports all internal and external processes. As a result, many tools that function independently, such as Load Calculator offering simple functions, become one of the functions of a larger system such as SeaRates.

2.5 *User Friendly*

User-friendly describes a hardware software interface that is easy to use. It is "friendly" to the user, meaning and it's not difficult to learn or understand. A user-friendly interface is not overly complex, but instead is straightforward, providing quick access to common features or commands. While "user-friendly" is a subjective term, the following are several common criteria found in user-friendly interfaces.

- Simple to install—Installation is the first point of contact for users, so it should be a friendly process. It doesn't matter whether it's an operating system or a single-client user application, the installation should be simple and well documented.
- Easy to update—As with the installation, an application's update process should be easy. Updates need to be simple enough to ensure that users continue to benefit from the hard work of the creators of the software. When users don't update, thus exposing issues, the software becomes less and less reliable and secure (as well as missing out on new features).
- Intuitive—In order to be user-friendly, an interface must be make sense to the average user and should require minimal explanation for how to use it.
- Efficient—not only should a piece of software work as expected, it should also be efficient. It should be optimized for specific architecture, it should have all memory leaks plugged, and it should work seamlessly with underlying structures and subsystems. From the users' point of view, the software should be an efficient means to completing their jobs. Software should not get in the way of completing a task, nor should it set up any roadblocks for users. The efficiency of a piece of software is tied up with its intuitiveness.
- Clean—a good user interface is well-organized, making it easy to locate different tools and options.
- Reliable—an unreliable product is not user-friendly, since it will cause undue frustration for the user. A user-friendly product is reliable and does not malfunction or crash.

The goal of a user-friendly product is to provide a good user experience. This may look different depending on the end user for whom the product is designed. For example, a user-friendly shipper's software will have a different interface than a terminal operator. However, the rules above apply to both types of software. Even if a program has many advanced features, it is still possible to make it user-friendly by designing a simple, clean, and intuitive interface.

2.6 *Mobile Application*

Due to the constantly developing mobile technology, which enables quick and easy access to information or tools, an increasing number of IT tool providers also offer a mobile version of software. Mobile applications support the software regardless of its basic functionality, it can be an ERP system as well as a SAP system, or a Freight Exchange, such as TimoCom or Trans.eu. Mobile devices are everywhere, making supercomputing accessible anytime—often as cloud services and applications—and making high levels of security a standard in this mobile environment. While this mobile trend offers tremendous opportunities, it also increases the pressure on IT providers improve their products. A lot of companies, are extending a subset of existing applications to mobile devices today. In many cases, these are micro apps that offer a subset of the features found within PC applications. Examples of micro apps include approvals, expense reporting, and time tracking.

3 Summary

Selected and characterized technologies of ICT supporting the management of the intermodal supply chain are examples of currently used tools, the selection and implementation of which depends on the range of business and the level of cooperation with other participants in the logistics process. It's important that individual links in the supply chain are characterized by a high level of integration. This significantly influences the application possibilities of the chosen technology and facilitating contacts with business partners and clients. The development of new distribution channels and the creation of products along with the development of ICT technologies are becoming the driving force for creating more and more effective innovative solutions, thus determining comprehensive approaches to supply chain management. The application of the technologies discussed may have a significant impact on the functioning of the entire supply chain by streamlining its operation, which allows obtaining benefits in the form of achieving synergy effects.

Increasing number of ICT tools providers, are focus on offering, complex solutions, that allow efficient data exchange, and process integration, not only inside the enterprise, but also between members of the supply chain. Actual many solutions are offered in the cloud, or with using electronic data exchange platforms. It's also important, that the software, need to be customize to user needs, and use of it, should be user friendly.

Today, we are all witnessing a fourth industrial revolution, for which the most characteristic feature is the disappearance of the barrier between people and machines and the use of the Internet of Things and cloud computing. Industry 4.0 is about integrating systems and creating networks and integrating people with digitally controlled machines that use the internet and information technologies—unifying the world of machines and the virtual world of the Internet (including the Internet of Things) and information technology. Virtual reality, the Internet of Things and 5G networks are the trends and technologies that will have the greatest impact on the transformation of the telecommunications sector in future. We can expect changes related to the development of standards and testing of 5G systems and the networking of an increasing number of devices and sensors within the Internet of Things ecosystem. Implementation of "software-defined networks" technologies SDN (Software-Defined Networking) i SD-WAN (Software Defined WAN) will bring about the most important market transformations in the enterprise segment and network architectures for many years. Therefore, when discussing current ICT solutions, and recommendations for information system developers it is important to take into account actual trends: Industry 4.0, cybersecurity, Intelligence Automation/Data Science, Analytical Methods Big Data, Cloud Computing, API, Internet of Things, Blockchain.

References

Curbera F, Duftler M, Khalaf R, Nagy W, Mukhi N, Weerawarana S (2002) Unraveling the web services web: an introduction to SOAP, WSDL, and UDDI. IEEE Internet Comput 6(2):86–93

Debicki T, Kolinski A (2018) Influence of EDI approach for complexity of information flow in global supply chains. Bus Logistics Mod Manage 18:683–694

Engel R, Pichler C, Zapletal M, Krathu W, Werthner H (2012) From en-coded EDIFACT messages to business concepts using semantic annotations. In: Commerce and enterprise computing (CEC), IEEE 14th international conference on, pp 17–25

He S, Li J, Gao X, Luo L (2015) Application of FTP in flaw detection of rail web. Optik-Int J Light Electron Opt 126(2):187–190

Herfurth M, Weiß P (2010) Conceptual design of service procurement for col-laborative service networks. Working conference on virtual enterprises. Springer, Berlin, Heidelberg, pp 435–442

Hodge R (2013) New data standards: challenge or opportunity? Superfunds Mag 381:42

Kohout J, Komárek T, Čech P, Bodnár J, Lokoč J (2018) Learning communication patterns for malware discovery in HTTPs data. Expert Syst Appl 101:129–142

Kolinski A, Jaskolska E (2018) Analysis of the information flow efficiency in the intermodal supply chain-research results. Bus Logistics Mod Manag 18:135–155

NSB CoRe (2018). https://ec.europa.eu/transport/themes/infrastructure/north-sea-baltic_en

Osmolski W, Kolinski A, Dujak D (2018) Methodology of implementing e-freight solutions in terms of information flow efficiency. In: Interdisciplinary management research XIV-IMR 2018, p 306–325

Saha D, Syamsunder M, Chakraborty S (2016) The SAP OEE add-on com-ponent of ERP. In Manufacturing performance management using SAP OEE, Apress, Berkeley, CA, pp 23–90

Sherif MH (2016) Protocols for secure electronic commerce, CRC press

Storz C (2007) Compliance with international standards: the EDIFACT and ISO 9000 standards in Japan. Soc Sci Jpn J 10(2):217–241

Zabinski J (2017) American national standards institute (ANSI). In: Encyclopedia of computer science and technology, p 130

IT Solutions Supporting Information Exchange in Intermodal Transport

Bartosz Guszczak and Roberto Mencarelli

Abstract The dynamic growth of trade in goods in Polish ports recorded in recent years, as well as the growing requirements of exporters and importers concerning efficiency, time and reduction of cargo handling costs and coordination of logistic port processes, create a number of requirements for fast and modern information circulation and safe transmission of documents. This chapter focuses on the complexity of information exchange in port operations. Digitalisation of processes for a more efficient exchange of information can be a solution. The digitisation challenge here is to develop and implement the Port Community System as an integrated maritime economy customer service platform, involving all stakeholders in port and border processes.

Keywords Intermodal transport · PCS_port community system

1 Introduction

Transport is nowadays one of the key sectors of the European economy, providing jobs for more than 11 million people and a major contributor to economic growth (its share of total gross value added for 28 EU Member States is 4.8%, or €548 billion). The European Commission aims to shape and support a transport policy that offers efficient, safe and sustainable solutions to create conditions for a competitive industry that generates employment and prosperity (EU transport policy).

The aim of the European transport policy is to help establish a system that lays the foundations for economic progress in Europe, enhances competitiveness and offers high quality mobility services while managing resources more efficiently. In practice, this means that the transport sector must consume less energy in a greener way—

B. Guszczak (✉)
Institute of Logistics and Warehousing, Estkowskiego 6, 61-755 Poznan, Poland
e-mail: bartosz.guszczak@ilim.poznan.pl

R. Mencarelli
RAM Logistica Infrastrutture e Trasporti Spa, Via Nomentana, 2, 00161 Rome, Italy
e-mail: rmencarelli@ramspa.it

A. Kolinski et al. (eds.), *Integration of Information Flow for Greening Supply Chain Management*, Environmental Issues in Logistics and Manufacturing, https://doi.org/10.1007/978-3-030-24355-5_9

making better use of modern infrastructure and reducing negative impacts on the environment and the most important natural resources. Solutions must be promoted that allow more goods to be moved using the most efficient means of transport or a combination of such means. It is necessary to develop ICT (Information and Communication Technologies) in-format tools enabling simpler and more reliable transport.

Polish seaports and the container terminals operating within them play an extremely important role in the intercontinental flow of container cargo and the creation of gross domestic product, linking sea and land cargo supply chains. The dynamic growth of trade in goods in Polish ports recorded in recent years, as well as the growing requirements of exporters and importers concerning efficiency, time and reduction of cargo handling costs and coordination of logistic port processes, create a number of requirements for fast and modern information circulation and safe transmission of documents.

2 Information Exchange, Current Situation

Digitisation is one of the most rapid changes taking place in our times, providing impetus for innovation and growth for most business sectors (Wronka 2011; Kolinski and Jaskolska 2018). Digital transformation has recently become a challenge for supply chains, public administration and the entire economies. The complex port environment also faces the challenge (Fechner 2007; Mindur 2002). The digitisation challenge here is to develop and implement the Port Community as an integrated maritime economy customer service platform, involving all stakeholders in port and border processes.

Polish ports play a vital role in intercontinental supply chains, combining sea and land transport. Combined with a consistent growth in international trade volumes, it unlocks significant growth potential for Polish ports, They process more cargo, not only containerised, and thus a growing number of ships. Nevertheless, the global cargo flows market shows growing competition for customers. We have witnessed rising requirements from exporters and importer for passenger, ship and cargo efficiency, time and cost reduction, and coordination of port services and companies. This calls for efficient support and effective information processing, ever shorter duration of service, clearance and inspection of ships or individual shipments. Disappointingly, the prospects are not bright. Polish ports are increasingly more acutely experiencing the absence of appropriate logistics infrastructure and effective data exchange systems, which increases traffic congestion and undermines the quality of logistics services at ports. The situation is even more complex as multiple port and border process stakeholders are involved, including maritime offices, border guard, customs services and border inspectors, port authorities, loading entities, ship-owners, liners, handling terminals, forwarders, logistics operators and carriers. The multitude of stakeholders also means there is an array of IT systems, communication standards, etc. Information is currently sent in many ways: over electronic mail, electronic data interface (EDI), fax, traditional mail or by phone. The variety of formats used to exchange information, time loss caused by administration of hard-copy documents, double data entry, low data security, risk of data loss, a huge amount of unnecessary operations—all these reduce the efficiency of port operations. Integration and optimisation pose a formidable challenge in this context. Stringent administrative procedures in terms of border and port operations, and strict data and information security requirements make the situation even more challenging and call for efficient document and information exchange. A prerequisite and the first step to success is full awareness of the Polish port community that the change to embrace digitisation is absolutely necessary. Representatives of public administration and business involved in border and port operations emphasise the need to implement a system in Poland modelled on European and global solutions that would ensure coordination of port services and operators as well as fast and modern information flow and secure document sending functionalities.

There are many problems to be solved, such as:

- growing trade volume in the context of limitations resulting from inability to grow territorially and absence of adequate logistics infrastructure in Polish ports
- increasing traffic congestion and the resultant poorer quality of port logistics services
- requirements from cargo owners (exporters/importers), sea carriers and operators pushing for shorter turnaround times and reduced logistics costs
- no unified and integrated customer service system assisted by IT tools and providing better service at modern sea ports,
- less competitive ports and outflow of cargo to German and Northern Adriatic ports—based on estimates:
  - German ports—ca. 600,000 TEU of Polish cargo
  - Northern Adriatic ports—ca. 80,000 TEU of Polish cargo
- many domain-specific customer service systems (maritime and safety, sea and land transport, forwarding, commercial, customs and border control), data duplication, longer turnaround times and reduced port efficiency
- existing legal requirements generate barriers to process and document digitisation—e.g. original copies or hard copies of documents required:
  - at international voyage registration, original copies of the following documents must be submitted to the harbourmaster on request:
    ship certificate,
    international measurements certificate,
    safety, loading line certificates,
    MARPOL certificate, financial security certificate,
    crew and passenger list,
    cargo manifest.

The scope of information exchange in port operations is very broad and involves very many stakeholders. The figure below shows examples of information exchange, which currently takes place over various channels (hard copy, e-mail, fax, EDI, internal IT systems, etc.).

The variety of standards for the exchange of messages among the respective stakeholders generates chaos in communication, and the requirement on the sender to modify the message multiple times to send it to many recipients. No standardised data sending standard means that the message must be converted into many data formats, depending on the format required by the recipient. Multiple data file modification often generates errors, resulting in the sending of distorted, incorrect or incomplete data.

The attempt to eliminate errors and streamline data transmission integrates the many stakeholders involved in the cargo flow at ports using a single communication standard.

3 Information Exchange Processes in Port Operations

This paper seeks to develop and improve services provided by entities involved in the cargo workflow at the port. There is an ongoing process of information exchange between stakeholders from the start of the ship handling process at the port until release of the cargo, or the ship's departure from the port. The information exchange is by multiple communication channels, depending on the standard used by the interacting entities. The data may be exchanged by:

- E-mail.
- Fax.
- Phone.
- Hard copy documents.
- EDI.

Since there is no unified standard for information exchange between the stakeholders, it is often necessary to send a single message in different formats and by different channels, to multiple stakeholders, and to receive feedback in different formats, often prone to errors. A non-standardised information exchange process is time-consuming and falls way short of the stakeholders' expectations, and therefore the services covered by this project envisage:

- Launch of a single standard or reduction in the number of message exchange standards.
- Automatic sending of messages to every stakeholder (single window).
- Launch of the electronic document workflow.

The services covered by the project were divided into categories based on specific activities involved in each, then specific services were selected of relevance to the respective operations which are to be standardised. Table 1 presents the summary of the main service group categories, and the summary of activities included in each group.

Table 1 Services originating from the shipowner/liner and provided to other cargo workflow stakeholders

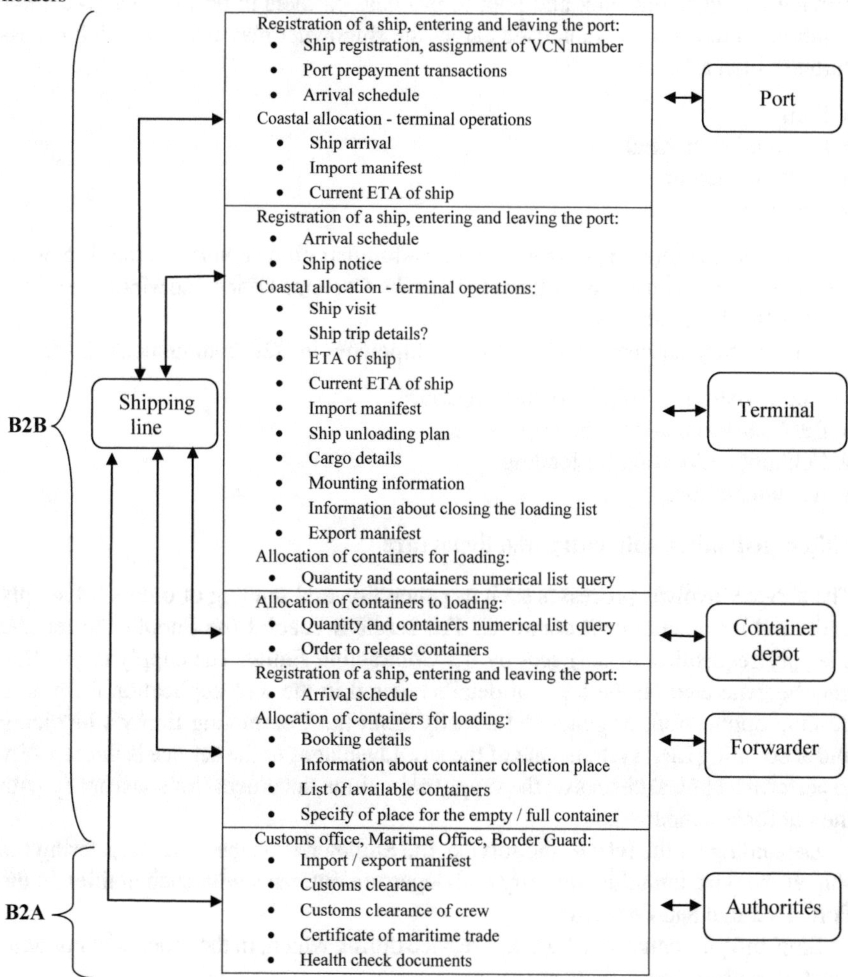

The services covered by the project are services provided among multiple stakeholders, often overlapping in their activities and scopes. They include:

- Ship registration, ship entry and departure.
- Berth allocation—terminal operations.
- Container allocation for loading.
- Transport contracting.
- Special orders.
- Cargo handling.
- Cargo transport.
- Terminal operations.

The existing situation was used as the starting point to determine a set of services provided by the shipowner and liner to business partners in terms of cargo administration. In terms of containerised cargo, the shipowner may interact with business partners such as:

- Port.
- Intermodal terminal.
- Container depot.
- Forwarder.

In terms of communication with public administration authorities, the shipowner may interact with business partners such as the Customs Office, the Maritime Office and the Border Guard.

The primary services provided by the shipowner in B2B communication are:

- Ship registration, ship entry and departure.
- Berth allocation—terminal operations.
- Container allocation for loading.
- Administration.

Ship registration, ship entry and departure

The service involves processes such as generation and sending of orders for supply under public procurement contracts. The services enables the Supplier to receive a supply requisition (order) sent by the Contracting Entity. The supply requisition may be generated on the basis of details entered in the web application form, in a desktop application, or generated directly from the Contracting Entity's invoicing and accounting ERP system. One of the key advantages of the service is the capacity to perform technical checks on the supply requisition document (for consistency with the Platform standards).

Depending on the type of the service, the Shipowner, as the entity registering the ship at the port, including its entry and departure, interacts with such entities as the Port, Terminal and Forwarder.

Ship and port entry registration—service during which, in the process of communication with partners, there is:

- Notification.
- Assignment of VCN number.
- Transmission of the schedule of calls.

In the course of ship registration, there are also services such as port prepayment transactions, which enable admission of the ship into the port. The Shipowner or the Liner is the owner of the ship entry registration service, and its recipient is the port entity.

In the process of ship registration at the port, the Shipowner sends the following messages to the respective recipients:

- Forwarder.
- Sea terminal.

Table 2 Documents exchange between stakeholders in port operations

Liner/liner agent	Forwarder	Schedule of calls	E-mail, SMDG	Schedule of calls
Shipowner/ship agent	Sea terminal	Pre-arrival notification	Email	Definition of mooring conditions
Liner/liner agent	Sea Terminal	Schedule of calls	E-mail, SMDG	Notification of ship call

The ship-owner transmits the schedule of calls information to two recipients—the Forwarder and the Sea Terminal. Both messages are transmitted at maturity level, through the SMDG message, or by-email, as an xls file attached to the message. Depending on the recipient of the message, the various schedule of calls information is transmitted, for instance—in communication with the Forwarder, the schedule of calls is provided as a schedule of ship entries to and departures from a given port, whereas in the case of information provided from the Ship-owner/Liner to the Sea Terminal, the schedule of calls information contains details of the anticipated ship call enabling the setting aside of resources and planning for terminal operations to handle the ship (Table 2).

Berth allocation—terminal operations

Services covered by the terminal operations category include messages exchanged between the Ship-owner—Liner and Port entities, the Harbourmaster and the Port Authority, as well as the Sea Terminal personnel.

As part of Berth allocation, the following message content is exchanged between the Liner and the Port personnel:

- Port calls—information provided to the Shipmaster, containing cargo details. The information is provided by e-mail or, where possible, by EDI IFTSAI.
- Import manifest—information transmitted by e-mail, to the attention of the Harbourmaster's office, used as the basis for the ship unloading plan and for any specific precautions that need to be taken while handling the ship.
- Current ship ETA—information submitted to the dispatcher at the Port Authority, by e-mail or by phone. It includes specific dates of ship departures, quantity and composition of cargo.

It should be noted that in terms of the exchange of berth allocation information with port entities, the Liner communicates with at least three different entities using various communications (e-mail, phone, EDI), providing a different set of data on ship departure to each. However, all the messages convey, in different formats and level of elaboration, the same information about cargo carried by the ship.

Berth allocation information exchanged between the Liner and the Sea Terminal includes:

- Ship call information—information provided over EDI IFTSAI on the ship arrival planned by the Liner, the departure schedule and the cargo details. The same information is provided to the Shipmaster. It is used by the Sea Terminal personnel to plan for ship handling in the terminal.

- Voyage details—information transmitted by e-mail with details of ship arrival, route of transport and cargo carried by the ship.
- Ship ETA—information transmitted by e-mail, specifying the expected time of the ship's arrival at the Terminal. Using this information, the Terminal staff check and plan for berth availability to enable entry and handling of the ship.
- Current ship ETA—precise details of current ship position and estimated time of port entry and arrival at the terminal. Using the current ship position details, the Liner provides information about any delays in the entry, to enable adjustments to the Sea Terminal operating routines. Details about the current ship ETA are provided, depending on the stakeholders' capacity, by e-mail, phone or EDI IFTSAI.
- Import manifest, cargo details—information about the ship's cargo, transmitted to the Sea Terminal as EDI COPRAR message, SMDG messages, or as an .xml file sent by e-mail. It is a document containing details of all batches of cargo onboard and identification details for such cargo. The import manifest quotes numbers of bills of lading, quantities of cargo units, names of loading personnel, cargo weights, freight rates, and other. Cargo details submitted to the Sea Terminal—the import manifest, enables the creation of the ship handling plan at the terminal, planning for, where necessary, special cargo handling, or planning for special precautions in ship handling.
- Ship unloading plan—cargo unloading plan for the ship prepared by the Liner, transmitted to the Sea Terminal by e-mail or as EDI BAPTILE message. It specifies in detail the location of the cargo on the ship, unloading and loading sequence. The cargo unloading plan for the ship depends on the information provided in the import manifest, and further route of cargo transport by sea as determined by the Liner.

If export goods are loaded on the sip at the Sea Terminal, the berth allocation and terminal handling service involves the exchange of additional messages between the Liner and the Sea Terminal, and includes:

- Fixing information—specification of requirements for cargo fixing on the ship, provided by the Liner by e-mail to the Sea Terminal, determines how the terminal handling staff secures the cargo on the ship.
- Cargo list closing information—transferred by e-mail to the Sea Terminal, enables detailed planning of loading works between the yard and the ship by the Sea Terminal staff. Using the ship's cargo list submitted by the Liner, the Sea Terminal checks the availability of the cargo at the yard, notifies loading readiness of the Liner when the full cargo set is ready for the loading.
- Export manifest (loading)—similarly to the import manifest, contains information on the cargo planned for transport by sea. Transmitted to the Sea Terminal by the EDI COPRAR message, SMDG messages, or as an xml file send by e-mail, it contains the details of all batches of cargo to be onboard, including identification details. The manifest quotes numbers of bills of lading, quantities of cargo units,

names of loading personnel, cargo weights, freight rates, and other. Cargo details submitted to the Sea Terminal—the export manifest, enables the creation of the ship handling plan at the terminal, planning for, where necessary, special cargo handling, or planning for special precautions in ship handling, or loading sequence.

When analysing the activities and messages falling into terminal handling services being part of berth allocation, it is evident that the exchange of messages in the same service activities takes place multiple times, among the various stakeholders. In the case of terminal operations, this applies to messages such as ship call information, import manifest, ship ETA and its update. Messages containing the same data are sent in different formats to different recipients, which translates into protracted operation time.

Container allocation for loading

The liner, in providing container allocation for loading services, cooperates with the Forwarder as the entity directly requesting making a container available for customer needs, and with the container terminal which is a depot for a given shipowner's containers. It should be noted that shipowner depots may be found both at intermodal sea and inland container terminals (Wagener 2017), they may be autonomous terminals which, other than container storage services, fail to provide container handling service between the various types of transport. Both in terms of shipowner depots at intermodal terminals and independent depots, cooperation with the liner in the area of container allocation for loading is the same and comprises the following activities at the Liner end:

- Query about the quantity and numerical list of the containers—a message sent by e-mail to the container terminals. The liner, in the reply message, expects that the terminal allocates containers having the required specifications, size and condition, and that will comply with the requirements of the Liner's customer in terms of cargo loading. The query may also be made by phone.
- Container issue order—a message with data of the container previously allocated by the terminal, and details of forwarding/person to report directly to the container terminal to receive it from its storage location. The container issue order is sent by e-mail.

In terms of container allocation for loading, the Liner also interacts with the Forwarder, to which it makes the container available, and with which it exchanges the following messages:

- Booking details—specifying the number of the allocated container, its number, ship name, voyage number, port of loading, unloading and destination, weight, seal numbers, booking number, forwarder name, forwarding name, type and size of the container. The booking details may be provided by the Liner over EDI COPRAR or by e-mail.
- Container pickup location information, and list of allocated containers—details of the container terminal from which the Forwarder may pick up the container

allocated by the Liner, transmitted by e-mail. The information is submitted concurrently with the list of the available containers, determined as the list of numbers or quantity.
- Container dropoff location details—specification of the terminal (depot) where the forwarder should drop off the empty container after unloading of import goods, or where the loaded container should be placed after loading to prepare it for further transport. The container dropoff location details are provided by the Liner by e-mail.

Administration

In the cargo handling process, cooperation of the Liner with public administration authorities, such as the Customs Office, the Maritime Office, the Border Guard and phytosanitary services, is also required. The entity with which the Liner interacts in terms of the A2B service, such as admin support of the ship and cargo, is the Customs Office. Messages transmitted by the Liner to public administration authorities cover:

- Customs clearance of the cargo.
- Border clearance of the crew.
- Submission of the import/export manifest for the cargo.
- Submission of the sea trade certificate.
- Provision of health inspection documents.

Ship administration takes place through the Customs Office, and all documents are submitted to the Customs Office, which further dispatches the relevant instructions and data to the relevant administration authorities over the electronic system.

Services originating from the Forwarder to other port workflow participants

The Forwarder conducts a number of activities in the cargo transport process to deliver the customer's order, cooperating with both public administration authorities, taking actions comprised in the B2A service, and with business partners to which the Forwarder provides B2B services.

Business partners with which the Forwarder may cooperate in delivering the forwarding order are primarily:

- Shipowner—Liner.
- Intermodal terminal—sea and land terminal, depending on the complexity of the transport route and using various types of transport.
- Container depot.
- Intermodal operator.
- Road carrier.
- Exporter/Importer—customer contracting the transport.

In terms of B2A administration, the Forwarder interacts with Customs Agencies, over which it conducts the administrative, customs and phytosanitary activities at the Customs Office.

Primary activities performed by the Forwarder for business partners are activities covered by the following services:

- Transport contracting.
- Cargo handling.
- Special orders.
- Cargo transport.
- Customs clearance.
- Terminal operations.

Transport contracting

The transport contract is made between the Forwarder and collaborating entities, from which the Forwarder contracts tasks on the cargo transport route. Depending on the type of the business partner, the activities involved in transport contracting, as the first stage of cooperation, may differ. If the Forwarder cooperates with the Shipowner—Liner, its activities include:

- Transport orders—based on data provided in the request for quotation, submitted by the Forwarder to the Liner by e-mail. They include the transport fee, preliminary cargo details, place of dispatch and destination of the cargo.
- Booking requests—on the basis of the transport order, the Forwarder submits a booking request to the Liner, specifying the number of the allocated container, name of the ship, voyage number, loading and destination ports, weight, seal numbers, booking number, forwarder name, forwarding number, type and size of the container. On that basis, the Liner receives details about a requirement for a container for loading, and books a place on the ship to transport the reported cargo. The booking request is submitted by e-mail, phone, or over EDI message, depending on the availability of the respective systems at the Forwarder end.

In communicating with the Intermodal Terminal and the Container Depot with respect to transport contracting, the Forwarder exchanges the same messages which, depending on the type of terminal used in the transport, are submitted to the sea or land terminal, or to the container depot. The type and place of the terminal used may be determined by the order of the customer, requirements of the Liner, or capacity of the Intermodal Operator. In delivery of the transport contracting service, the Forwarder cooperates with the terminals by performing the following activities:

- Handling order—request to change the means of cargo transport (container) submitted by the Forwarder to the Terminal by e-mail or by entry of order details to the terminal system (if feasible for the Forwarder).
- Dropoff pre-notification—information submitted by e-mail about the date of container dropoff, with numerical container data, full/empty indication, details of cargo transported in the container, including name of the goods, weight, markings, special instructions. The pre-notification also contains details of the carrier and vehicle bringing and placing a given container at the terminal. Instructions for further processing of the cargo and container are provided along with the pre-notification (booking data).
- Pickup notification—submitted by e-mail, by manual entry of data in the terminal system, or by EDI COPINO, information from the Forwarder providing numerical

date previously obtained from the Shipowner or the Customer (Importer/Exporter) or container pickup number. The pickup notification also includes details of the vehicle and person reporting to the terminal to pick up the container. In the case of empty containers at the container depot, the pickup notification submitted by the Forwarder to the Terminal is verified and matched with the notification received from the Shipowner—Liner. The cargo (container) is issued to the notified person only if the two notifications match.

- Repair or special service order—an additional order for special services not covered by the transport order, which may cover orders for goods handling, fixing, fumigation, storage, weighing, or other. Depending on the range of extra services provided by the Terminal, the Forwarder may also order container repair, or container adaptation to cargo transport—opening, cleaning, floor repair, etc. Orders for additional services are submitted by e-mail or in the terminal system as manually added orders.

The Forwarder may also cooperate with the Intermodal Operator in the provision of the Transport contracting service. Activities it performs in that case for the Intermodal Operator include:

- Forwarding order—in which the Forwarder specifies cargo details, transport route, transport fee, order delivery date and conditions for the order. The forwarding order is sent by e-mail as an Excel or PDF file, or over EDI.
- Cargo pickup availability notification—is submitted electronically or over the phone, and specifies when cargo covered by a transport order is available for the Intermodal Operator, that is when inspection, customs or loading is completed. Based on the pickup availability information, the Intermodal Operator may plan cargo transport between terminals entrusted to it.

In the case of cooperation with a Road carrier, the primary activities involved in the Transport Contracting are:

- Transport order—order with cargo details, carrier details, freight rate. The transport order, sent by the Forwarder electronically, is used by the Carrier to proceed with the transport.
- Container and cargo pickup order—specifies details of the transport order, pickup location, container and cargo dropoff location.
- Bill of lading—typically includes address of the seller, destination address, reference number of the recipient and supplier (e.g. order number or contract code), type of transported cargo and its quantity and unit price. The bill of lading is often the basis for the customs services to inspect and collect duties on the value of the transported goods.

In terms of transport contracting by the Forwarder, various types of information are exchanged with other participants in the port workflow. Cargo details are however provided on multiple occasions: in the case of imported goods complete details of the goods are provided already at the start of the transport contracting process, and in the case of exported goods, the message with the details is provided on multiple occasions to the same entity. This results from the requirement to update the details after loading. Updating details of the goods is required for Forwarder—Ship-owner, Forwarder—Intermodal Operator, Forwarder—Terminal communication. In that case, multiple sending of updated details in various formats to various recipients markedly extends the cargo handling process and may result in errors in data transmission.

Cargo transport

Cargo road transport is the key task of the Forwarder in its relations with other port workflow participants. All actions taken by the Forwarder pertain to cargo transport and handling processes, however the process of physically transporting the cargo takes place between the Forwarder and the Terminal and between the Forwarder and the Importer or Exporter. The following services are provided in the process of cargo transport between the respective entities:

- Forwarder notification—service provided by the Forwarder to the Terminal, involving the Forwarder's transmission of information about the planned container drop-off. Depending on the technical capacity of the partners, the information is provided as EDI-COPINO message, by e-mail or electronically by directly entering the notification to the Terminal system by the Forwarder. The Forwarder notification includes container details—its number and characteristics, and transported cargo details (weight, type of goods, quantity, particular terms of carriage), details of the drop-off method (details of the driver, road or rail carrier) as well as booking details and sea transport details.
- Cargo readiness notification—service provided by the Forwarder to the Terminal involving the update of the anticipated readiness to pick up or deliver the cargo or container to/from the terminal. The forwarder notifies the Terminal by e-mail of the updated cargo arrival time to the terminal and container drop-off.

In terms of the Forwarder—Exporter/Import relation and the cargo transport service, the Forwarder provides to the business partner the vehicle arrival notification service and the transport time notification service.

- Vehicle arrival notification—is the information provided by e-mail or phone, where the Forwarder, using data obtained from its subcontractor, specifies the exact time of vehicle arrival at the customer's location. The vehicle arrival notification contains carrier and driver details, such as first and last name of the driver, phone number, personal ID card number, registration number of the tractor and trailer, and number of the arriving container. Proper arrival notification to the Importer or Exporter predetermines the cargo unloading/loading capacity. The notification is verified against the actual details of the driver, vehicle and container when the vehicle arrives at the location of the customer (Importer/Exporter).
- Transport time notification—information about tentative time of travel from the export cargo pickup location to the destination or, in the import operation, the time of travel from the pickup terminal to the target customer. The service, provided by the Forwarder to the Exporter or Importer, transmitted by phone or e-mail on request, is used to plan for the distribution of labour at the customer's location to

smoothly perform the cargo unloading and loading, and to assess, on an ongoing basis, the options for further cargo transport coordination, e.g. if loading on the ship is contemplated.

Terminal operations

Terminal operations include the scope of internal services provided at sea and land container terminals, which are initiated by external services resulting from the cooperation between the Forwarder and the Container Terminal or between the Forwarder and the intermodal operator operating on the terminal's premises, or forwarding orders to the cooperating container terminal. In terms of activities closely linked with the terminal operations services, the Forwarder cooperates with the Terminal and the intermodal operator with respect to the following:

- Forwarder notification—an item of services included in Terminal Operations, being an activity performed by the Forwarder for the Terminal. Depending on the partners' capacity, the information is provided by e-mail, as an attached excel file, as an EDI EDIFACT message, or directly in the Terminal system by the Forwarder. The Forwarder notification includes container details—its number and characteristics, and transported cargo details (weight, type of goods, quantity, particular terms of carriage), details of the dropoff method (details of the driver, road or rail carrier) as well as booking details and sea transport details.
- Revised notification—service involving the Forwarder's provision to the Terminal of details updating the shipped cargo details. The service is provided if the details provided in the forwarder notification are changed, which may result from the actual weight being different than notified, or from a change in the cargo transport route, or booking or from ship notification changes. The revised notification is transmitted just as the forwarder notification—by e-mail, EDI message, or as a change to the notification directly in the terminal system.
- Container loading list—if the cargo is handled by an intermodal operator as an agent in the transport by rail, the Forwarder provides the Intermodal Operator with a lost of container covered by the loading and transport order. The transmission of the list of containers for loading is made by e-mail, as an attached xls file, with container number details. The provision of the container loading list enables the Intermodal Operator to verify cargo details and loading container availability at the terminal from which the Operator picks up the cargo for rail transport.
- Preliminary cargo details—similarly to the container loading list, the preliminary cargo details are provided in the Forwarder—Intermodal Operator relation. The provision of cargo details may take place concurrently with the provision of the container loading list, or later. The list of preliminary cargo details includes the Forwarder-projected weight and type of the cargo. The details are obtained by the Forwarder from its customers—the Importer or Exporter; these are typically details which are verified only after the loading of the cargo in the export relation. Transmission of the preliminary cargo details is necessary for the intermodal operator to plan for the transport.

Customs clearance

The customs clearance process is a service in the administration-to-business (A2B) relation, however it affects the relation and information flow in the cargo handling process among the various process stakeholders. In terms of services provided by the Forwarder to other stakeholders in the process, the customs clearance information flow takes place between the Forwarder and the Terminal, and between the Forwarder and the Customs Agency. Activities involved in the customs clearance and performed by the Forwarder for the Terminal include:

- Transmission of customs status information.
- Transmission of customs clearance completion information.
- Transmission of revision requirement.

In the B2A relation, the Forwarder interacts with public administration authorities, such as customs offices that place further orders with other public administration authorities (phytosanitary). The Forwarder may interact directly with the Customs Office, but also, which is commonplace, may interact with public administration authorities through the Customs Agent. In terms of collaboration with Customs Agencies/Customs Office, the Forwarder reports the customs clearance requirement. The customs clearance order—takes place between the Forwarder and the Customs Office, and may also involve the Customs Agent.

4 Conclusions

The scope of information exchange in port operations is very broad and involves very many stakeholders. The variety of standards for the exchange of messages among the respective stakeholders generates chaos in communication, and the requirement on the sender to modify the message multiple times to send it to many recipients. No standardised data sending standard means that the message must be converted into many data formats, depending on the format required by the recipient. Multiple data file modification often generates errors, resulting in the sending of distorted, incorrect or incomplete data. The solution can be development and implementation the Port Community System as an integrated maritime economy customer service platform, involving all stakeholders in port and border processes.

PCS is an element of a coherent system of communication between Polish and world seaports and numerous port stakeholders, railway junctions or logistic centres. PCS is an integrated platform for customer service in the maritime economy with the participation of all stakeholders in port processes. It is an electronic platform thanks to which both private and public entities can exchange data and documents, making mutual cooperation more transparent and effective.

References

Contemporary transport technologies (2002) In: Mindur L (ed) Technical University of Radom, Radom

Creation of conditions for functioning and development of intermodal logistic network in Poland (2011) In: Mindur L, Krzyżaniak S (eds) Methodological aspects. Institute of Logistics and Warehousing, Poznań

EU Transport Policy, European Union. http://europa.eu/pol/trans/index_pl.htm

Fechner I (2007) Supply chain management. WSL, Poznań

Kolinski A, Jaskolska E (2018) Analysis of the information flow efficiency in the intermodal supply chain-research results. Bus Logist Mod Manag, 135–155

Mindur L (2002) The concept of combined transport. In: Mindur L (ed) Modern transport-we technologies. Radom University of Technology, Radom

Rydzkowski W, Wojewódzka-Król K (ed) (2017) Transport. Wydawnictwo Naukowe PWN, Warszawa, 2005, nr 337

Wagener N (2017) Intermodal logistics centres and freight corridors—concepts and trends. LogForum 13(3):273–283

Wronka J (2011) Basic barriers to the development of intermodal transport in Poland. In: Mindur L, Krzyżaniak S (eds) Creating conditions for the functioning and development of the intermodal logistics network in Poland. Institute of Logistics and Warehousing, Poznań

Wronka A, Inteligentne łańcuchy dostaw, Studia Ekonomiczne. Zeszyty Naukowe Uniwersytetu Ekonomicznego w Katowicach

Determining the Digitalization Degree of Information Flow in the Context of Industry 4.0 Using the Value Added Heat Map

Dagmar Piotr Tomanek, Christine Hufnagl and Jürgen Schröder

Abstract Supply chain management concentrates mainly on the material flow. The information flow plays a lesser role in terms of value creation. Digitalization in the context of Industry 4.0 promises a higher efficiency in logistics and manufacturing. To move away from a paper-based to a digital production planning and control is also a step forward to a green production system. At the same time a highly focused Industry 4.0 approach can cause plenty of media disruptions in a process. Disruptions in the information transfer leads to a reduction of the value creation. In the course of digitalization the ingratiation of the information flow and the material flow became a key factor. An integration of the information flow requires, as a first step, an analysis of the value of an information. The Value Added Heat Map (VAHM) is a modelling tool that categorizes and visualize the information flow depending on the digitalization degree. The implementation to processes in manufacturing showed that the VAHM method can contribute to a better transparency of the often not fully documented internal information flows. The Value Added Heap Map uncovers media breaks via colour transitions and helps to improve the supply chain within the context of Industry 4.0 by identifying and eliminating possible sources of wastage. The Value Added Heat Map is an innovative method that can contributes to an integration of the information flow in supply chain management processes.

D. P. Tomanek (✉) · C. Hufnagl
Technische Hochschule Ingolstadt, Zentrum für Angewandte Forschung (Research Center) and Business School, Esplanade 10,
85049 Ingolstadt, Germany
e-mail: dagmar.tomanek@thi.de

J. Schröder
Technische Hochschule Ingolstadt, Business School, Esplanade 10, 85049 Ingolstadt, Germany

A. Kolinski et al. (eds.), *Integration of Information Flow for Greening Supply Chain Management*, Environmental Issues in Logistics and Manufacturing, https://doi.org/10.1007/978-3-030-24355-5_10

1 Introduction in the Topic of Industry 4.0

Digitalization represents a megatrend of the future. Following Naisbitt (1986), a megatrend has a duration up to 50 years. Megatrends have a global character and affect to all living areas e.g. politics, consume or industry. Digitalization in the context of the industrial sector is summarized in the following under the term Industry 4.0.

Industry 4.0 stands for the 4th industrial revolution. The impact of Industry 4.0 is comparable with the initial industrial revolution at the end of 18th century, which based on the introduction of water- and steam-powered mechanical manufacturing facilities. The second major industrial revolution took place at the beginning of the 20th century and based on the introduction of electrical-powered mass production. The information flow during the term of so called Industry 1.0 and 2.0 is characterized by analogue communication (Günthner et al. 2017). This type of information flow is expressed by verbal or written exchange of information and a paper-based documentation (n.p. 2012).

With computerisation and connectivity in industry based on the introduction of electronics and IT began the era of the third industrial revolution (n.p 2012). During the development of Industry 3.0 the information flow changed from analogue to electronic communication. Since 1970 the exchange of information is characterized increasingly by the Internet and the Intranet (2017).

Today the industry is at the beginning of the fourth industrial revolution. Drivers of Industry 4.0 are Cyber-Physical-Systems (n.p. 2012). The information flow is characterized by digital communication in the form of an Internet of Things (Günthner et al. 2017).

During the evolution of industry from 1.0 to 4.0 the complexity increased. In parallel to this, the transformation from analogue to digital communication increase the information flow complexity (see Fig. 1).

Today almost every industrial sector is disrupted radically by Cyber-Physical-Systems. A Cyber-Physical System (also CPS) represents the integration of computation with physical processes (Günthner et al. 2017). They are "systems with embedded software, which:

- record physical data and affect physical processes
- evaluate and save recorded data
- interact both with the physical and digital world
- connected with one another and in global networks
- use globally available data and services
- have a series of dedicated, multimodal human-machine interfaces (Lee and Seshia 2017)".

In summary, Cyber-Physical Systems are interconnecting the physical and digital worlds. This means that physical things are no longer cut from the virtual world.

The Internet of Things (also IoT) represents the integration of real objects in a digital network on the basis of Cyber-Physical-Systems. Through the Internet of Things physical objects are able to act and to communicate in a virtual network. In summary, the IoT is also called as a system of systems.

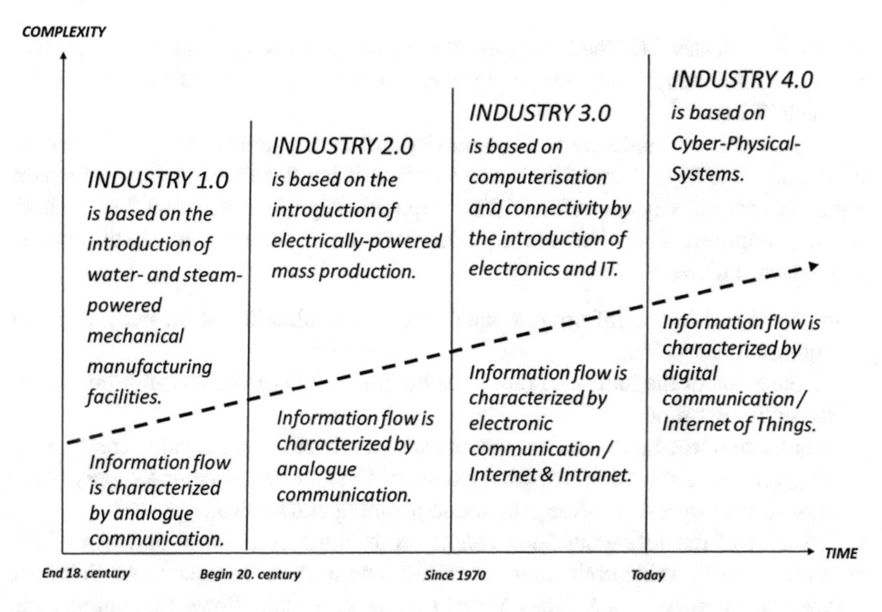

Fig. 1 Evolution of industry and transformation of information flow (Günthner et al. 2017)

The implementation of the Internet of Things and Cyber-Physical-Systems in production stands also for a communication of human and machines in real-time. An exchange of information in real-time is the basis of a self-organized factory within the meaning of Industry 4.0 (Tomanek and Schröder 2017; Kocsi and Oláh 2017).

Industry 4.0 is not an end in itself but serve to achieve monetary cost saving (see Fig. 2). Following Bauernhansl, Industry 4.0 promise cost saving in manufacturing, logistics and quality up to 20%. Inventory and maintenance costs could by reduced up

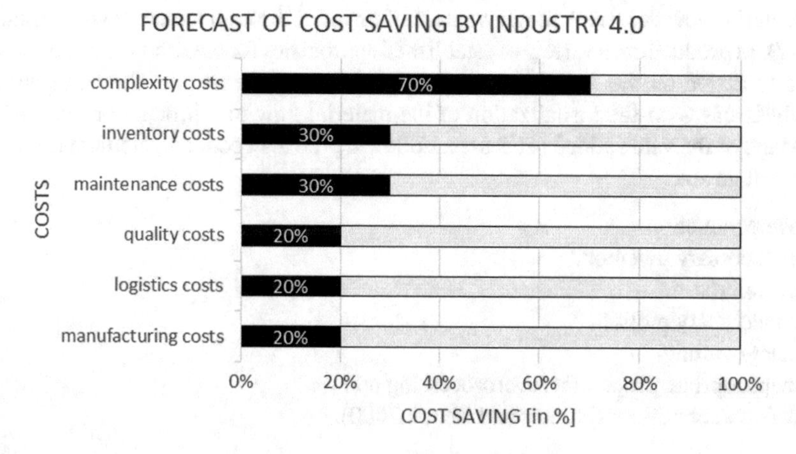

Fig. 2 Forecast of cost saving by industry 4.0 (Bauernhansl 2017)

to 30% by Industry 4.0. The largest impact in cost saving is identified in complexity. Bauernhansl forecasts that complexity costs of 70% can be saved by Industry 4.0 (acatech 2011).

An increasing complexity and complexity cost savings are no irreconcilable contradictions (see Figs. 1 and 2). On the contrary, it implies that Industry 4.0 makes a growing complexity controllable. This requires a digital information flow without media disruptions. Therefore, the requirements to an information flow in the context of Industry 4.0 are:

- Improvement of the information quality by an availability of all necessary and required information.
- Acceleration of the industrial processes by availability of decision-relevant information in real-time.
- Avoidance of media disruption, which cause a reduction of the value creation.
- Integration of a digital information exchange to enhance a greening supply chain management by a paper-free production planning and control.
  The goal of the following book chapter is to emphasize the importance of the information flow in production processes. The authors Tomanek and Schröder developed a method, the Value Added Heat Map, which allows a layout-specific analysis of the information flow. Furthermore, the presented method also allows the determination of a digitalization degree and a degree of compaction. This article also contains instructions for increasing the digitalization degree of information flow presented by an application example. The benefits of a high digitalization in production processes are put in context of greening supply chain management.

2 Theoretical Basis: Information Flow

A material flow describes the movement of material through a factory (Bauernhansl 2017). In production, a variety of established approaches for value added optimization lead to action on the analysis of the material flow. For example, the value stream analysis bases on the visualization of the material flow to eliminate or minimalize wastage of the value added processes. Following the lean concept, elementary forms of wastage are:

- overproduction,
- unnecessary inventory,
- transportation,
- unnecessary motion,
- delay/waiting,
- inappropriate processing/overprocessing and
- defects/corrections (Rother and Shook 2000).

Overproduction represents a production of more parts than are needed for use or sell. Overproduction cause inventory of semi-finished or finished goods than required.

Due to overproduction and unnecessary inventory additional not-value added transportation of goods and materials is necessary. Unnecessary motion of people, such as walking or searching represents wastage. Any kind of waiting, employees as well as machines, between the end of one activity and the start of the next process is not-value added. Overprocessing describes wastage in form of putting more energy or activity than is actually needed to fulfil a product. Defects result in rework or disposal of the product. Both is wastage and should be avoided in production (Ohno 1993).

Verrier et al. postulates a positive correlation between lean and green production. This means that an elimination or minimization of non-value added activities also leads to a sustainable production (Ohno 1993).

Compared to the material flow, the importance of the information flow has still a marginal meaning in the era of Industry 4.0. In the opinion of the authors, an isolated examination of the material flow and the information flow is not-value added. A flow of information is a mandatory requirement to enable the material flow. An information flow uses data and documents to describe the communication between production and controlling processes (Verrier et al. 2015). An inadequate internal communication or defective information transfer has been already recognized as an additional form of wastage (Erlach 2010). Consequently, a value added information flow is not only a contribution to a lean production but also leads to a greening manufacturing and in general to a sustainable supply chain.

2.1 *Information Value*

Information technologies promises a competitive advantage in management practice (Schröder and Tomanek 2012). To date, there are several approaches to define the information value. An approach that base on the derivative nature of the information flow is to link the information value with the information benefit (Porter and Millar 1985). This means that the value of an information can be deducted from the impact on the material flow.

Research on the bullwhip effect demonstrates in a vivid manner how insufficient or incorrect exchange of information lead to wastage in the value chain. The bullwhip effect describes a phenomena that fluctuations in demand occurs more strongly at the end of a value chain than at the beginning. Causes of the bullwhip effect are excessive inventory investment, poor customer service, lost revenues, misguided capacity plans, ineffective transportation, and missed production schedules (Reese 2016). A measure for the avoidance or minimization of the bullwhip effect is, inter alia, the continuous improvement of the information flow (Lee et al. 1997).

It follows that information value is created by information, which are transmitted correctly, complete and in a timely manner (Reese 2016). Disturbance of the information flow leads to non-value added wastage. Disturbances in information flow can occur e.g. because of media disruption. Media disruptions describe media changes during an information transfer, which can cause incorrect, incomplete or delayed information.

2.2 Media Disruptions

A media disruption is defined as a change of medium during the transmission of information within the transmission chain (Reese 2016). For example, a transmission of an electronic information into an oral form constitutes a media break. In general, if a received electronic information is further developed into another form, this development may create communication mistakes. Mistakes in the exchange of information can lead to errors in the value chain. Moreover, media disruptions can cause redundancy and additional effort. This reflects often in non-value added activities like printing, copying or scanning.

In Fig. 3 an example of order processing is sketched. This process incudes following steps:

(1) A purchase order is received online from the customer.
(2) An order list is created.
(3) The order list is printed out.
(4) Remarks of the order fulfilment are recorded in written form in the order list.
(5) The order list is scanned for documentation. The scanned copy is saved in the data base. The paper copy is disposed.

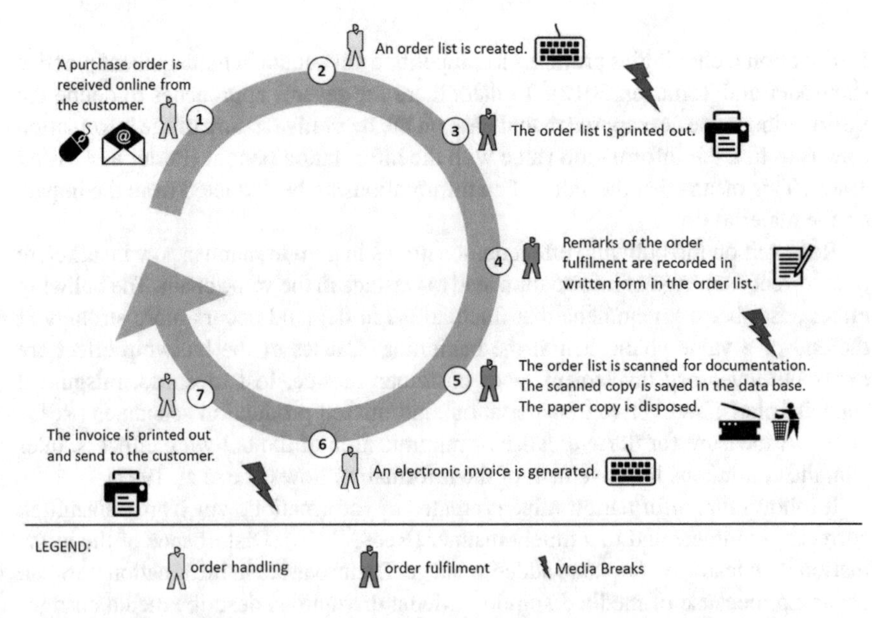

Fig. 3 Example: media disruptions within order processing (Springer Gabler Verlag 2015)

(6) An electronic invoice is generated.
(7) The invoice is printed out and send to the customer.

The analysis of the information flow in the shown example contains four media disruptions. Between step 2 and 3 as well as between step 6 and 7 the medium change from electronic to paper-based. Between step 4 and 5, in turn, the medium change from paper-based to electronic.

From an added-value perspective, media disruptions should be eliminated. If this is not possible media disruption should be at least minimized.

The identification of media disruptions in practice is often complex. A challenge lies in the lack of transparency in the information flow within the company. Another problem lies in the absence of adequate forms of design. In the literature, to date, no independent symbol exists for the visualization of media disruptions.

2.3 Diagram Options

In literature, data visualization methods in application are often related to big data (Tomanek and Schröder 2018). There are only a few approaches for visualizing an information flow (Gorodov and Gubarev 2013). Visualizing tools like value stream analysis or Sankey diagram focus stronger on the material flow than on the information flow.

3 Value Stream Analysis

Value stream analysis is a known method for identifying and avoiding wastage within production processes. Value stream analysis is linked with the Toyota Production System, Today, value stream analysis is used in many industries for process improvement. In this method, different types of arrows are used as symbols for the information flow (Table 1). A jagged arrow represents an electronic information, e.g. a master over the ERP system. A straight arrow visualizes a manual information transfer. A simple example of a manual information is giving a worker an oral instructions. Transmission of information, in particular in the production sector can also be mapped through a go-and-see planning, a levelled production planning or a Kanban symbol (Günthner and Schneider 2011).

The value stream method visualizes in an outstanding way the material flow. This visualization includes also an integration of the information flow in one diagram. But the potentials of the information flow are not visible at first appearance. Stock, latency or lay time get negative influenced by missing, sluggish and incomplete information flow and consequently they extend generally the process time. A simple example of a Value Steam Analysis is shown in Fig. 4.

Table 1 Value stream analysis—symbols used for visualization of an information flow (Balsliemke 2015)

Symbols for the information flow	Meaning
	Manual information flow
	Electronic information flow
6-σ	Electronic information flow
OXOX	Levelled production planningtt
	Route of a kanban card

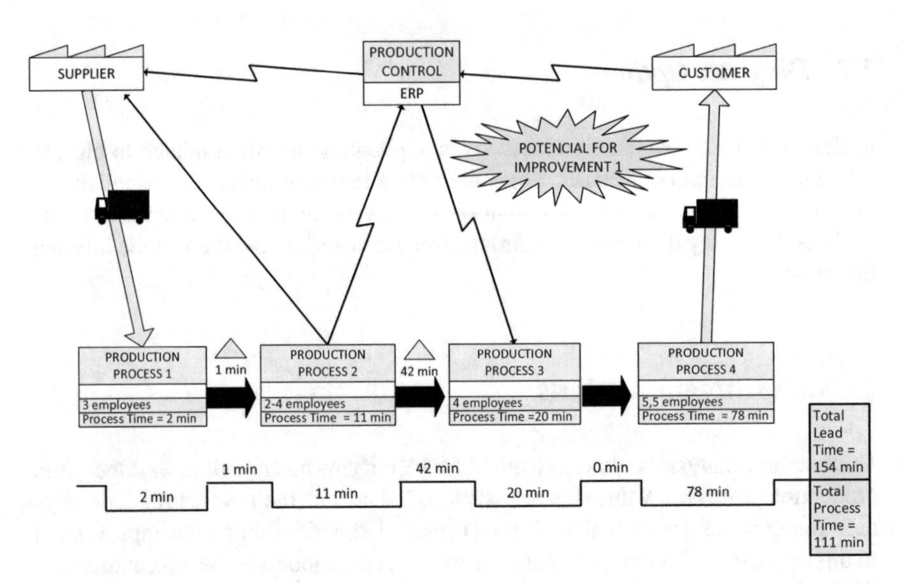

Fig. 4 Example: value steam analysis

4 Sankey Diagram

Sankey diagrams has originally been used for thermodynamic systems. Sankey diagrams were used for an identification of heat losses. Analogous to the value stream analysis, Step by step Sankey diagrams have been successfully applied to other disciplines (Balsliemke 2015). Today the Sankey diagram is applied also to production shop floors. In production, this method can be used to visualize the flows of materials or costs. This tool helps to identify material losses, for instance by production faults or inefficient processes. The main components of a Sankey diagram are arrows. Arrows interlink the individual process steps and they define the direction of a flow.

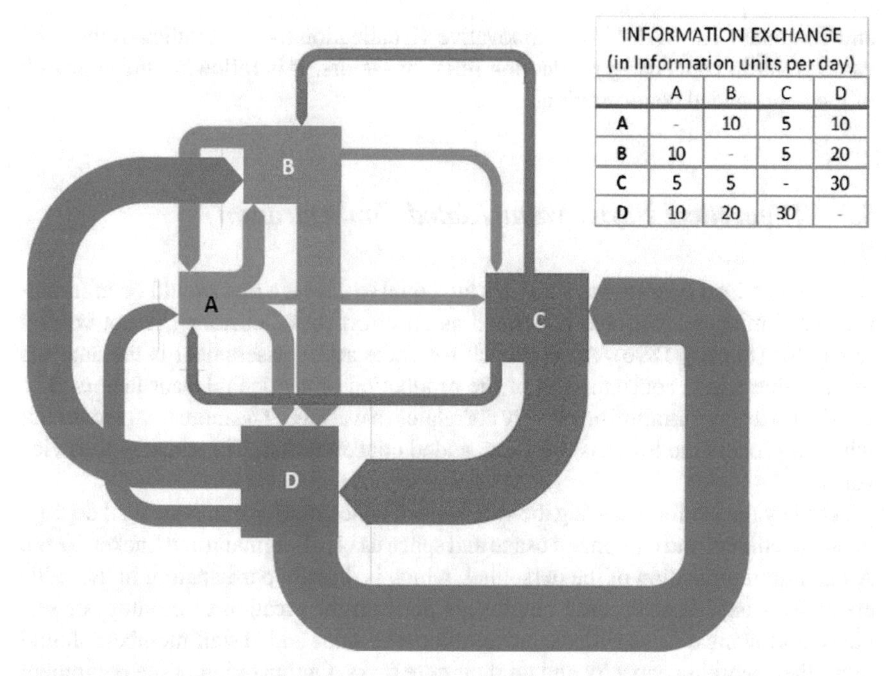

INFORMATION EXCHANGE
(in Information units per day)

	A	B	C	D
A	-	10	5	10
B	10	-	5	20
C	5	5	-	30
D	10	20	30	-

Fig. 5 Example: Sankey diagram

The thickness of an arrows represents the quantity of the substance, which occurs in the flow (Sankey 1896).

The visualization of a Sankey diagram can be combined with a factory layout. The drafting of this diagram starts always with recording the flows of the occurring substance within the analyzed subsystem. The result is a simple, model-like diagram of the process steps and the material flows. The results allows a practical assessment of the analyzed system. Within a production the visualization through a Sankey diagram can e.g. display crossing material flows, which cause production backlogs. Transport bottlenecks or material loops are further examples that can be visualized by this approach.

A Sankey diagram is not only limited to the material flow in production. This method can also be used to visualize the flows of information. An example of an information exchange visualized by a Sankey diagram is shown in Fig. 5.

5 Value Added Heat Map

Methods like value stream analysis or Sankey diagram are used to visualize an information flow. But they are not able to determine the value of an information flow. For that reason the Value Added Heat Map has been developed by Schröder and Tomanek. The VAHM is an innovative visualization tool that indicates the level value creation concerning production relevant factors. It is following the approach of the value added concentration.

5.1 *Theoretical Basis: Value Added Concentration*

Non-value added operational activities are equal to wastage that should be minimalized or eliminated. Wastage is defined as an effort that a customer is not willing to pay for (Sankey 1896). An approach for value added assessment is the analysis of the value added concentration of the production or service relevant factors. The value added concentration negatively correlates to wastage. The more wastage occurs within a process the lower is the value added concentration. The same applies vice versa.

The key factors for assessing the value added concentration are: personnel deployment, machinery and equipment usage and space usage (Bergmann and Lacker 2009). A maximum utilization of the personnel, which is directly participating in the value creation, is expedient because employees perform the creation of products or services. Following a value added concentration, the value added staff members should focus their working capacity and on their core tasks. Optimization of the equipment or machines should also be pursued to ensure the maximum concentration of benefit. Spaces within the shop floor are usually a highly limited resource. Unused or reserved space create no or only limited value. The main aim of value added concentration is to reduce spaces that do not create value, to ensure that sufficient space for the actual value adding process is available (see Fig. 6).

An innovating method for the analysis and visualization of the value added concentration in a shop floor is the Value Added Heat Map invented by Schröder and Tomanek and patented at German Patent and Trademark Office (Tomanek and Schröder 2018).

5.2 *Methodology*

The Value Added Heat Map base on the methodically of a thermal heat map camera. It is an innovative tool that indicates the level of value creation concerning production relevant factors. The Value Added Heat Map has been originally used to visualize the usage of production space with regard to their value. Tomanek and Schröder classified the space consumption in shop floors according to their value. For example, spaces in a factory reserved for production lines are maximum value added. They are directly used for value creation. In a shop floor, you can also find spaces that are necessary for operating, e.g. staging areas for required materials, spaces for intermediates and finished goods the plants. These spaces have limited added value contribution. On the contrary, areas for empties, waste and blocked defective parts do not contribute

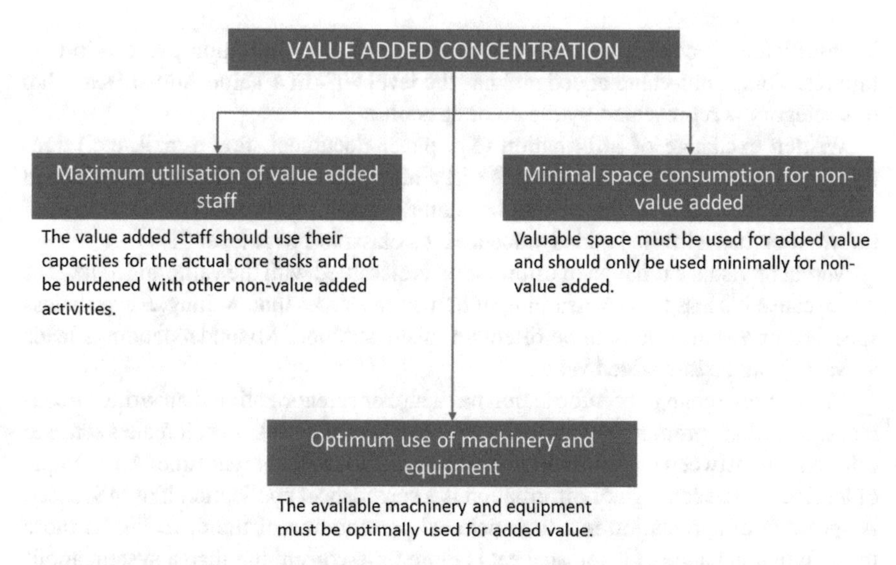

Fig. 6 Value added concentration (Tomanek and Schröder 2018)

to value creation. Space that are not used are not-value added. In a Value Added Heat Map each square meter of an analyzed shop floor is evaluated with regard to a value level. The Value Added Heat Map supports the optimization of a production layout. The aims is to generate smaller area with the same value or the same area with a higher value (Schröder and Tomanek 2017).

In general the Value Added Heat Map is a tool to visualize the value added level of production relevant factors by using color scaling. In addition to the visualization, the VAHM develops conclusive key performance indicators. The graphical result of the analysis resembles a thermal image. Therefore, it is called Value Added Heat Map. By using this method potentials for improvement can be recognized easily. The Value Added Heat Map is not limited to improve the usage of available space. It can be applied to further production relevant factors like exchange of information.

6 Evaluation Scale for Information Flow

Regarding the implementation of the Value Added Heat Map to information flow, it is important to note that information contribute differently to the added value. The authors classified the information flow in six value added levels, namely from zero to five. As a scale for the categorization serves the effort that is related to the information exchange. The lower the effort, the higher the value added level and vice versa.

Insufficient, incorrect or unnecessary exchange of information cause effort by failures. This is not-value added and has the level "0". In a Value Added Heat Map this category is represented by the color dark blue.

Written exchange of information (e.g. paper document, fax, e-mail, etc.) contributes to value creation. But there is still a high potential for improving the added value. The time needed for writing consumes human resources. This exchange of information has a limited added value and it's classified with level "1".

Verbal or visual exchange of information is classified with the value added level of "2" because it's less time-consuming of human resources than writing. Nevertheless speeches or visuals cues can be often get misunderstood. Misunderstandings leads to wastage and a low added value.

Electronic exchange of information has a higher categorization than written documents or verbal communication. In the case of electronic exchange, it makes sense to differentiate between an information flow in real-time and not real-time. An example of an electronic exchange of information is a spreadsheet application like MS Excel. A spreadsheet application is a time-delayed presentation of the data. Furthermore the data maintenance of a spreadsheet is more time-consuming than a system application. This is why an electronic information flow not in real-time has a lower value added level, namely "3", than an information exchange in real-time. An information flow in real-time has a value added level of "4".

The highest value added level of information flow is represented by digitalization. Implementing Internet of Things and Cyber-Physical-Systems in a shop floor enable a communication of employees and machines in a network in real-time. A digital exchange of information has the maximum value added level of "5" (see Table 2).

To draft a Value Added Heat Map a current layout of the analyzed shop floor is required. The factory layout serves as the basis for the assessment of the information flow. The information flow is represented by arrows. In a Value Added Heat Map,

Table 2 Value added heat map—evaluation scale for information flow (Tomanek and Schröder 2016)

Categorization	Value Added Level	Dimension of Information Flow	Scale
No Added Value	0	Insufficient, incorrect or unnecessary exchange of information	Effort
Limited Added Value	1	Written exchange of information (e.g. paper document, fax, e-mail, etc.)	
	2	Verbal or visual exchange of information	
	3	Electronical exchange of information not real-time (e.g. by spreadsheet application)	
	4	Electronical exchange of information real-time (e.g. by system-application)	
Maximum Added Value	5	Digital exchange of information real-time (e.g. by Internet of Things and Cyber-Physical-Systems)	

the arrow's thickness represents the quantity of the substance, which occurs in the flow. In the case of visualizing the information exchange, the thickness refers to the amount of information units. The color of the arrows represents the value added level of information exchange.

To identify the information flow of a process it is helpful to visualize and quantify the material flow first. It is important to investigate, which information enhance the material flow. It is also relevant to quantify the number of information units, which are generated and transferred by the process. It is advisable to adjust the dimension of the information flow to a time unit. The authors recommend a shift or a day as a suitable time unit.

7 Layout-Specific Digitalization Degree and Degree of Compaction

In a Value Added Heat Map, each information transfer has to be determined according to an evaluation scale for information flow (see Table 2). The targeted information flow corresponds to a fully digital exchange of information in real-time without media disruptions. Consequently, the value added levels of an analyzed information exchange determinates a digitalization degree. The significance of a Value Added Heat Map for information flow is compressed by a key performance indicator, namely the "layout-specific digitalization degree". The layout-specific digitalization degree is calculated from the quotient of the sum of each transferred information units multiplied with the corresponding value added level and the amount of transferred information units multiplied with the highest possible value added level (see Eq. 1). Following the evaluation scale for information flow, the highest possible value added level is "5". The layout-specific digitalization degree indicates, which percentage the degree of information flow promotes added value.

Key Performance Indicator "(Layout-Specific) Digitalization Degree" (Tomanek and Schröder 2017)

$$\begin{aligned} &(Layout - Specific)\ DIGITALIZATION\ DEGREE\ [\%] \\ &= \frac{\sum_{i=1}^{N} (Information\ Transfer \times Value\ Added\ Level)_i}{N \times 5} \times 100; \end{aligned} \tag{1}$$

Information Transfer $i = 1, \ldots, N$;
N = *Amount of transferred information per time unit*
Value Added Level $= 0, \ldots, 5$;

In addition to the layout-specific digitalization degree, a degree of compaction indicates the theoretical potential of improvement of the information flow in percentage. The degree of compaction is calculated by the difference between 1 and the layout-specific digitalization degree (see Eq. 2).

Key Performance Indicator "Degree of Compaction" (Tomanek and Schröder 2018)

$$\begin{aligned} &\mathit{DEGREE\ OF\ COMPACTION}\ [\%] \\ &\quad = 100 - (Layout - Specific)\ \mathit{DIGITALIZATION\ DEGREE}; \end{aligned} \tag{2}$$

8 Example of Application: Manufacturing

For determining the digitalization degree of an information flow it is useful to analyse the material flow first. The interaction of material flow and information flow represents the value added flow. In the following application example the authors selected an area at a production facility of an automotive supplier.

8.1 Analysis of Material Flow

A prerequisite for a holistic analysis of the information flow is an analysis of the material flow. In the following application example, the selected production line contains six working stations. At each working station the finished goods are check for quality reasons. Good parts (OK-parts) are transported directly from a working station to the distribution warehouse. Bad parts (NOK-parts) have to be reworked at two special rework stations. After a quality check at the end of rework, good parts are transported to the distribution warehouse. Reworked bad parts are disposed. Because of the high rework effort, NOK-parts from the working stations can't be moved directly to a rework stations. For that reason, parts for rework has to be temporary stored in an additional space in the production area. Based on data from a multi-moment-recording the analyzed material flow consists of a stream of 2.012 good parts and 130 bad parts for disposal. This is a process output of in sum 2.142 good and bad parts per day. In addition, the material flow of the rework process amount to 858 parts per day (see Fig. 7).

In detail, at six working stations 1.284 parts per day are good parts. They are transported from each working station directly to the distribution warehouse. In average, the rework at each working station is 143 parts per day. In sum, 858 NOK-parts per day has to be stored in a temporary warehouse for rework. In the shown example, the rework rate equates to 67%.

At the first rework station 198 parts per day are reworked. 164 of these parts are ok and 34 parts has to be disposed. At the second rework station 660 parts per day

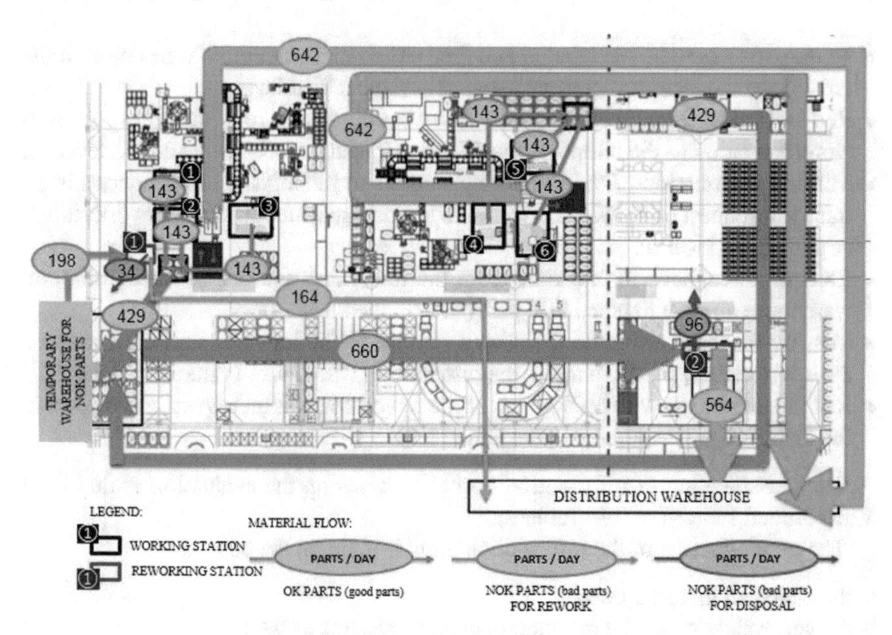

Fig. 7 Application Example: Sankey diagram of material flow (Tomanek and Schröder 2018)

are reworked. 564 of these parts are ok and 96 parts has to be disposed. The detailed material flow is shown by a Sankey diagram in Fig. 7.

8.2 Analysis of Information Flow

The information flow of the production process described above is analyzed using the Value Added Heat Map by Schröder and Tomanek. For that purpose the quantity and quality (according to the value added level, see Table 2) of information units is determined. A single information unit is defined as a production relevant content, e.g. confirmation of a finished order. In practice, the identification of information unis is difficult. A workable approach to determine the information flow in manufacturing is following the material flow.

Building on the analyzed material flow in Fig. 7 the corresponding information exchange is identified, qualified (classified) and quantified. The compounded information exchange forms the information flow. The visualization of the information flow includes all information necessary for the steering of the material flow. In the application example following information exchanges were identified and classified:

- Electronic feedback in real-time from the working stations to the distribution warehouse about a finished OK-part (value added level 4).
- Missing feedback from the working stations to the logistics staff concerning a transport request of OK-parts to the distribution warehouse (value added level 0).

- NOK-parts has to be red labeled and then stored in a quarantine warehouse in the area of the working station (paper-based—value added level 1).
- Red labeled boxes with NOK-parts at the working station request a transport from working station to the temporary warehouse (paper-based—value added level 1).
- Missing information of the reworking stations to the logistics staff concerning a transport request of NOK-parts from the temporary warehouse to a rework station (value added level 0).
- Electronic feedback in real-time from the reworking stations to the distribution warehouse about a finished OK-part (value added level 4).
- Missing feedback from the reworking stations to the logistics staff concerning a transport request of OK-parts to the distribution warehouse (value added level 0).
- Red labeled boxes with NOK-parts at the reworking station request a transport to the disposal (paper-based—value added level 1).

The classification of information exchange bases on the evaluation scale of the Value Added Heat Map (see Table 2).

The quantification of the information units is done on the basis of

- the analyzed material flow,
- further evaluation of the production control systems as well as
- supplementary observations made in the shop floor of the factory (multi-moment analysis).

The analyzed material flow consists an amount of 2.012 good parts per day, 130 bad parts per for disposal and 858 rework parts per day (see Fig. 7). The amount of material movements is not necessarily equal to the amount of information units.

Taking into account aspects like product type and the related product cage pallet for transportation the information exchange in given example contains 1.472 identified information units per day. The visualization of the information flow by a Value Added Heat Map is shown in Fig. 8. The determination of the information units is described in the following paragraphs in detail.

During the analyzed period 200 orders were electronically confirmed in real-time. 128 orders were confirmed directly from six working stations. 72 orders were confirmed after rework at two rework stations. In the shown use case, an order feedback is equivalent to a confirmed cage pallets with good parts. A cage pallet can contain from 6 to 12 parts depending on the product type. In average, a cage pallet in the application example includes an approximate amount of 10 good parts. Each order feedback transmitted to the production control system represents an information unit. 200 information units match to 13% of the registered information flow. According to the evaluation scale in Table 2 a real-time electronical exchange of information has a value added level of “4”. In a Value Added Heat Map a digitalization degree of “4” is represented by orange arrows.

At six working stations, an end of line test screens out not-OK-parts for rework. Each single not-OK-part is labeled by a printed out sticker and stored in a cage pallet. Following the material flow in Fig. 7, an amount of 858 parts for rework is labeled by a written document. This is equivalent to 858 information units. In addition,

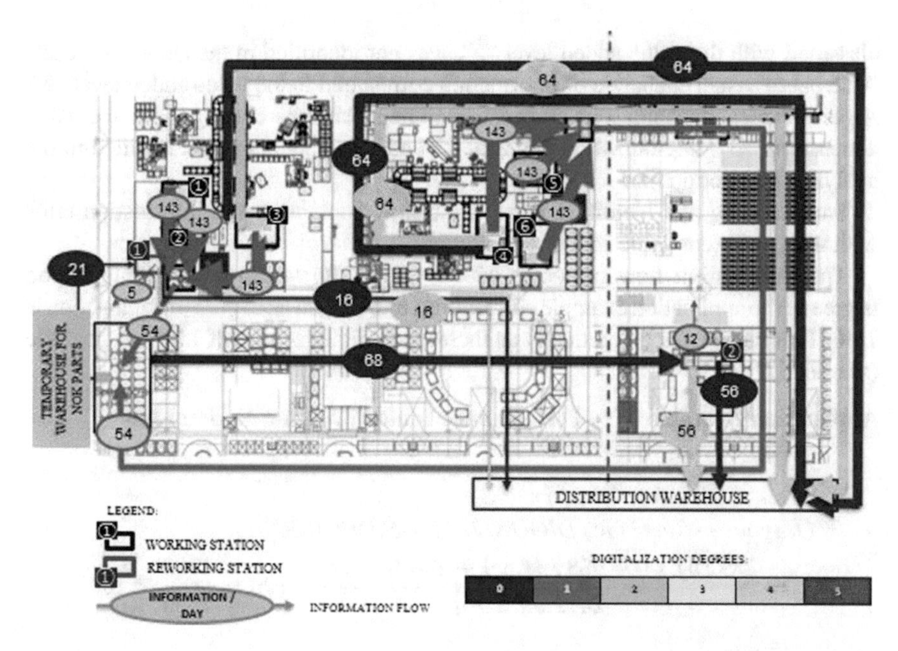

Fig. 8 Application Example: value added heat map of information flow (Tomanek and Schröder 2018)

each cage pallet with parts for rework is marked by a printed out written document (equivalent to an information units). In the case study, 108 cage pallets for rework were transported from working stations to a temporary warehouse for NOK-parts and then to two rework stations. After rework 17 cage pallets are still NOK-parts. They are marked for as parts for disposal by a printed out written document. In summary, 983 information units per day are written notification. This corresponds to 67% of the analyzed information flow. Written exchange of information is classified with the value added level "1". For visualization is used the color light blue.

An insufficient exchange of information caused non-value-adding wastage like unnecessary searches and handling for the forklift operators, stock keeping in the production line, and empty journeys of the forklift. In the application example, this level of information was identified in the form of missing notification about the physical transport of goods. Effects were observed during the multi-model record, when the forklift operators were searching for the boxes with OK-parts and not-OK-parts for transport. Specifically, 200 transport orders to distribution warehouse and 89 transport orders to rework stations were not transmitted to forklift drivers. In sum 289 information units were not-value added, what corresponds to 20% of the analyzed information flow. Consequently, these information units were classified with the value added level "0". According to the evaluation scale in Table 2 information with the level "0" are colored in dark blue in a Value Added Heat.

In the application example, the level of 4 was the highest recorded digitalization degree. This means that a digital exchanges of information in real-time, which is classified with the value added level “5”, was not identified in the shown use case. A verbal or visual exchange of information correspond to the value added level “2”. An electronical exchanges of information in non-real-time correspond to the value added level “3”. Digitalization degrees “2” and “3” were also not identified in the analyzed production line.

Summarizing, the identified information flow contains 1.472 information units with three different value added levels.

The information flow promotes added value. The layout-based digitalization degree of the application example corresponds to 24.2%. The degree of compaction is 76%. This means that theoretically an increase of digitalization of 75.8% is possible (see Eq. 3).

Application Example: Key Performance Indicators Information Flow

$$(Layout - Specific)\ \mathit{DIGITALIZATION\ DEGREE} = \frac{289\, Inf.x\ 0 + 983\, Inf.\, x\ 1 + 200\, Inf.\, x\ 4}{1472\ Inf.\ x\ 5} \times 100 = \frac{1783}{7360} \times 100 = 24.2\%$$

$$\mathit{DEGREE\ OF\ COMPACTION} = 100 - 24.2 = 75.8\% \tag{3}$$

A layout-based digitalization degree and the resulting degree of compaction in the shown example points out potential for improvement regarding the digitalization degree.

First of all, the information flow with the value added level “0” has to be eliminated immediately. In the application example the notification for the transport of OK-parts and not-OK-parts must be available to all forklift operators. A technical solution is to link the already existing electronic log-on system of finished goods with a request of a physical transport directly to the forklift operators.

In a further step, the information units with the value added level “1” have to be minimized. An up to 67 percent paper-based production line disclosure savings potentials. Through digitalization 983 written documents per day can be reduced. This reflects not only in cost saving, but also in time saving by reducing the writing effort. In general the intension of an information flow optimization through digitalization is an improvement of the value creation. An optimization of the added value includes also the minimization of media disruptions. Existing media disruptions in the analyzed example are visualized by color shifts of the information flow (see Fig. 8).

Summarizing, the application of the Value Added Heat Map can be used to optimize the information flow by pointing out a missing information exchange, determining the digitalization degree and visualizing media disruptions.

9 Summary

The Value Added Heat Map by Schröder and Tomanek has been already applied to production relevant factors like space usage, utilization of machinery and intralogistics traffic load. In this chapter this methodology has been successfully adapted to a further production relevant factor, namely the information flow.

The presented method shows how non-value added wastage caused by deficient or defective information can be visualized. It's an innovative holistic approach for optimizing the added value.

The Value Added Heat Map is also a method for determining the digitalization degree of the information flow in manufacturing companies. Determining the digitalization degree is a mandatory requirement in the context of Industry 4.0.

Moreover, a digitalization of the information flow can lead to a sustainable production without wastage. Furthermore, a digitalized information flow without media disruption is a contribution to a greening supply chain management.

The Value Added Heat Map method is not only limited for visualization. It is possible to use the VAHM to benchmark the information flow of different single production lines or even different production plants within one company. A visualization and optimization of the information flow among the whole supply chain, including suppliers and customers, can lead to an increase of the added value.

An industry-wide benchmark is also recommended to accelerate the value added optimization. Moreover, the Value Added Heat Map can be transferred to service companies to analyze and benchmark the information flow. In future, the benchmarking could also be carried out for cross-industry comparison of business service and production sectors.

There is a need for further research in improvement the value added approach of analyzing the information flow. The power of the presented case study is insufficient to quantify the impact of the Value Added Heat Map for greening the supply chain management. For that reason the authors suggest an adaption of the Value Added Heat Map on further companies and branches. From a scientific point of view, it is expedient to establish a higher sampling rate to confirm or refute the results of the described case study. In particular, initial research carried out by the authors showed promising results especially in service industry. Furthermore, it is also important to identify by addition research how the quality of data (especially master data) influence on the added value.

This book chapter discuss the importance of the information flow, determination of a layout-specific digitalization degree and gives recommendations for increasing the digitalization degree using an application example. Furthermore, digitalization in production is put in context of a greening supply chain management. Summarizing the presented research, the key recommendations from this work is to optimize the information flow for a lean and green supply chain.

References

acatech (2011) Cyber-physical systems—driving force for innovation in mobility, health, energy and production;

Balsliemke F (2015) Kostenorientierte Wertstromplanung: Prozessoptimierung in Produktion und Logistik. Gabler, Wiesbaden, p 9 as well as Rother M, Shook J (2000) Sehen lernen: Mit Wertstromdesign die Wertschöpfung erhöhen und Verschwendung beseitigen, 1. Aufl. LOG_X, Stuttgart, pp 100–101

Bauernhansl T (2017) Die Vierte Industrielle Revolution – Der Weg in ein wertschaffendes Produktionsparadigma. In: Vogel-Heuser B, Bauernhansl T, ten Hompel M (eds) Handbuch Industrie 4.0 Bd.4. Springer Reference Technik. Springer Vieweg, Berlin, Heidelberg, p 297 as well as n.p. (2012) Im Fokus: Das Zukunftsprojekt Industrie 4.0; Handlungsempfehlungen Zur Umsetzung; Bericht Der Promotorengruppe Kommunikation. Forschungsunion, p 31

Bergmann L, Lacker M (2009) Denken in Wertschöpfung und Verschwendung. In: Dombrowski U, Herrmann C, Lackerand T, Sonnentag S (eds) Modernisierung kleiner und mittlerer Unternehmen: Ein ganzheitliches Konzept. Springer, Berlin, p 161 as well as Bhasin S, Burcher P (2006) Lean viewed as a philosophy. J Manuf Technol Manag 17(1):56–72

Erlach K (2010) Wertstromdesign: Der Weg zur schlanken Fabrik. Springer, Berlin, pp 32–33 as well as Koch S (2015) Einführung in das Management von Geschäftsprozessen: Six Sigma, Kaizen und TQM, 2nd edn. Springer, Berlin, p 138

Gorodov EY, Gubarev VV (2013) Analytical review of data visualization methods in application to big data. J Electr Comput Eng 2013:3–7

Günthner WA, Schneider O (2011) Methode zur einfachen Aufnahme und intuitiven Visualisierung innerbetrieblicher logistischer Prozesse., Forschungsbericht des IGF-Vorhabens 16187 N der Bundesvereinigung Logistik (BVL) e.V., Lehrstuhl für Fördertechnik Materialfluss Logistik, Garching, p 32

Günthner W, Klenk E, Tenerowicz-Wirth P (2017) Adaptive Logistiksysteme als Wegbereiter der Industrie 4.0. In: Vogel-Heuser B, Bauernhansl T, ten Hompel M (eds) Handbuch Industrie 4.0 Bd.4. Springer Reference Technik. Springer Vieweg, Berlin, Heidelberg, p 297

Günthner W, Klenk E, Tenerowicz-Wirth P (2017) Adaptive Logistiksysteme als Wegbereiter der Industrie 4.0. In: Vogel-Heuser B, Bauernhansl T, ten Hompel M (eds) Handbuch Industrie 4.0 Bd.4. Springer Reference Technik. Springer Vieweg, Berlin, Heidelberg, p. 297 as well as n.p. (2012) Im Fokus: Das Zukunftsprojekt Industrie 4.0; Handlungsempfehlungen Zur Umsetzung; Bericht Der Promotorengruppe Kommunikation. Forschungsunion, p 13

Kocsi B, Oláh J (2017) Potential connections of unique manufacturing and industry. LogForum 13(4):389–400

Lee EA, Seshia SA (2017) Introduction to embedded systems. A cyber-physical systems approach, 2nd edn. MIT Press, Cambridge, MA, p 1

Lee H, Padmanabhan V, Whang S (1997) The bullwhip effect in supply chains. In: Sloan management review. Spring, p 93

Naisbitt J (1986) Megatrends. 10 Perspektiven, die unser Leben verändern werden [Vorhersagen für morgen], 6th edn. Bayreuth, Hestia

n.p. (2012) Im Fokus: Das Zukunftsprojekt Industrie 4.0; Handlungsempfehlungen Zur Umsetzung; Bericht Der Promotorengruppe Kommunikation. Forschungsunion, p 13

Ohno T (1993) Das Toyota-Produktionssystem, 1st edn, Campus-Verlag Frankfurt am Main, p 46 f

Porter ME, Millar VE (1985) How information gives you competitive advantage. Harvard Bus Rev 63(4):149. Retrieved from https://search.proquest.com/docview/1296463701?accountid=135362

Reese J (2016) Management von Wertschöpfungsketten. Unternehmenskooperation ohne Märkte. Verlag Franz Vahlen, München, p 80

Reese J (2016b) Management von Wertschöpfungsketten. Verlag Franz Vahlen. Unternehmenskooperation ohne Märkte, München, p 28

Rother M, Shook J (2000) Sehen lernen: Mit Wertstromdesign die Wertschöpfung erhöhen und Verschwendung beseitigen, 1. Aufl. LOG_X, Stuttgart, pp 100–101

Sankey HR (1896) The thermal efficiency of steam-engines (including appendixes). In: Minutes of the proceedings 125, no 1896, pp 182–212 as well as Koch S (2015) Einführung in das Management von Geschäftsprozessen: Six Sigma, Kaizen und TQM, 2nd edn. Springer, Berlin, pp 82–94

Schröder J, Tomanek DP (2012) Wertschöpfungsmanagement: Grundlagen und Verschwendung. Arbeiten der Hochschule Ingolstadt Nr. 24:17

Schröder J, Tomanek DP (2017) Layoutbezogenes Wertschöpfungsprogramm Value Added Heat Map nach Schröder und Tomanek, registration at German Patent and Trademark Office: 15.05.2017, DE, registration number 30 2017 012 325, 08.08.2017

Springer Gabler Verlag (eds) (2015) Gabler Wirtschaftslexikon. Medienbruch, Stichwort. Available at: http://wirtschaftslexikon.gabler.de/Archiv/77699/medienbruch-v9.html. Access 12 Nov 2015

Tomanek DP, Schröder J (2017a) Analysing the value of information flow by value added heat map. In: Dujak D (Hrsg) Proceedings of the 17th international scientific conference business logistics in modern management, Faculty of Economics in Osijek, p 86

Tomanek DP, Schröder J (2017b) Analysing the value of information flow by value added heat map. In: Dujak D (Hrsg) Proceedings of the 17th international scientific conference business logistics in modern management, Faculty of Economics in Osijek, p 88

Tomanek DP, Schröder J (2018a) Value Added Heat Map. Eine Methode zur Visualisierung von Wertschöpfung. Springer Fachmedien Wiesbaden, Wiesbaden, p 74

Tomanek DP, Schröder J (2018b) Value Added Heat Map. Eine Methode zur Visualisierung von Wertschöpfung. Springer Fachmedien Wiesbaden, Wiesbaden, p 11

Tomanek DP, Schröder J (2018c) Value added heat map. Eine Methode zur Visualisierung von Wertschöpfung. Springer Fachmedien Wiesbaden, Wiesbaden, p 69

Tomanek DP, Schröder J (2018d) Value added heat map. Eine Methode zur Visualisierung von Wertschöpfung. Springer Fachmedien Wiesbaden, Wiesbaden, p 70

Tomanek DP, Schröder J (2018e) Value added heat map. Eine Methode zur Visualisierung von Wertschöpfung. Springer Fachmedien Wiesbaden, Wiesbaden, p 71

Tomanek DP, Schröder J, Wirz M (2016) Value added heat map—a new method for the optimization of production space. In: 13th international conference on industrial logistics (ICIL 2016), Sawicz T (ed) Alnus Sp. z o.o, 2016, Krakow, pp 315–323

Verrier B, Rose B, Caillaud E (2015) Lean and Green strategy: the Lean and Green House and maturity deployment model. J Cleaner Prod, p 153

Green Supply Chain Management in Retailing Based on Internet of Things

Jelena Končar, Sonja Vučenović and Radenko Marić

Abstract Retail sector is about to face changes in the functioning of supply chains. The biggest challenge for retail management is how to achieve sustainable (green) supply chains and at the same time be efficient and profitable in the market. Production and distribution are already adapting to the trends of reducing energy consumption and thus affecting the environment. The key to success is the rationalization of operations, the reduction of costs and, on this basis, the provision of a more efficient supply chain, which enables competitiveness in the market. Such a retail supply chain requires information monitoring, flow, consolidation and management that implies the implementation of the Internet of Things. The aim of this chapter is to analyse the justified introduction of the Internet of Things to the retail sector. An empirical study was conducted covering two strata. The first stratum consisted of 30 managers of retail chains from the Western Balkans region. It was tested whether the introduction of the Internet of Things contributed to the improvement of indicators such as customer information, profitability, rationalization of costs, more efficient storage and logistics activities, packaging and waste disposal, product returns and customer safety and health. The second stratum covered 150 customers who were tested about the significance of these indicators for their perception of eco-quality (green quality). On the basis of the obtained results, regression models are defined which show the significance of each of the analysed indicators. The results of the research can serve as a guide to the management of retail chains on which indicators

J. Končar · S. Vučenović · R. Marić (✉)
Department of Trade, Marketing and Logistics, Faculty of Economics in Subotica, University of Novi Sad, Novi Sad, Serbia
e-mail: radenko.maric@ef.uns.ac.rs
URL: http://www.ef.uns.ac.rs/ofakultetu/osnovni-podaci/kadrovi/maric-radenko.php

J. Končar
e-mail: koncarj@ef.uns.ac.rs
URL: http://www.ef.uns.ac.rs/ofakultetu/osnovni-podaci/kadrovi/koncar-jelena.php

S. Vučenović
e-mail: sonjavucenovic1@ef.uns.ac.rs
URL: http://www.ef.uns.ac.rs/ofakultetu/osnovni-podaci/kadrovi/lekovic-sonja.php

A. Kolinski et al. (eds.), *Integration of Information Flow for Greening Supply Chain Management*, Environmental Issues in Logistics and Manufacturing, https://doi.org/10.1007/978-3-030-24355-5_11

they should develop and optimize through the Internet of Things. The limitations of the existing research and suggestions for future research are outlined in the paper.

Keywords Retail · Supply chain · Internet of things · 4.0 technologies · Retailers · Customers

1 Introduction

In recent years, a growing number of academic research (Tan et al. 2018; Wortmann and Flüchter 2015; Xia et al. 2012; Kopetz 2011) is dedicated to the problem of implementing the Internet of Things (IoT) in production and service processes and activities aimed at rationalizing costs and fostering sustainable development. The data show that many large European companies are active in the implementation of modern (digital) technologies, collecting 20–30% of revenues based on the use of consolidated data and applications supported by the Internet of Things (Uckelmann et al. 2011; Cedeño et al. 2018). In particular, the application of the Internet of Things is expanding in supply chains and retail sector. In the global market, the amount of business data and information is constantly increasing, which in the retail sector implies tracking each individual customer, each customer's purchase and storing of data. Consequently, retailers have a little room for decision-making on their action in relation to a particular market and final customers. The Internet of things, as such, in the retail sector allows tracking of each individual product from the supplier to the process of consumption by the final customers, which, in economic terms, should finally cause a decrease in the total cost of business operation (Liao and Cheung 2001).

In most organizations, information travels along well-known routes. Protected information is entered into databases and analysed, reports are being created, and thus the efficiency of management chains increases. Information also comes from external sources and from public sources on the Internet or purchased from information providers (Chui et al. 2010). However, predictable information paths change because the physical world becomes more of a type of information system. In such a system, disposing of right information is crucial, especially in the service sector, where reaction has to be immediate and in line with the ever-growing and changing demands of the market and customers. The placement of right information is possible only through the implementation of modern technology (4.0 technology), which will through wired and wireless networks, often using the Internet protocol (IP), connect all elements in one business system by installing sensors in products and physical objects. These networks place information and data from the manufacturer, through the distributor, to the final place of consumption, as well as feedback on product disposal, return of unused or damaged goods, etc. Huge amounts of information are converted into databases used for analysis. Revolutionary aspect of all of this is that these physical information systems start to be used in all spheres of business and mostly without human intervention (Whitmore et al. 2015; Chui et al. 2010).

Studies (Pang et al. 2015; Yan and Huang 2009) confirm that thanks to the Internet of things, or sensors placed into products, it is nowadays possible to track the products, i.e. production and service processes along the entire supply chain from the procurement of raw materials, the process of production and assembly of products, transport, distribution, conditions and manner of storage, transport to retail facility, arrangement, purchase, consumption and disposal. In this way, e.g. food producers, transport and logistics companies, and retail chains have a completely different perspective on business processes and supply chains. When sensors are placed in products (or packaging), companies can track the movement of these products, primarily monitor their interactions. Business models can be fine tuned to take advantage of these behavioural data. For example, in agreement with manufacturers and retailers, some insurance companies install location sensors in purchased cars. This allows insurance companies to define their insurance costs on the basis of how and in what way the car is driven and how and where it travels. Through the Internet of things, prices can be adapted to actual safety risks, instead of being based on an estimate of driver's gender, age, place of residence, etc. (Chui et al. 2010).

In supply chains, the implementation of the Internet of things is of particular importance as a means of promoting green management and sustainable development. In recent years, China has experienced the expansion of green cities and green management based on the Internet of things that connects all objects and population into a single system and places information on behaviour, consumption, service speed, critical points, etc. (Vlacheas et al. 2013).

Basically, Internet of things contributes to green supply chain management by providing accurate information on: (1) *consumption of natural resources*—which is important for the sustainability principle, (2) *method of production* of a specific product—e.g., the crucial information for food products is the composition or whether products contain emulsifiers, sweeteners, artificial colours and aromas, whether they were treated with artificial fertilizer, whether packaging is biodegradable, etc., (3) *manner and conditions of transport*—CO_2 emissions, delivery speed and safety, (4) *storage/manipulation operations*—these are information on electricity consumption, temperature and humidity, lighting, etc., (5) *retail*—information on consumption, size of turnover of a certain product group, manner of goods handling, shelving, and stock status, and finally, (6) *return*—information on how to return and how to deal with unused products or expired products, how to dispose of packaging and packaging waste, etc. are particularly important. In this way, the Internet of things provides information that is, on the one hand, useful for sustainable behaviour along the entire supply chain, and at the same time, on the other hand, enables all the chain participants to make their business more efficient and reduce cost by performing timely business activities on the basis of the obtained data.

Bearing in mind the presented aspects, *the subject of this paper* is explaining the need for the implementation of the Internet of Things in retail and its contribution to the green supply chain management. *The aim of the paper* is to analyse the justification of the implementation of the Internet of Things in the retail sector and to assess the importance of the Internet of Things to the development of indicators such as customer information, profitability, reduced costs, storage and logistics activities, disposal of packaging and waste, product returns and customer health and safety.

In this context, a theoretical research was carried out for emphasizing the significance of the implementation of the green supply chains for the retail sector, the importance of the introduction and the basic characteristics of the Internet of Things in retail and the contribution of 4.0 technology to green management. Empirical research included two samples, retail chain management, and customers who evaluated the importance of the Internet of Things for business activities.

The chapter comprises five sections. Following the Introduction, the *Review of Literature* presented the most important theoretical views of the significance of the green concept or the sustainability concept in supply and retail chains, and it defined the Internet of things, its basic characteristics and its contribution to retail. The third section presented the *Methodology* along with the research hypotheses, the method used in the research, the sample, the procedure and the methods of testing the hypotheses. The section with *Research Results* presented the results of the conducted tests, while the section of *Discussion* proposed a model that will contribute to more effective implementation of the Internet of Things and point to the importance of implementation for certain indicators. The final section summarized the conclusions of the most important results and provided suggestions for future research.

2 Literature Review

2.1 Implementation of Green Supply Chain in Retailing

Retail plays a role in supply chains through irreplaceable contact with the end customer. Since customers' demands are continually changing and increasingly reflect the need for organic or green products and eco-quality, retailers must meet such requirements and adjust their supply policy to green trends on the market. Implementing a sustainable or green supply chain in retail can improve the retailer's reputation, since socially responsible behaviour is important to modern customers and becomes crucial in achieving competitiveness. The key to success is the rationalization of operations, the reduction of costs and, on this basis, the provision of a more efficient supply chain, which enables competitiveness in the market (Buddress 2014; Sarkis 2003).

In the conditions of the modern globalization process, all the participants in the supply chain have significantly influenced the natural cycles and the state of the natural environment by performing their activities, which largely restricts unhindered technical and technological progress and production process. As environmental problems are bigger, manufacturers and all participants in the supply chain face bigger challenges in creating quality eco-products. These challenges involve large investments in research, exploitation of raw materials, elimination of harmful substances, as well as effective placement, advertising and promotion of quality of ecological production, distribution and placement in relation to traditional processes. The authors (Clark et al. 2009; Waage 2007; James 1997) confirm that end customers are increasingly showing positive reactions to green technologies and eco-products. They value not only the final product as such but the entire product development path through all stages of the life cycle. In addition to their functional properties, they are increasingly focused and oriented towards the purchase of products for ecological (sustainable) reasons. Customers place value not only on functional and structural components of quality, nor the quality of service, supply or delivery, but they expand it to the field of eco-quality. From the aspect of green management, this means that the quality and usability of products must now be considered as a unity of sustainable development, technical and technological determinants and environmental aspects. Some studies conducted on global market show that global demand for green products is growing at an annual rate of as much as 20% (Kopić et al. 2008), while it is estimated that retail sales of organic products annually exceed 100 billion US dollars (US$), with the trend that 93% of global customers expect on the market exclusively those products that meet social and environmental requirements (RILA 2018).

All of this points to the need to implement green supply chain management in retail sector in order to control the complete process of making products, from the quality of raw materials and reproduction material, the composition of products, procedure of packaging and composition of packaging, transport and storage, to the placement of the finished product into retail facilities and final consumption and subsequent disposal. This development path of the product to the final customer implies the availability of a large amount of information that is almost impossible to filter, store and process without the use of modern information technology and product digitization (4.0 technology).

The necessity of implementing a green supply chain in retail based on information and data integrated in the Internet of things, implies certain specificities in retailers' operations. The technology itself requires certain adjustments to retail operations, but also requires retailers to change their way of doing business, primarily in the area of contact with customers. As a result, in addition to lowering the costs of operations in the retail supply chain, competitiveness is achieved and is visible to the customers. Thanks to the development of a green supply chain, retailers build their competitiveness on non-price forms of competition, i.e. through socially responsible behaviour, which is increasingly important for modern customers.

2.2 Characteristics of IoT in Retailing

In the past few years, global market witnessed development and implementation of advanced technologies, and above all digital technologies within production, distribution and placement processes. To be competitive and to survive on the market, companies must follow the development and modernize and automate their supply chains. The fundamental changes bring digital technologies such as: the introduction of Internet of things, including open software platforms, open communications, open databases with strong embedded processors so that the networked business activities of all participants in the supply chain become transparent and easily accessible (Gomez et al. 2019; Karabegović and Karabegović 2018). The trend of computer application is ubiquitous in all areas, and the use of data from all sources of business implies the application of Internet of things. Today, every mobile phone becomes a personal computer through which it accesses the Internet and is intensively used in the buying process. The Internet of things does not exist only in the data collection segment, but it becomes crucial in the analysis of data and decision-making based on the collected data, therefore, and thus it is a useful tool in the entire supply chain. It becomes the imperative of modern business, and consequently, more and more academic and scientific research has been dedicated to the issue of the application and importance of IoT in production, placement and consumption.

Authors define the Internet of things (IoT) as a network of objects equipped with radio frequency chips, sensors, RFID tags and other similar technologies, so that these objects can communicate with one another, as well as with their users (customers). IoT represents the next evolution of the Internet, which shows tremendous progress and ability to collect, analyse and distribute data that we can convert into information, data and knowledge (Machado and Shah 2016). In this context, the literature emphasizes that the application of IoT implies the unity of three paradigms: Internet orientation, implementation of sensors and semantic orientation or knowledge (Gubbi et al. 2013). IoT is actually a network of interconnected objects uniquely addressable on the basis of standard communication protocols.

When it comes to the retail sector, the implementation of such a network that connects all participants in supply chains becomes an imperative in the modern global market (Končar and Leković 2013). Internet of things (IoT) provides new opportunities to retailers by connecting via the Internet each final customer with a retail facility and activities that are performed in the purchasing process, thus combining data from the physical and virtual world.

Digitalization is becoming a growing trend in the retail sector, which has always been a labour-intensive activity, and today the human factor is increasingly replaced by technology. Thanks to the Internet, retailers have the ability to carry out their operations more efficiently, but this also makes it easier to perform in an increasing number of marketing channels. The IoT movement provides retailers with opportunities in three critical areas: customer experience, supply chain and new revenue channels and streams. Retailers, who are reluctant to develop and execute the IoT strategy, risk being pushed out of the market, because if competitors implement the IoT, it is highly probable that those who hesitate to do so will be pushed out of the market.

The basic advantages of IoT in retail (Thanh and Thanh 2018) are reflected in the more efficient flow of information along the entire supply chain, precise information on the condition of the product in transport, warehouse and in stock, information on customers' habits, purchasing method, their needs and purchasing motives, information on how to display products, expiration date, the manner of keeping and handling the products, the condition of the packaging and especially information on the disposal of packaging and packaging waste, the return of unsold (unused) products, product breaking, scrap, power consumption, water consumption, carbon dioxide emissions, etc. All of these data, as the foundation of IoT infrastructure and architecture in retailing at the start of the IoT application, were provided by RFID tags, which were placed on the packaging of the product and were used to transmit information about the product through radio frequencies. RFID (Radio Frequency Identification) technology is a system of remote data sending and receiving between RFID tags, chips or transmitters and RFID readers, i.e. terminals for decoding, validating and sending information to a computer system (Alsinglawi et al. 2019). RFID tag is a data carrier for identifying a product consisting of a microchip placed on a small antenna that can be separate or integrated with an RFID sticker (Chu and Lisitsa 2018; Ugarak and Cice 2013). RFID sticker is placed on the desired packaging (product, transport packaging, pallets, containers, etc.). In this way, it is possible to track the movement of products throughout the entire supply chain.

Modern technology has led to evolution in IoT infrastructure so that today, besides RFID, communication protocols for constrained devices, (mobile) sensors and sensors networks are increasingly used, as well as interface, mobile applications, robotics, 4.0. technologies, etc. (Wang et al. 2013). There is a particular trend that the entire IoT infrastructure is less controlled and monitored by the human factor.

Today, the amount of business data on the global market is constantly increasing. It is interesting that some projections show that at global level there will be 40 zettabytes of useful data, which is more than the estimated amount of all the grains of sand on all the beaches on earth. In retail it means tracking each individual customer, its every purchase and storage of data, giving retailers room for decision-making on the placement policy and market performance. Apart from that, the key factor in introducing Internet of things in retail is cost reduction, optimization of business operations, more efficient fulfilment of the end customers' demands, and thus achieving a higher volume of operating income.

Furthermore, naturally, the IoT implementation greatly contributes to green management, sustainable development and achieving eco-quality (Kelly et al. 2013). Having obtained timely and accurate information, all participants in the supply chain as well as the final customers are familiar with final product they will use, its composition, the method of production and the entire development path to the retail facility. IoT enables all participants in the supply chain to define points critical to the natural environment and sustainable development in the chain itself. Furthermore, IoT enables more efficient tracking of customer data from the loyalty program, adjusting the offer to the selected market segment, introducing the logistics concept of a quick response, etc.

However, on the other hand, the application of IoT in the retail sector has some negative aspects. Namely, the introduction of IoT technology implies high initial costs of installing sensors, maintaining, storing data, collecting data, as well as processing data that requires the software, as well as advanced knowledge in this field. Therefore, regardless of the pressure on the digitization of retail, small retailers of a local character will have no interest in introducing the Internet of things because of the above mentioned costs. This is considered one of the biggest negative features of IoT implementation that threatens to further deepen the gap between big retail chains and small independent retailers. It is also known that there are restrictions on the introduction of electronic payments with retailers operating with only one or a few retail facilities and performing locally. Certainly, it is very important that retailers before the implementation of Internet of things thoroughly evaluate the actual effects of this concept on their business. Data management and security (safety) of customer data is also a big threat.

2.3 Green Supply Chain Management Based on Internet of Things

Real and virtual worlds are rapidly evolving and creating the Internet of things (IoT). In fact, IoT has encouraged companies to focus their business and activities on the fourth industrial revolution called Industry 4.0. The basic characteristic of modern business in the context of Industry 4.0 will be the flexibility in the scope of production and the adjustment to the end market (Shrouf et al. 2014). It will reflect in a wider integration between manufacturers, suppliers, trading companies and customers, and above all in sustainability along the entire supply chain.

Supply chain management is the coordination and management of a complex network of activities involved in bringing finished products to final customers. All product life cycle phases affect the ecological suitability of the supply chain, from resource extraction to production, use and reuse, final recycling or disposal. This precisely reflects the green supply chain management (GSCM), which is defined as green procurement + green manufacturing + green distribution + reverse logistics (Ninlawan et al. 2010). The idea of GSCM refers to the rational use of natural resources for the production of new products (concept of sustainability), production that is safe for customers' health (products without chemicals, GMOs, additives, emulsifiers E102, E103, E104, etc.), rational use of fuels, electricity and water consumption, reduced emissions of CO_2, elimination or minimization of waste along the supply chain, proper disposal of packaging and packaging waste, reverse logistics, controlled use of pesticides, artificial fertilizers, etc. Such a green supply chain can be organized and operationalized only through networked sensors, data and information through the Internet of Things.

The literature (Pang et al. 2015; Gubbi et al. 2013; Ugarak and Cice 2013; Kopić et al. 2008) points out the following as the most important indicators in the retail sector in direct correlation with the implementation of IoT and green supply chains: customer information, profitability, rationalization of costs, more efficient storage and logistics activities, disposal of packaging and waste, product returns and customer safety and health.

Informing customers—through the IoT platform, all product data are collected in a single database, where the necessary information are then filtered and sent to final customers via mobile phones and mobile applications (Lee and Lee 2015). In this way, customers who prefer eco-products know whether and to what extent the entire path of bringing a product to a retail facility is actually in line with the principles of sustainable development and green production, procurement and distribution (Grubor 2008).

Profitability—thanks to the implemented IoT, retailers as well as other participants in the supply chain can fully adapt their business to the logistics concept of the Quick Response (QR). This means that the entire supply chain should generate higher revenues on the basis of more flexible production in terms of what and how much should be placed on the market, for example: whether the market needs eco-products, to what extent, whether they have a seasonal character, if they are modern, etc. IoT enables significant revenue growth through flexibility, which implies that the volume of production is dramatically increased or decreased during the various demand fluctuations, i.e. that the change of production from one product to another is fast, timely and efficient.

Rationalization of costs—refers to the obligation of the supply chain participants to properly channel the obtained information on the consumption of fuel, electricity, gas, water, etc., and, on the basis of the obtained data, direct their efforts towards rationalization of the use of energy and energy products in order to adapt business activities to the concept of sustainability.

Storage and logistics activities—are one of the most important indicators directly influenced by IoT because they are transparent information about the storage state and conditions of products in stock (temperature, light, moisture), the way the product is handled and manipulation activities in the warehouses (inventory replenishment, cost rationalization), transport speed, product safety and security during transport, CO_2 emissions, etc.

Disposal of packaging and packaging waste—IoT completes information on the selection of adequate packaging materials for specific product groups, which implies a good knowledge of the functional characteristics of the product being packaged, knowledge of the characteristics of packaging materials (durability, resistance, ease of use, functionality) and above all, the perception of the ecological status of packaging, packaging material and packaging waste.

Product return—the implemented IoT allows for more efficient tracking and grouping of data on the product expiration date and perishability. In this way, retailers are timely informed about the time period when the product due to the expiration date has to be withdrawn from the shelves, from the storage, warehouse, etc. It is especially important to monitor how it is further disposed of as waste, reproduction material or raw material.

Customer safety and health—through the IoT network, information are grouped and then sent to customers, but only those related to the safety of products for the health of final customers in terms of the content of allergens, organic substances, the treatment of products with pesticides, etc. This is particularly important for the placement of food products such as: meat and meat products, milk and dairy products, fruits and vegetables, confectionery products, etc.

Bearing in mind the presented aspects, the aim of this chapter is to analyse the justification of introducing green supply chain management on the basis of IoT by establishing the statistical correlation and significance of IoT for the mentioned indicators.

3 Methodology

3.1 Aim and Hypothesis

Given the defined theoretical significance of the implementation of IoT for green supply chains in the retail sector, the main goal of the empirical research is to determine a statistically significant correlation between IoT and green indicators, such as: customer information, profitability, cost rationalization, storage and logistics activities, disposal of packaging and waste, product return and customer safety and health. The goal and subject of the research are operationalized through the following research hypotheses:

- $\mathbf{H_1}$ reads: the implementation of the Internet of things in the retail sector is statistically significantly correlated with green indicators, such as customer information, profitability, cost rationalization, storage and logistics activities, disposal of packaging and waste, product return and customer safety and health. Alternative hypothesis $\mathbf{H_{1a}}$ reads that these differences are coincidental and that the implementation of the Internet of things does not significantly affect the improvement of the observed indicators.
- $\mathbf{H_2}$ reads: green sustainability indicators like customer information, cost rationalization, storage and logistics activities, disposal of packaging and waste, product return and customer safety and health statistically significantly affect customer satisfaction in the selection of finished products and their perception of eco-quality in supply chains. Contrary to that, the second alternative hypothesis $\mathbf{H_{2a}}$ implies that this correlation is not statistically significant and that these indicators do not have a significant effect on the behaviour of customers.

3.2 Variables

The presented research and conducted analyses included different variables in order to as precisely as possible confirm or reject the hypotheses. Final customers and retailers are taken as independent grouping variable. In other words, respondents from the Western Balkans market are divided into two independent groups, as final customers of finished products and as managers of retail facilities participating in the implementation of IoT. Independent variables of interval type of measurement are assessments of respondents' satisfaction with indicators: customer information, prof-

Table 1 Research sample

Market	Managers		Final customers	
	n	%	n	%
Croatia	8	27	37	25
Serbia	14	47	42	28
Bosnia and Herzegovina	4	13	36	24
Montenegro	4	13	35	23
TOTAL	30	100	150	100

Source author's calculation

itability, cost rationalization, storage and logistics activities, disposal of packaging and waste, product return and customer safety and health. Each of these indicators has been operationalized through three-five Likert-type items (1–5 points). A dependent variable is the implementation of green supply chain management in the WB retail that is based on the Internet of things.

3.3 Research Sample

The survey covered a total of 180 respondents from selected markets of the Western Balkans (Croatia, Serbia, Bosnia and Herzegovina, Montenegro). The sample is gender-balanced, i.e. 52% of female respondents and 48% of male respondents. In terms of age, most respondents had between 40 and 60 years of age (55%), there was 30% of respondents aged between 20 and 40, and 15% older than 60. The research sample consists of two strata. The first stratum was made by retail facility managers who are involved in the implementation of IoT, are familiar with the results and use IoT. The second stratum represented final customers of finished products. The number of respondents in the first stratum was 30, representing the middle level of managers (e.g. Heads of warehouses, Managers of sales facilities, Heads of distribution and transport, persons responsible for receiving goods, etc.). They work on the WB market in companies that have organized their green supply chains on the IoT basis, such as: Delhaize Maxi, Mercator, Spar, Voli Trade, Delta Transport Logistics, etc. The second stratum had 150 respondents, representing the users of final products who make purchases in retail chains on the WB market. Customers were equally represented in the analysed WB countries (markets). The representation of respondents in the sample is given in the Table 1.

3.4 Measurement Procedure and Instruments

The research of the participants' opinions was carried out during August and September 2018 in the Western Balkans region. The survey was conducted through a questionnaire, that was e-mailed to employees and customers of the largest WB retail chains (Delhaize Maxi, Mercator, Spar, Voli Trade, etc.). Respondents were divided into two strata: managers of retail facilities (Stratum 1) and final customers (Stratum 2). The task was for the respondents to express their agreement with the following statements on a five-point scale, such as: (a) *for managers*—on a rating scale of 1–5, rate each indicator individually, how the implementation of IoT is associated with the improvement of green indicators, (b) *for customers*—on a rating scale of 1–5, rate how the indicators are related to your satisfaction with the purchase and selection of eco-products.

3.5 Data Analysis

The data collected in the research were processed and the necessary tests were performed on the basis of the statistical package for social sciences SPSS 20. Respondents' answers from both strata were described by descriptive statistics. The mean values for each of the analysed indicators were presented, as well as the deviation for each dimension. The accuracy of the set hypotheses was examined by the statistical method—Multiple regression analysis.

4 Research Results

Descriptive statistics describe the extent to which respondents from both strata agree with the statement that there is a statistically significant correlation between these indicators and the implementation of IoT in the green supply chain management in the Western Balkans retail sector. Rating 1 expresses the lowest, while rating 5 expresses the highest agreement. Ranking results for each indicator are provided in Table 2. First, the results and necessary data will be presented for the testing of the hypothesis for stratum of managers, and then for the final customers' stratum.

The above table shows that respondents from the first stratum show the highest agreement with the indicator of Storage and logistic/manipulation activities ($M = 4.68$), and the lowest with Cost rationalization ($M = 3.11$). In addition to these rankings of each individual indicators and sub-items, descriptive statistics is presented in the following Table 3.

Managers are of the opinion that the implemented IoT in the retail supply chains has the greatest impact on the indicator of Storage and logistics/manipulation operations in the sense that in this way it ensures a better and timely connection between

Table 2 Ranking results for the stratum of managers

No.	Criterion	Sub-criterion	Rating
1	Customer information	Timeliness	4.6
		Accuracy	4.7
		Use of mobile applications	4.9
		Use of RFID tags	3
		Mutual exchange of information	3.9
2	Profitability	Increase in turnover size	4
		Lowering costs	3
		Flexible production	4.8
		More efficient business	4.9
3	Cost rationalization	Electricity consumption	3
		Gas consumption	3.2
		Fuel consumption	4.2
		Water consumption	2
4	Storage and logistics activities	Concept of Quick Response (QR)	4.8
		Fast and efficient transport	4.6
		GPS support	4.7
		Storage conditions (temperature, humidity, light, etc.)	4.6
5	Disposal of packaging and waste	Environmental status of packaging	4
		Transportation of packaging waste	4.5
		Possibility of recycling	3.9
6	Product return	Information on the product expiration date	5
		Efficient monitoring of product returns and disposal	3
7	Safety and health	Product composition	4.9
		Method of manufacturing and treatment of raw materials	4
		Use of artificial additives	4.2
		Perishability	4.5
		Amount of allergens	4
		Use of GMO	4.8

Source author's calculation

Table 3 Descriptive statistics for examined indicators

Indicators	Min	Max.	Mean	St. deviation
Customer information	2.63	4.97	4.2200	1.17253
Profitability	2.00	4.67	4.1750	1.36797
Cost rationalization	1.33	4.25	3.1000	0.77569
Storage and logistics activities	2.00	5.00	4.6750	0.613810
Disposal of packaging and waste	1.67	4.67	4.4000	0.96842
Product return	1.00	5.00	4.0000	1.08754
Safety and health	2.33	4.97	4.4166	0.814879

Source author's calculation

manufacturers, distributors and retailers. The development of the Quick Response concept is especially emphasized, as a timely reaction to changes in the requirements of the final market. The second ranked indicator in terms of significance is Customer safety and health (M = 4.41). Use of IoT provides customers with the opportunity to check the organic composition of products and the content of all substances and raw materials (natural or artificial) which may be potentially harmful to health. Great significance is attached to the Disposal of packaging and waste (M = 4.4), and especially to information on the ecological status of packaging and the packaging waste transport routes. Timeliness and accuracy of information is in fourth place (M = 4.22), while a somewhat lower significance is attributed to Profitability (M = 4.175) probably due to major initial investments in the implementation of IoT, as well as to the indicator of Energy cost savings (M = 3.1).

A multiple regression analysis was applied to examine the impact and correlation of the mentioned group of indicators on the concept of IoT-based green supply chain. First, the overall stratum of managers of retail chains will be tested, and then each indicator individually. In the first stratum, the Enter method was applied, in which all independent variables were included together in order to predict the dependent variable. The obtained results indicate that the regression model is statistically significant ($F = 5.29$, $p < 0.001$). In other words, this means that a set of tested indicators statistically significantly predicts the implementation of IoT in the green supply chain management in the retail sector. It describes 67.4% of the variance of the criteria. In addition to the overall contribution of the set of predictors, the contribution of individual predictors was also examined. Their contribution is provided in the following Table 4.

Observing the contributions of individual indicators, it can be concluded that indicator of Storage and logistics/manipulation activity mostly contributes to the description of criterion variance (B = 1.438). This means that in the opinion of managers, the implemented IoT increases the efficiency of transaction processing, ensures more secure transportation of goods and services purchased, rationalizes the number of staff in the warehouse, increases the speed and quality of supply, reduces inventory levels, reduces the number of faults and transport risks, ensures better atmospheric

Table 4 Contribution of individual predictors to the description of the criterion variable

Model	Unstandardized coefficients		Standardized coefficients	T	Significance
	B	Std. error	Beta		
1 (constant)	4.283	1.853		4.842	0.000
Customer information	0.749*	0.545	0.431	3.714	0.001
Profitability	0.228	0.478	0.011	0.0127	0.051
Cost rationalization	0.177	0.420	0.140	0.8742	0.073
Storage and logistics activities	1.438**	0.798	0.235	0.0217	0.000
Disposal of packaging and waste	0.819**	0.137	0.358	2.461	0.000
Product return	0.313*	0.392	0.285	1.529	0.001
Safety and health	0.923**	0.251	0.470	0.9843	0.000

Source author's calculation
* Significant at the level 5%, ** significant at the level 1%

storage conditions, etc. The following indicators also made a statistically significant correlation with IoT: Disposal of packaging and waste (B = 0.819) and Customer safety and health (B = 0.923). Given the implemented IoT, it is possible to more efficiently monitor the environmental status of packaging and controlled disposal of packaging waste. On the other hand, information on the composition of the product and the process itself are available to the final customers through mobile applications in a transparent manner. A slightly lower statistical significance is attached to Customer information (B = 0.749) and Product return (B = 0.313). According to opinion of the management, there is still no statistically significant correlation between IoT and Profitability (B = 0.228) and Cost rationalization (0.177).

Based on the conducted testing, it can be concluded that a statistically significant correlation exists between the observed indicators: (a) green indicators: Customer information, Profitability, Cost rationalization, Storage and logistics activities, Disposal of packaging and waste, Product return and Customer safety and health and (b) implemented IoT in the retail sector. This means that the changes of one or the other indicators statistically significantly affect their mutual relationship, i.e. the growth of the value of one indicator influences the growth of the value of the second indicator and vice versa. This conclusion confirms the first research hypothesis $\mathbf{H_1}$ in the paper, while the alternative hypothesis $\mathbf{H_{1a}}$ is rejected.

In the second research stratum, it was tested whether and how the implementation of IoT in green supply chains in the retail sector is important to final customers, and how they evaluate (rate) these indicators in terms of their individual significance to making a final decision on the purchase of eco-products. As in the case of the first stratum, the descriptive statistics described the extent to which respondents agree with the statement that there is a statistically significant correlation between

Table 5 Ranking results for the stratum of final customers

No.	Criterion	Sub-criterion	Rating
1	Customer information	Timeliness	4.6
		Accuracy	4.9
		Use of mobile applications	4.7
		Use of RFID tags	3.5
		Mutual exchange of information	4.1
2	Cost rationalization	Electricity consumption	3.5
		Gas consumption	2
		Fuel consumption	4
		Water consumption	2.1
3	Storage and logistics activities	Concept of Quick Response (QR)	4.6
		Fast and efficient transport	4.8
		GPS support	3
		Storage conditions (temperature, humidity, light, etc.)	4.9
4	Disposal of packaging and waste	Environmental status of packaging	4.6
		Transportation of packaging waste	4.4
		Possibility of recycling	4.7
5	Product return	Information on the product expiration date	4.8
		Efficient monitoring of product returns and disposal	3.5
6	Safety and health	Product composition	5
		Method of manufacturing and treatment of raw materials	4.5
		Use of artificial additives	4.8
		Perishability	4
		Amount of allergens	4.5
		Use of GMO	4.7

Source author's calculation

the mentioned indicators of the implementation of IoT and their decision on the purchase and selection of finished products in a retail facility. The following table (Table 5) presents the rating results for each indicator individually as well as for the sub-criteria offered to the final customers.

When testing the respondents in this stratum, indicator of Profitability is omitted, since final customers do not have an insight into the volume of consumption and realized sales turnover. The highest average rating, as well as the highest individual rank by sub-criteria, were assigned to Customer safety and health (M = 4.58). Final customers expect from the implementation of IoT in the green supply chain to provide

Table 6 Descriptive statistics for examined indicators

Indicators	Min	Max.	Mean	St. deviation
Customer information	2.67	4.67	4.3600	1.28541
Cost rationalization	1.00	3.33	2.2000	0.88679
Storage and logistics activities	1.67	4.33	4.3250	0.78412
Disposal of packaging and waste	1.67	4.97	4.5667	1.17428
Product return	1.33	4.67	4.1500	1.19845
Safety and health	2.33	4.97	4.5833	0.97423

Source author's calculation

them with accurate information about the composition of products, raw materials and manufacturing method. Almost equally, environmentally-oriented customers expect that implemented IoT provide them with the information on the environmental status of packaging, proper disposal of waste, and the possibility and options for recycling packaging materials (M = 4.56). More detailed descriptive indicators and descriptive statistics are shown in Table 6.

In addition to the aforementioned indicators concerning Customer safety and health and Disposal of packaging and packaging waste, the largest average ratings in this stratum was achieved by the possibility of timely information and data transparency (M = 4.36), and Storage and logistics/manipulation activities (M = 4.35), where final customers particularly value information provided by IoT regarding storage conditions (temperature, humidity, light, etc.) and speed and accuracy of transport. Other indicators have a lower average significance rank.

As in the case of the first stratum for the testing of the second research hypothesis, a multiple regression analysis was applied to examine how significant the implementation of IoT is in green supply chains. In this context, the correlation was tested between the mentioned group of indicators and the final customers' decision on the selection of eco-products. The obtained results indicate that the regression model is statistically significant ($F = 10.964$; $p < 0.001$), i.e. according to the opinion of final customers, the set of tested indicators statistically significantly predicts the implementation of IoT in the green supply chain management in the retail sector. The contributions of individual indicators are illustrated in the Table 7.

Based on the data provided in the table, it can be concluded that the criterion variable is statistically significantly predicted by the indicators of Customer safety and health (B = 1.271) and Disposal of packaging and waste (B = 1.012). Final customers see the correlation between the implementation of IoT and green supply chain management in the mentioned indicators, which is the availability of information that will convince customers of the origin and composition of products, the ways of eco-packaging and environmental protection in the form of proper disposal of packaging waste. The next rating in terms of importance is achieved by Customer information (B = 0.938), where with the help of modern mobile applications based on IoT, customers can filter and select the most important information in the supply

Table 7 Contribution of individual predictors to the description of the criterion variable

Model	Unstandardized coefficients		Standardized coefficients	T	Significance
	B	Std. error	Beta		
1 (constant)	4.002	0.975		4.598	0.000
Customer information	0.938**	0.655	0.542	2.306	0.000
Cost rationalization	0.371	0.330	0.247	0.8742	0.061
Storage and logistics activities	0.727*	0.584	0.346	0.0217	0.001
Disposal of packaging and waste	1.012**	0.247	0.249	2.542	0.000
Product return	0.617	0.483	0.487	1.529	0.005
Safety and health	1.271**	0.362	0.574	0.8759	0.000

Source author's calculation
*Significant at the level 5%, ** significant at the level 1%

chain. Statistically significant correlation is also evident for the indicators of Storage and logistics/manipulation activities (B = 0.727), which based on IoT can provide customers with information about the conditions of storage of products, the manner and conditions of transport, inventory replenishing, product breaking, scrap, etc.

According to the results of the conducted testing in Table 7, it can be concluded that the analysed green indicators: Customer information, Cost rationalization, Storage and logistics activities, Disposal of packaging and waste, Product return and Customer safety and health statistically significantly influence customer satisfaction in the selection and purchase of products and their perception of eco-quality and the need for the implementation of IoT in supply chains. Therefore, the second research hypothesis $\mathbf{H_2}$ is accepted, while the alternative hypothesis $\mathbf{H_{2a}}$ is rejected.

5 Discussion

The significance of the observed indicators in introducing IoT into the green supply chain management system was analysed (assessed) from two different points of view: from the perspective of retail management and final customers. The reason for such a two-way analysis is the fact that the concrete research indicates advantages (as well as the shortcomings) in the application of IoT for the retail sector and supply chains, on the one hand, and the targeted segments of the market (final customers), on the other hand. Based on summarized responses and conducted tests, it is possible to perform two regression models that show the impact of each individual indicator on the implementation of IoT, viewed from the aspect of managers (model 1) and the aspect of final customers (model 2).

$$y = 4.28 + 0.75x_1 + 0.23x_2 + 0.18x_3 + 1.44x_4 + 0.82x_5 + 0.31x_6 + 0.92x_7 \quad (1)$$

$$y = 4 + 0.94x_1 + 0.37x_3 + 0.73x_4 + 1.01x_5 + 0.62x_6 + 1.27x_7 \quad (2)$$

In the above models, y is the dependent variable (implementation of IoT), while x is independent variable, such as: x_1—Customer information; x_2—Profitability; x_3—Cost rationalization; x_4—Storage and logistics activities; x_5—Disposal of packaging and waste; x_6—Product return and x_7—Customer safety and health.

The first model represents an equation that shows the importance and interaction between the introduction of IoT into the green supply chain management and each observed indicator. The second model presents the final customers' opinion on how much IoT affects the indicators that are primary to the customers for sustainable behaviour and consumption, as well as the selection of final products. The results obtained coincide with the results of related studies (Pang et al. 2015; Gubbi et al. 2013; Yan and Huang 2009), which demonstrate that the application of IoT enables more efficient monitoring of the product, i.e. the process of manufacturing and process of servicing along the entire supply chain, from the procurement of raw materials, the manufacturing itself and the assembly of products, through transport and distribution, to the conditions and manner of storage and shelving and sale of products to final customers.

When observed by models, respondents' responses match in terms of information provided by IoT regarding customer safety and health, environmental status of packaging, storage and logistics/manipulation operations, and transparent and two-way information. According to the opinion of the respondents, IoT has a lower importance in terms of profitability and rationalization of costs, which is contrary to studies carried out on the global market (Whitmore et al. 2015; Chui et al. 2010; Liao and Cheung 2001), which show that the introduction of IoT significantly reduces labour costs, affects the growth of business revenues and rationalizes the service itself. However, the Western Balkans market is still underdeveloped and the introduction of IoT requires high initial investment, as well as high costs of employee training, sensor installation, product and packaging labelling, networking of facilities and devices, etc. All of the above, in this initial period of implementation of IoT, makes it difficult to estimate its actual impact on expenditure savings and turnover growth.

Apart from this, the mentioned models can also have practical significance because they show to the green supply chain management which indicators are most influenced by IoT, which indicators attention should be paid to and should be optimized, and how final customers look at the introduction of IoT and which indicators should be more promoted and developed through IoT, in order to influence their satisfaction and thereby build long-term relationships (CRM).

6 Conclusion and Further Research

The need to research the implementation of the IoT concept in green supply chain management in the retail sector has arisen as a result of the fact that it is a segment of the service process that is increasingly emerging as a key competitive advantage in the global market. Observing the evolution of new technologies in retail, electronic retailing dominated by the middle of the second decade of the 21st century. However, technologies applicable in retail facilities will have the increasing influence in further retail development.

Digitalization is becoming a growing trend in the retail sector, which has always been a labour-intensive activity, and today the human factor is increasingly replaced by technology. For retailers, the growth of end-customer data and their eco-habits and preferences is especially enabled by IoT. In many cases, the application of IoT structure and applications will be a necessary condition for maintaining competition, as well as being able to manage data and transparency of information, protecting customer safety and health, more effectively monitoring logistics/manipulation operations, etc.

In other words, the application of IoT becomes the future of technology development by retailers, who through technology find ways to become competitive. Thanks to IoT, they are simply trying to improve their purchasing experience.

In this context, empirical research was conducted with the aim of determining which indicators in the retail sector are directly influenced by IoT and which indicators are most valued by final customers when making a purchase decision. Confirmed research hypotheses support the fact that statistically significant correlation exists between the indicators: Customer information, Profitability, Cost rationalization, Storage and logistics activities, Disposal of packaging and waste, Product return and Customer safety and health, and the application of IoT in the green supply chain management in the retail sector, and that their interaction cannot be excluded or ignored. Based on the multiple regression analysis, two regression models were presented, which in the form of mathematical equations show the intensity and direction of this relationship.

The shortcomings of the conducted research are reflected in the territorial limitation of the research and the size of the sample. The research is limited exclusively to the four countries of the Western Balkans, and the objective reason for this choice of geographical location of the research is the authors' familiarity with the mechanisms of the retail sector functioning and customers' habits in the regional market. Regarding the size of the sample, the second stratum made of 150 final customers is sizeable, while the first stratum of 30 managers is limited for objective reasons that relate to the unavailability of respondents from the management hierarchy to participate in this type of research. In addition, seven indicators were tested, which can lead to simplified conclusions. Objectively, there is a need for the analysis to be extended to certain subgroups within the mentioned indicators.

Therefore, as part of the guidelines for future research, it is necessary to: (1) include a larger sample of respondents (e.g. 100 managers and 300 final customers),

(2) focus on the analysis and testing of subgroups within the mentioned indicators and how their impact is reflected in the application of IoT, and (3) expand testing to the global market, e.g. to test supply chains in the region of South Eastern Europe and make comparisons between countries that are EU members and countries that are in the EU accession process (transition countries).

References

Alsinglawi BS, Nguyen QV, Gunawardana U, Simoff S, Maeder A, Elkhodr M, Alshehri MD (2019) Passive RFID localization in the internet of things. In: Recent trends and advances in wireless and IoT-enabled networks. Springer, Cham, pp 73–81
Buddress L (2014) Integration: a supply chain view. Strateg Manage 19(3):3–9
Cedeño JMV, Papinniemi J, Hannola L, Donoghue I (2018) Developing smart services by internet of things in manufacturing business. LogForum 14(1):59–71
Clark G, Kosoris J, Hong LN, Crul M (2009) Design for sustainability: current trends in sustainable product design and development. Sustainability 1(3):409–424
Chui M, Löffler M, Roberts R (2010) The internet of things. McKinsey Q 2:1–9
Chu G, Lisitsa A (2018) Penetration testing for internet of things and its automation. In: 2018 IEEE 20th international conference on high performance computing and communications; IEEE 16th international conference on smart city; IEEE 4th international conference on data science and systems (HPCC/SmartCity/DSS), pp 1479–1484
Gomez C, Chessa S, Fleury A, Roussos G, Preuveneers D (2019) Internet of Things for enabling smart environments: a technology-centric perspective. J Ambient Intell Smart Environ 11(1):23–43
Gubbi J, Buyya R, Marusic S, Palaniswami M (2013) Internet of things (IoT): a vision, architectural elements, and future directions. Future Gener Comput Syst 29(7):1645–1660
Grubor A (2008) Marketing mix instruments combining in the function of the sustainable development. Ann Fac Econ Subotica 20:149–157
James P (1997) The sustainability cycle: a new tool for product development and design. J Sustain Prod Des, 52–57
Kopić M, Cerjak M, Mesić Ž (2008) Consumer satisfaction with the offer of organic products in Zagreb. IN: 43rd Croatian and 3rd international symposium on agriculture, Opatija, Proceedings, pp 259–265
Končar J, Leković S (2013) Trends of modernization and structural transformation of trade as a function of sustainable development. Ann Fac Econ Subotica 29:407–419
Karabegović IV, Karabegović EM (2018) Implementation of Industry 4.0 by application of robots and digital technologies in production processes in China. Tehnika 73(2):225–231
Kopetz H (2011) Internet of things. In: Real-time systems. Springer, Boston, MA, pp 307–323
Kelly SDT, Suryadevara NK, Mukhopadhyay SC (2013) Towards the implementation of IoT for environmental condition monitoring in homes. IEEE Sens J 13(10):3846–3853
Lee I, Lee K (2015) The Internet of Things (IoT): applications, investments, and challenges for enterprises. Bus Horiz 58(4):431–440
Liao Z, Cheung MT (2001) Internet-based e-shopping and consumer attitudes: an empirical study. Inf Manag 38(5):299–306
Machado H, Shah K (2016) Internet of Things (IoT) impacts on supply chain. Retrieved, 1–19
Ninlawan C, Seksan P, Tossapol K, Pilada W (2010) The implementation of green supply chain management practices in electronics industry. Proc Int Multiconference Eng Comput Sci 3(1):17–19
Pang Z, Chen Q, Han W, Zheng L (2015) Value-centric design of the internet-of-things solution for food supply chain: value creation, sensor portfolio and information fusion. Inf Syst Front 17(2):289–319

Retail Industry Leaders Association (RILA), Available at: http://www.retailcrc.org/sustainability/Lists/Briefings/Attachments/14/RILA%20Issue%20Brief%20-20The%20Value%20of%20Sustainability%20in%20Retail%20Marketing.pdf. 18 Sept 2018

Sarkis J (2003) A strategic decision framework for green supply chain management. J Clean Prod 11(4):397–409

Shrouf F, Ordieres J, Miragliotta G (2014) Smart factories in Industry 4.0: A review of the concept and of energy management approached in production based on the Internet of Things paradigm. Ind Eng Eng Manage (IEEM) 2014:697–701

Tan C, Sun Y, Li G, Jiang G, Chen D, Liu H (2018) Research on gesture recognition of smart data fusion features in the IoT. Neural Comput Appl, 1–13

Thanh HT, Thanh TD (2018) Integrating scan mobile with electronic signage solution in supermarket and retail store. Sci Technol Dev J-Econ -Law Manag 2(1):98–110

Uckelmann D, Harrison M, Michahelles F (2011) An architectural approach towards the future internet of things. In: Architecting the internet of things. Springer, Berlin, pp 1–24

Ugarak J, Cice J (2013) Use of RFID technology in distributive companies. In: 40th National conferences of quality, Kragujevac, pp 82–90

Vlacheas P, Giaffreda R, Stavroulaki V, Kelaidonis D, Foteinos V, Poulios G, Moessner K (2013) Enabling smart cities through a cognitive management framework for the internet of things. IEEE Commun Mag 51(6):102–111

Wortmann F, Flüchter K (2015) Internet of things. Bus Inf Syst Eng 57(3):221–224

Whitmore A, Agarwal A, Da Xu L (2015) The Internet of Things—a survey of topics and trends. Inf Syst Front 17(2):261–274

Waage SA (2007) Re-considering product design: a practical "road-map" for integration of sustainability issues. J Clean Prod 15(7):638–649

Wang W, De Suparna Gilbert C, Moessner K (2013) Knowledge representation in the internet of things: semantic modelling and its applications. Automatika 54(4):388–400

Xia F, Yang LT, Wang L, Vinel A (2012) Internet of things. Int J Commun Syst 25(9):1101–1102

Yan B, Huang G (2009) Supply chain information transmission based on RFID and internet of things. In: Computing, communication, control, and management—ISECS international colloquium, vol 4, pp 166–169

Role of IoT Solutions in Reducing CO_2 Emission and Road Safety in Car Rental and Car Sharing Market

Michał Adamczak, Rebeka Kovacic-Lukman, Adrianna Tobołа, Maciej Tórz and Piotr Cyplik

Abstract It is required by increased traffic on European roads to undertake actions not only to reduce CO_2 emission but also to enhance traffic safety. A driver's driving style is an important aspect that influences CO_2 emission and driving safety. Considerable challenges are to train drivers on a massive scale and convince them to change their driving style. Nevertheless, the authors notice two favourable circumstances that are a popularity increase in a short-term car sharing and the Internet of Things technology development. It would be possible to educate drivers on a massive scale with no additional costs included due to systems that analyse the drivers' driving style in the short-term car rental service performance. This analysis would be based on telematics data and its processing and transfer to the drivers as hints oriented to eco driving and safe driving. The authors suggest that such solutions would significantly influence the driver's driving style improvement. In the authors' view, cars use more fuel (and thereby, emit more CO_2) than declared by their producers in the short-term car rental service performance. At the same time, the drivers cause more damage than the mean of the entire Polish market. The chapter aims at verifying 3 research hypotheses. One of them regards fuel consumption (and CO_2 emission) and two of them regard drivers' safety in the short-term car rental service performance. The research was conducted in July 2018 based on a sample of 699 cars that were part of the fleet of an international car rental company operates on Polish market and specialises in short-term car rental services. The conducted research results made it

R. Kovacic-Lukman
Faculty of Logistics, University of Maribor, Mariborska cesta 7, 3000 Celje, Slovenia
e-mail: rebeka.kovacic@um.si

M. Adamczak · A. Tobołа · M. Tórz · P. Cyplik (✉)
Poznan School of Logistics, Estkowskiego 6, 61-755 Poznan, Poland
e-mail: piotr.cyplik@wsl.com.pl

M. Adamczak
e-mail: michal.adamczak@wsl.com.pl

A. Tobołа
e-mail: adrianna.tobola@wsl.com.pl

M. Tórz
e-mail: maciej.torz@wp.pl

A. Kolinski et al. (eds.), *Integration of Information Flow for Greening Supply Chain Management*, Environmental Issues in Logistics and Manufacturing, https://doi.org/10.1007/978-3-030-24355-5_12

possible to state that car consume much more fuel (and thereby emit more CO_2 to the atmosphere) than declared by their producers. This is referred to the short-term car rental service performance. The drivers, who use this service, cause traffic damage more frequently than the average for all cars in Poland and the damage results are larger than the mean (measured by the average damage value).

Keywords CO_2 emission · Eco driving · Safe driving · Car rental · Car sharing

1 Introduction

The decrease in carbon dioxide emission and the increase in road safety might be achieved by using a few methods. One of the most spectacular methods is to develop road infrastructure that enables safe and thereby, accident risk decrease. The next method is to develop combustion, electrical engines and electronic systems that support drivers in dangerous situations. Although these elements are significant, they cannot ensure road safety and will not lead to reducing CO_2 to a massive extent. Presently, the key role is played by drivers who make decisions about driving on the road and have a given driving style. Therefore, it is so significant for drivers to be conscious about how their driving style influences safety and CO_2 emission. The drivers' education is obviously possible but it is not certainly the most effective solution. It is much more efficient to train drivers while they are driving cars but such training have several confinements. They are not so common, paid and drivers need to have time for them. Having considered a more and more rapid development of the short-term car rental and sharing services, one might make use of the above tendencies to get drivers trained.

Contemporary telematics solutions make it possible to make a real-time transfer of a number of car parameters to a data analysis system working in a cloud. Such solutions based on the Internet-of-Things (IoT) technology make it possible to make up-to-date assessments of the drivers' driving style and send feedback about what behaviours a particular driver should correct to drive their car more economically and environmentally friendly. Having put such facts as the increase in the car rental popularity and driving style analysis technologies, one should figure that the cars, which are used to perform the short-term car rental service, might be perfect tools for the sake of eco-driving and safe driving training. This solution will make it possible to perform low-free massive trainings and have drivers gradually change their driving style.

While tending to fulfill the above concept, the authors initially focused on assessing the present situation, i.e. comparing CO_2 emission and traffic damages caused by drivers in the short-term car rental service performance. The chapter aims at verifying the following three research hypotheses.

H1: The average fuel consumption and CO_2 emission by cars that perform the short-term car rental service (in Poland) is larger than the consumption rate declared by the producer.

H2: While performing the short-term car rental service (in Poland), drivers cause traffic damages and suffer damage more frequently than an average car in Poland.
H3: While performing the short-term car rental service (in Poland), drivers cause traffic damages which value is larger than a whole car population in Poland.

2 The Importance of Eco-Driving and Safe-Driving Issues Related to the EU Sustainable Development Requirements

The sustainable development essence is predominately related to undertaking such actions that do not endanger the natural environment and do not cause its destruction or degradation. It is required by the sustainable development strategy to predominately integrate ecological and economic issues which is implied by the care for the environment and also for production efficiency (Szołtysek 2009). The EU-defined term "sustainable development" embraces the ability to meet the needs of one generation so that it is possible to fulfill the needs of future EU inhabitants. Thus, the sustainable development relies on such integration of economic, social, environmental materiality that enables their mutual progress (https://ec.europa.eu, 19.09.2018).

As part of the research conducted in this article, it is necessary to appropriately identify the terms notions of eco driving and safe driving.

Eco-driving is a "driving technique improvement due to economic, ecological and operating vehicle benefits" (Caban et al. 2017). Eco driving comes from Sweden (Zakrzewski and Zbyszyński 2014), where an economical driving method contributed not only to making road traffic more fluent but also to considerably decreasing exhaust gas emission. Eco driving creates conditions to diminish energy use and eliminate the negative exhaust gas influence on the environment. The enforcement of economical driving rules should predominantly contribute to minimising the degree of the CO_2 harmful influence on climate and to diminishing the fuel combustion by cars and, consequently, to diminishing their usage costs. It is required by an effective eco driving enforcement to predominately possess vehicles that meet exhaust gas emission norms that are decisive about having vehicles enter large cities. In Tables 1 and 2, there are all present EURO exhaust gas emission norms with regard to harmful substances emitted by cars with petrol and diesel engines.

The exhaust gas components in the table consecutively stand for:

- CO—carbon monoxide,
- HC—hydrocarbon,
- NO_x—nitric oxide,
- PM—particulate matters.

The European Union keeps modifying the exhaust gas values included in the EURO 6 norm. At this place, one might distinguish EURO 6b, 6c, 6d variations. This is implied by discrepancies between Diesel engine emission measures in laboratory

Table 1 EURO emission standards for vehicles with gasoline engines

EURO Emissions Standards	The standard is valid from	Components of exhaust (g/km)				
		CO	HC	NO_x	HC + NO_x	PM
EURO 1	01.07.1992	2.72	–	–	0.97	–
EURO 2	01.10.1996	2.20			0.50	–
EURO 3	01.10.2001	2.30	0.20	0.15	–	–
EURO 4	01.10.2006	1.00	0.10	0.08	–	–
EURO 5	01.10.2009	1.00	0.10	0.06	–	0.005
EURO 6	01.01.2014	1.00	0.10	0.06	–	0.005

Source Caban et al. (2017)

Table 2 EURO emission standards for vehicles with diesel engines

EURO Emissions Standards	The standard is valid from	Components of exhaust (g/km)				
		CO	HC	NO_x	HC + NO_x	PM
EURO 1	01.07.1992	3.16	–	–	1.13	0.14
EURO 2	01.10.1996	1.00	0.15	0.55	0.70	0.08
EURO 3	01.10.2001	0.64	0.06	0.50	0.56	0.05
EURO 4	01.10.2006	0.50	0.05	0.25	0.30	0.009
EURO 5	01.10.2009	0.50	0.05	0.18	0.23	0.005
EURO 6	01.01.2014	0.50	0.05	0.08	0.17	0.005

Source Caban et al. (2017)

settings and the measures taken in real conditions. The European Commission will probably introduce the EURO 7 norm in 2020 (European Commission 2016, 2017, https://www.fleeteurope.com, 24.09.2018). If these norms are failed to be fulfilled, this might result in having conventionally driven cars, especially with diesel engines, be prohibited to enter large cities.

The increase in driving fuel efficiency and environmental friendliness might be influenced by significant factors that are elements fully dependent on the vehicle user, e.g. driving dynamics (aggressive acceleration and braking), too late gear change, the end-of-life state of consumables, e.g. tyres, brakes, air conditioning usage frequency, no use of braking by gear reduction, excessive heating use, no prediction of road situations (forcing rapid braking) and the road state (Lempart and Malik 2013; Caban et al. 2017). Thus, the change of drivers' behaviours seem to be of key importance in the trend to the eco driving development, which means to make them acquire and consistently use the new ecological and economic driving standards and get them appropriately motivated to change their present driving style. This might lead to decreasing fuel combustion even by 10% and confine the gas exhaust from 5 to 20% (Kuiken and Twisk 2001; Barkenbus 2010; Stillwater and Kurani 2012).

As regards to the driving fuel efficiency improvement, it is necessary to refer to innovative solutions used by producers of leading car brands. These solutions contribute to increasing numerous drivers' fuel efficiency. Among such solutions, there are a tyre pressure control system, start-stop system, systems indicating the necessary to gear up/kickdown or computer systems that analyse a driver's style and indicate information on e.g. average fuel combustion (Caban et al. 2017).

Solutions to improve environmental performance are not just about vehicles and drivers. The works are carried out in many other fields, such as improving the flow of information aimed at reducing mistakes and the need to implement additional transports (Osmólski et al. 2018).

The necessity to follow eco-driving rules also depends on the traffic fluency. It is easier to apply eco driving rules if most road users fulfill their standards. The more fluent traffic, the smaller fuel combustion and emission of greenhouse gases (Qian and Chung 2011). The most significant eco driving benefits are:

- full integration of the environment, society and economy which is a sustainable development objective,
- limitation of car operating costs and usage pace,
- decrease in the level of emitting exhaust gases and harmful substances to the atmosphere and
- increase in the road safety level (Lempart and Malik 2013).

As to eco-driving, one should also refer to the safe driving policy. If an aggressive driving style is given up, it positively influences not only the natural environment but is predominately related to limiting the accident risk due to fluent driving on the road. The ecological driving style application makes it possible not only to make considerable fuel savings but, above all, to make more efficient maneuvers (e.g. rapid braking) than in the case of aggressive driving. Thereby, the terms eco-driving and safe driving should be considered together as compatible elements. However, one should notice that such factors as age, driver's experience, proper habits, etc. are of high importance when it comes to the car passengers' safety. Although contemporary vehicles offered by contemporary automobile concerns have modern safety systems, there is a continuous increase in the maximum speed to be achieved by a driver. This becomes a sort of factor that has an impact on decreasing safety on European roads.

Although European countries had been trying to implement the eco driving policy for about 30 years, the European Union noticed that it was necessary to apply this policy in all member states to an equal degree. This will result in fulfilling the sustainable development programme and diminishing the environment pollution (Caban et al. 2017). In order to enhance the driving environmental friendliness, the European Commonwealth imposed directives on member states. The directive should determine the procedures to limit gas emission and passengers' safety. The directives include:

- The directive no. 2010/40/UE of the European Parliament and Council that determines the ITS (Intelligent Transportation Systems) implementation framework. As these systems are innovative applications, they aim at improving security, fuel efficiency, environmental friendliness and road transport comfort. Their task is to predominately integrate the system users, vehicles and infrastructure (Shaheen and Finson 2013; Zakrzewski and Zbyszyński 2014). The ITS systems make it possible to predominately select an optimal track that ensures a minimum number of kilometers. This track avoids increased traffic. This results in a smaller number of substances harmful to the atmosphere.

- The EU Green Paper issued is mainly related to road transport in large cities that are concentrations of a very large number of cars somewhat unaccustomed to the exhaust gas emission norms in force. The Green Paper assumptions are predominantly referred to difficulties in controlling exhaust gas amounts emitted to the atmosphere. The European commission mentions an inappropriate driving style of numerous drivers as one of the factors that have a giant influence on climate changes. Therefore, it is necessary for European drivers to be obedient to eco driving and safe driving rules as it makes it possible to fulfill the assumptions to reduce greenhouse gases by 20% to 2020.
- Similarly to the Green Paper, the White Paper was issued by the European Council and is predominantly focused on reducing the use of petroleum as a pollution source as well as high costs. It is assumed in the White Paper that the emission of harmful greenhouse gases should be diminished by the transport economy of the entire European Union by at least 60% by 2050. It is estimated that both costs, exhaust gas emissions and the number of accidents will rise if the society does not get accustomed to the eco and driving rules. These conditions are necessary to be met but need to be supported by modern technologies (European Commission 2011; Zakrzewski and Zbyszyński 2014).
- The "Europe 2020" strategy (in other words The strategy for intelligent and sustainable development favouring social inclusion) (European Commission 2010). As regards to eco driving and safe driving, the main strategy assumption is to predominantly make effective use of resources in entire Europe by fulfilling economic assumptions. They envisage the emission of harmful substances to be diminished and the transport economy energy efficiency to be increased. This might be achieved by fulfilling the assumptions of ecological, economical and safe driving (Zakrzewski and Zbyszyński 2014).

The EU actions are predominately focused on limiting the transport impact harmfulness on the natural environment and road safety. In order to meet the EU-imposed conditions, it is predominantly necessary to integrate modern technologies that enable emission of smaller amounts of exhaust gases to the atmosphere. This should be accompanied by the simultaneous tendency to increase drivers' awareness about the necessity to change their driving style.

3 The Car Rental Role in Increasing the Society Mobility and Congestion Reduction

A continuously increasing number of private motor vehicles used by inhabitants of most large cities have caused serious problems related to more and more traffic jams, the infrastructure resulting in no free parking places in city centres or an increasing environment pollution level. No actions for enhancing passenger mobility in large cities and metropolis throughput cause more and more institutions that offer cars lent to passengers even for few-minute periods of time. These passengers get around the city nearly on a daily basis (Fiedler et al. 2017). Analysis of the root causes of this state of affairs allows to find an answer to the question what should be done to reduce the number of cars in cities and thus the emission of harmful substances into the atmosphere (Wieczerniak et al. 2017). Due to the digital revolution, the availability of mobile car rental places keeps growing. Such solutions make it possible not only for decreasing car operating costs (e.g. insurances, servicing, repairs) and thereby, to diminish not only the amount of exhaust gases emitted to the atmosphere but also traffic intensity on European roads (Bondorová and Archer 2017). When choosing a car, an average driver frequently purchases a vehicle that exceeds their needs. The capabilities of cars with large engine capacities are unused in most cases. This happens because the traffic of various intensity makes it impossible to effectively use a car that burns huge amounts of petrol or diesel oil to drive long distances (with more or less constant speed). These vehicles are getting unprofitable to users mainly due to operating costs but typically town cars, i.e. with engines of low capacities are much better adjusted to city conditions (frequent starts and stops). Therefore, the purchase of large cars is frequently unprofitable and does not enable complete usage of the vehicle capacities. Car rental places would make it feasible to comfortably travel around the country. Such car rental places which offer a wide variety of cars make it feasible for their users to quickly and cost effectively travel distances within cities and comfortably travel around the country (Car-Sharing fact sheet 2009).

One might observe a new trend of using car rental places in streets in large cities. This trend enables vehicle use for travelling every distance in the city. A wide variety of cars located within one city area is offered by car sharing rental places that are institutions dealing with car rentals for dozen or so minutes. When one merely has an application in the mobile phone, one might rent a car and leave it in any location in a city or sometimes in a country. This is accompanied only by bearing costs of travelling the distance from point A to point B. It is proven by the research on the efficiency of using both private and rented cars that private cars are not used by their owners by almost 90% of time (during their work, shopping or stay at home). Insurance and parking costs are generated by cars that hold parking places most of the daytime. It is proven by the research conducted in the USA and Portugal that the guarantee of an appropriate fleet of cars rent in a mobile way would make it possible decrease the number of private cars in large cities by 1/3 in the former case of insurance costs and by 10% in the latter case of parking costs as accompanied by the same level of the city inhabitants' mobility. The car use degree typical of car sharing depends on the city infrastructure and specificity. Furthermore, contemporarily used mobile rental places of motor vehicles do not reach their maximum efficiency due to using a very large number of private sector cars at the same time.

The owners of private cars predominantly tend to maintain the mobility that is required to be at an appropriate level. This is accompanied by the possibly lowest car operating cost. People who live in cities must therefore decide how to meet their need to move (more about decision making models in Hadas et al. 2011). In the case of a continuously increasing number of vehicles in streets, it is impossible to avoid high traffic intensity responsible for forming traffic jams that cause increasing car operating costs. Such a situation resembles a vicious circle that predominantly requires a radical decrease in the number of cars in large cities. Car rentals are placed large demands regarding their guarantee of the required or even a higher mobility level to all passengers that travel by rent cars. It is also necessary to form a proper infrastructure, e.g. parking places for rent cars only which discourage owners of private vehicles to use their own cars. These conditions are necessary to be met to make it possible to notice real benefits from using cars typical of car sharing (Fiedler et al. 2017).

It was proven within the research conducted by Zipcar, an American rental place and Berkeley's Transportation Sustainability Research Center that it was made feasible by the rent car use possibility to decrease in using private cars by 5% in the case of students and by 20% in the case of employed people. Furthermore, the CO_2 emission level decreased by 0.1–2.6% as a result of using Zipcar.

The research conducted in Milan proved that the introduction of mobile car rental places made it possible for the city inhabitants to use different forms of transport, e.g. combining a rent car with public transport. This lead to purchasing about 30,000 private vehicles less than in previous years and the greenhouse gas emission decrease by about 18%. A similar situation is developing in Paris where the city authorities started a project of mobile and electric car rental places at the same time in 2011. This has made it possible to diminish the CO_2 emission by 15,000 tones for 6 years but it is planned that the demand for private cars will have fallen by about 36,000 by 2023. As a result, the CO_2 emission decrease will fall by approximately 75,000 tones (Bondorová and Archer 2017).

It is indicated by the examples as presented in article that a number of European cities have obtained favourable results by introducing mobile rental places. In most cases, the users of private vehicles decided not to possess them in favour of renting cars. Thus, it was possible to considerably enlarge the city throughputs and, as a consequence, there was less exhaust gases emitted to the atmosphere. Car sharing has an influence not only on limiting the environment pollution, but also impacts the users' tendency to much more frequently use other means of transport, for instance public means of transport, bicycle, travel on foot, and therefore, has a significant impact on the metropolis inhabitants' mobility (Katzev 2003).

There are numerous benefits from giving up using one's own car in favour of using rental places:

- decrease in traffic intensity in large cities that results in increasing the city throughput, and consequently, shortening the travel duration period,
- the possibility to shutdown a number of parking places that enables obtainment of extra space in the city,
- the increase in the mobility level of numerous city inhabitants by making it feasible to combine various means of transport,
- decrease in emitting harmful substances to the atmosphere,

- fulfilling EU directives related to the air quality improvement,
- decrease in operating costs (insurance, parking fee, servicing),
- the possibility to decrease the car lending costs by using an economical driving style (due to feedback generated by applications),
- the possibility to choose a vehicle precisely adopted to customers' needs dependent on the situation (city travel, motorway travel, etc.),
- the possibility to use the car rental place by small companies that do not have their own fleet (Litman 2000; Car-Sharing fact sheet 2009).

The environment pollution degree will be diminished by giving up driving private cars that are used all day long. The cars typical of car sharing are mostly small, efficient and sometimes hybrid or electric vehicles. This causes the decrease in the level of emitting harmful substances to the atmosphere. It is estimated that one regularly rent car might replace even 15 private vehicles (Bondorová and Archer 2017) and it depends on several factors such as the infrastructure specifics and the number of inhabitants. This will certainly lead to diminishing congestion and thereby, will contribute to increasing passengers' mobility. Presently, it is nonetheless of key importance to gain support from state institutions that predominantly aim at ensuring an appropriate infrastructure in large cities. This is favourable to developing rental places. Another goal of the public institutions is to form proper campaigns that motivate inhabitants to give up using private cars. It should be noticed at this place that most city inhabitants, who have private cars, choose their own one due to the inadaptability of public transport to their own needs. The introduction of car rental places would make it possible to combine city transport with car sharing. Thereby, the passengers' mobility would be increased.

4 The Analysis of CO_2 Emission and Traffic damage Caused by Drivers Performing the Short-Term Car Rental Service

As part of the research on CO_2 emission, one analysed the combustion of 699 cars from segments A, B, C, D and delivery trucks (the cars were 88% of the investigated population and the delivery trucks 12% respectively). The vehicles were used to perform a short-term car rental service in one of the international companies that specialise in offering short-term car rentals. All the cars in the analysed company were equipped with telematics sensors and the parameter data of the vehicles were sent to the central system. However, these data are not passed on to drivers. Among the analyzed cars, 232 had diesel engines and 467 gasoline engines. Among 699 cars there were 102 different car types. The car type consisted of such elements as brand, model, engine type, type of gear box. The fuel consumption analysis was conducted in July 2018. In the investigated sample, there were 67 types of cars with petrol engines and 35 types of cars with diesel engines. The investigated cars drove more 2 300 000 kilometres, consumed more than 151,000 fuel litres and emitted more than

367 CO_2 tones (it was assumed in the calculations that 2.33 CO_2 and 2.64 CO_2 kg is generated in the case of consuming one litre of petrol in the former case, diesel gas in the latter case).

One compared the fuel consumption (thereby, the CO_2 emission) of the analysed company fleet and the declared fuel consumption according to the methodology presented in Fig. 1.

The conducted analysis first step was to select vehicles from the investigated company fleet to the research population. The selection was performed based on the following criteria:

- exclusively cars from segments A, B, C, D and vans,
- cars that drove more 200 km in the analysed period of time (July 2018).

In the case of the selected group of vehicles (with 699 cars as mentioned above), data on the driven distances and consumed fuel was downloaded from the telematics systems. This enabled calculation of an average fuel consumption rate for each vehicle type. One also conducted a collective analysis of a driving type defined as driving in the built-up area and outside it. In order to compare the fuel consumption rate in the analysed population with the fuel consumption rate declared by the producer, it is necessary to specify a mix mode structure.

In the selected population, one collected data on the combustion as declared by the producers for each type of cars. This was simultaneous to the calculation of the average fuel consumption by the vehicles. Based on the data, one calculated a declared consumption rate in the mix mode for each type of cars. The mix mode was

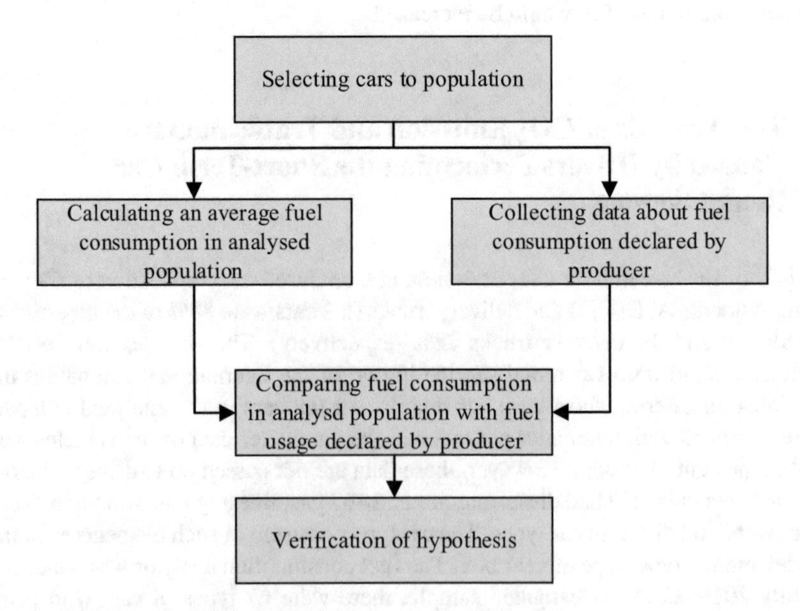

Fig. 1 Research methodology, own study

calculated based on the calculation proportion (consumption rate in cities 0.58 and consumption rate outside the cities 0.42).

At the next stage, one compiled the average fuel consumption of the cars in the investigated population with the fuel consumption rate declared by the producer (in the mix mode as calculated at the previous stage). This compilation made it possible to notice that the average fuel consumption in the analysed population was higher than declared by the producer. In order to make sure about this difference essence, one formulate a pair of statistical hypotheses (corresponding to the research hypothesis H1):

H_{0a}: *The average fuel consumption of cars in the analysed population equals the fuel consumption declared by the producer.*
H_{1a}: *The average fuel consumption of cars in the analysed population is different than the fuel consumption declared by the producer.*

This pair of statistical hypotheses were verified by means of Two-Sample T-test. ($\alpha = 0.05$). The test results are presented in Fig. 2.

The Two-Sample T-test results enable rejection of the null hypothesis (p-value $< \alpha$). Thus, it should be stated that the fuel consumption of cars included in the research population is statistically significantly different from the fuel consumption declared by the producer by about 1 litre per 100 km (exactly 0.9471 l/100 km, it is 19.7% more then fuel consumption declared by the producer). Thereby, it might concluded that the cars, which perform the short-term car rental service, consume more fuel (emit more CO_2 to the atmosphere) than declared by their producer.

Another investigated aspect was the cars damages. One compared the cars damages classified from investigated population with an average cars damages registered in Poland by the Polish Financial Supervision Authority (KNF 2018). As part of the cars damages analysis, one investigated two aspects: frequency of damage and

Two-Sample T-Test and CI: Average fuel consumtion [l/100km; Fuel consumtion declared by car producer

```
Two-sample T for Average fuel consumption [l/100km vs Fuel
consumption declared by car producer

                                        N   Mean StDev  SE Mean
Average fuel consumption              699  6,55  1,61    0,061
Fuel consumption declared by car producer  699  5,60  1,14    0,043

Difference = μ (Average fuel consumption [l/100km) - μ (Fuel
consumption declared by car producer)
Estimate for difference:  0,9471
95% CI for difference:  (0,8008; 1,0934)
T-Test of difference = 0 (vs ≠): T-Value = 12,70  P-Value = 0,000  DF
= 1255
```

Fig. 2 Two-Sample T-Test and CI for comparison average fuel consumption (l/100 km) versus fuel consumption declared by car producer, own study

average value of damage for a two types of damages: TP (third-party) and CC (comprehensive cover). The names of the damage types are taken from the car insurance name that covers the damage. The comparison results of the frequency of damage aspect are presented in Fig. 3.

This means that drivers during renting a car cause traffic damage and suffer damage much more frequently than an average car in Poland. Thereby, the research hypothesis H2 has been positively verified.

The average damage value comparison results are presented in Fig. 4.

Average value of TP damage in analyzed population is higher than average for a whole Poland This means that cars that perform the short-term rental service cause damage within the TP insurance with their value approximately twice higher than the market mean. It might be only suspected that the higher damage value equals a larger car damage and higher danger of the accident participants. The CC damage value in both groups is similar. Nevertheless, it is possible to positively verify the

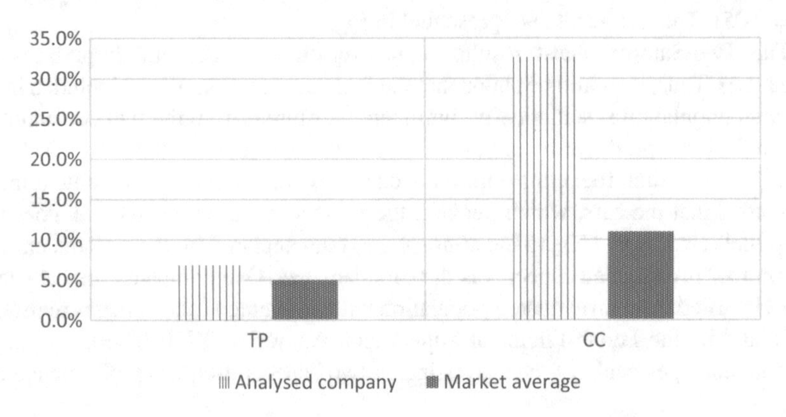

Fig. 3 Comparison of frequency of damage, own study based on (KNF 2018)

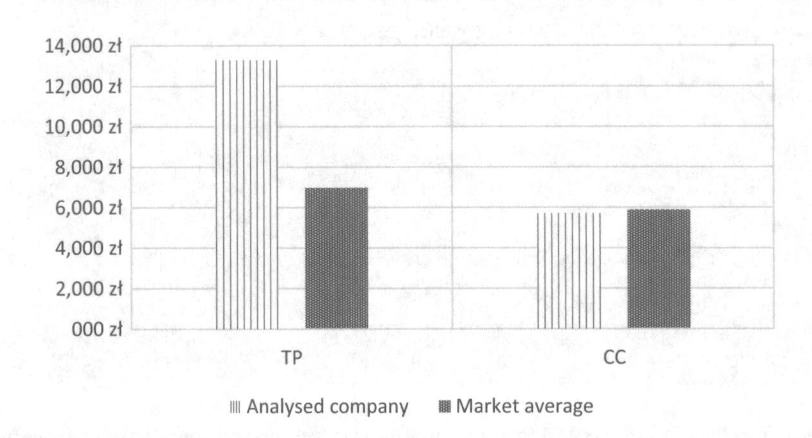

Fig. 4 Comparison of average value of damage, own study based on (KNF 2018)

research hypothesis H3 due to a large difference in the damage covered by the liability insurance. When performing the short-term car rental service, drivers cause damage which value is much higher than the global car population in Poland.

5 Business Benefits for Companies Offering a Short-Term Car Rental Service

It should be noticed that eco and safe driving will bring not only ecological but also economic benefits. The beneficiaries of such behaviours will be both drivers and companies specialising in short-term car rental services. As to the drivers, it is related to smaller fuel consumption (smaller fuel combustion by even 10% and exhaust gas emission limitation from 5 to 20%) (Kuiken and Twisk 2001; Barkenbus 2010; Stillwater and Kurani 2012). The economic benefits will be particularly accumulated as a smaller use of cars (better engine state, smaller frequency of exchanging brake blocks and other elements prone to be quickly used during aggressive driving). Eco and safe driving will directly lead to decreasing the harmfulness or rate of accidents of the used cars. If there is a crash, the damage results will be much smaller than in the case of a statistical crash. This will lead to the statistical damage value. Therefore, not only the fleet of cars will be in better technical condition, which will consequently result in diminishing the exhaust gas emission, but also the repair time and cost will be smaller. This will result in a smaller number of cars absolutely necessary to ensure the mobility. As referred to the work (Cudziło et al. 2018), it is necessary to analyse the influences of suggested solutions within the entire supply chain (including short-term car rental services). On balance, all the above elements have an impact on the ecological and economic aspect of a driver-user, but also on the financial parameters of short-term rental companies.

6 Conclusions

Summarizing the research presented in this chapter, it should be noted that as a result, three research hypotheses were positively verified. This means that the cars, which perform the short-term rental service, generated more CO_2 than declared by the producer. When the short-term rental service was being performed in July 2018, 699 analysed cars generated more CO_2 than indicated in the producer's declaration by more than 53.5 tones. It might be stated by using a simplified approximation that this difference will be more than 600 tones annually.

It is noticed with respect to safety that during the rental service, drivers cause damage much more frequently than a statistical driver and the damage is additionally much more destructive to other road users (the average damage covered by the third-party insurance will be twice as high as the mean damage in Poland).

One might also conclude that it is possible to limit operating cost related to the car exploitation. These costs are directly related to improving drivers' technique and behaviors.

Thereby, it might be concluded that it is possible to improve the drivers' style and make it closer to eco and safe driving. IoT technology based solutions are a helpful tool in improving the driving technique.

As part of further research, the team is planning to perform an analysis of the instances of breaking the speed limits and other dangerous behaviours by drivers who perform the short-term rental service. All these research actions should result in developing the system functionality and applications educating drivers and encouraging them to conform to the eco and safe driving style.

Acknowledgements This paper has been the result of the study conducted within the grant by the Ministry of Science and Higher Education entitled "Development of production-logistics systems" (project No. KSL 2/17) pursued at the Poznan School of Logistics.

References

Barkenbus JN (2010) Eco-driving: an overlooked climate change Initiative. Energy Policy (38)

Bondorová B, Archer G (2017) Does sharing cars really reduce car use? Transport & Environment, Brussels

Caban J, Sopoćko M, Ignaciuk P (2017). Eco-driving, przegląd stanu Zagadnienia. Autobusy, 6, Radom

Car-Sharing fact sheet (2009) Car-Sharing reduces the burden on both cities and the environment—the environmental impacts of Car-Sharing use. Car Sharing fact sheet number 3(3), Brussels

Cudziło M, Voronina R, Dujak D, Koliński A (2018) Analysing the efficiency of logistic actions in complex supply chains—conceptual and methodological assumptions of research. LogForum 14(2):171–184

European Commission (2010) Communication from the commission Europe 2020 A strategy for smart, sustainable and inclusive growth, KOM(2010), 2020 Brussels

European Commission (2011) White Paper "Roadmap to a Single European Transport Area—Towards a competitive and resource efficient transport system", COM(2011) 144, Brussels

European Commission (2016) Commission Regulation (EU) 2016/646 of 20 April 2016 amending Regulation (EC) No 692/2008 as regards emissions from light passenger and commercial vehicles (Euro 6), Brussels

European Commission (2017) Commission Recommendation (EU) 2017/948 of 31 May 2017 on the use of fuel consumption and CO_2 emission values type-approved and measured in accordance with the World Harmonised Light Vehicles Test Procedure when making information available for consumers pursuant to Directive 1999/94/EC of the European Parliament and of the Council, Brussels

Fiedler D, Cap M, Certicky M (2017) Impact of mobility-on-demand on traffic congestion: simulation-based study. In: 20th IEEE international conference on intelligent transportation systems (ITSC 2017)

Hadas L, Stachowiak A, Cyplik P (2011) Decision making model in integrated assessment of business-environment system—a case study. In: Information technologies in environmental engineering (New trends and challenges, Environmental Science and Engineering), pp 419–429

Katzev R (2003) Car sharing: a new approach to urban transportation problems. Anal Soc Issues Public Policy 3(1):65–86

Kuiken M, Twisk D (2001) Safe driving and the training of calibration. SWOV Institute for Road Safety Research, Leidschendam

Lempart M, Malik P (2013) Proste rozwiązania – wymierne korzyści czyli ekojazda w koncepcji równoważonego rozwoju (Simple Solutions—Measurable Benefits, That is Eco Driving in the Concept of Sustainable Development). Zeszyty Naukowe Politechniki Poznańskiej numer 60 Organizacja i Zarządzanie (Scientific Journal of Poznan University of Technology series of "Organization and Management. no. 60)", Poznań

Litman T (2000) Evaluating carsharing benefits. Transp Res Rec J Transp Res Board

Osmolski W, Kolinski A, Dujak D (2018) Methodology of implementing e-freight solutions in terms of information flow efficiency. Interdisciplinary Management Research, XIV IMR 2018, Opatija, pp 306–325

Polish Financial Supervision Authority. Quarterly Bulletin. The market of insurances. IV/2017 https://www.knf.gov.pl/?articleId=61185&p_id=18. Access 12 Dec 2018

Qian G, Chung E (2011) Evaluating effects of eco-driving at traffic intersections based on traffic micro-simulation. In: Australasian Transport Research Forum 2011 Proceedings, Adelaide

Shaheen S, Finson R (2013) Intelligent transportation systems, reference module in earth systems and environmental sciences. Elsevier, Amsterdam

Stillwater T, Kurani K (2012) In-vehicle ecodriving interface: theory, design and driver responses. In: 91st annual meeting of the transportation research board, Washington, DC

Szołtysek J (2009) Logistyka zwrotna (Reverse logistics). Institute of Logistics and Warehousing, Poznań

Wieczerniak S, Cyplik P, Milczarek J (2017) Root cause analysis methods as a tool of effective change. In: Proceedings of international scientific conference business logistics in modern management (Business logistics in modern management), pp 611–627

Zakrzewski B, Zbyszyński M (2014) Pojęcie "Ekojazdy" w aspekcie polityki i przepisów Unii Europejskiej. Czasopismo Logistyka (The term "ecodriving" in the light of the EU policy and regulations). J Logistics 4:1486–1502. https://ec.europa.eu/info/strategy/international-strategies/global-topics/sustainable-development-goals/eu-approach-sustainable-development_pl. 19 Sept 2018. https://www.fleeteurope.com/en/taxation-and-legislation/europe/features/do-you-know-your-euro-6-your-6c-and-6d-temp. 24 Sept 2018

Integrating Life Cycle Thinking, Ecolabels and Ecodesign Principles into Supply Chain Management

Matevž Obrecht

Abstract Population growth, increasing global production and consumption all cause increasing material, water, energy and land use and cause harmful impacts for the environment as well as for human health. Due to limited and scarce resources on planet Earth, environmental depletion, harmful emissions and climate changes our society is pushed one hand as well pulled on the other by increasing environmental awareness, available green technologies and developed concepts of environmental protection towards more environmentally sound future, necessary to enable further existence of human rase in its present extent. Environmentally sound economy such as circular economy, ecodesign and sustainable development will become not just a part of comparative advantage in achieving differentiation strategy but also a possible potential answer to forecasted socio-economic challenges in the forthcoming decades. Because companies are more and more connected and related to their supply chain partners, activities related with environmental improvements must also be focused on supply chain perspective and not just on one company representing just a small part of much broader supply chain or supply network. This chapter is therefore related on investigating green supply chain management, integrating life cycle thinking, eco-design principles and environmental labelling into supply chain management.

Keywords Green supply chain · Life cycle assessment · Ecodesign · Ecolabels

1 Introduction

Due to increasing number of population, standard of living and consequently increasing human activities and production, environmental concern is gaining importance. It is becoming clear that our planet can not restore itself anymore and that resources

---

M. Obrecht (✉)
Department of Sustainable Logistics and Transport, Faculty of Logistics,
University of Maribor, Mariborska 7, 3000 Celje, Slovenia
e-mail: matevz.obrecht@um.si
URL: http://fl.um.si/kontakt/obrecht-matevz/

A. Kolinski et al. (eds.), *Integration of Information Flow for Greening Supply Chain Management*, Environmental Issues in Logistics and Manufacturing, https://doi.org/10.1007/978-3-030-24355-5_13

are used unsustainably (Obrecht and Knez 2016). As human activities cause severe harmful local and global environmental impacts, economic activities are increasingly considering environmental aspects. It is believed that environmentally conscious and more sustainable oriented operation may give organizations a competitive advantage especially on a long run (Plouffe et al. 2011; Albino et al. 2009; Dangelico et al. 2017; Gerstlberger et al. 2014).

Numerous facts revealed that current linear economy is unsustainable. Population growth and increasing standard of living require ever increasing material extraction, food, water and energy use. Consequently prices of these materials are increasing, arable land and forest areas are disappearing, accessible clean water is questionable on a long term, biodiversity is changing rapidly etc. (The 2030 Water Resource Group 2009; Alexandratos and Bruinsma 2012; International Energy Agency 2017). Due to forecasted trends environmentally sound economy such as circular economy, life cycle based ecodesign and sustainable supply chains will become not just a part of comparative advantage in achieving differentiation strategy but also a possible potential answer to forecasted socio-economic challenges in the forthcoming decades (Bešter 2017) and a systematic solution for sustainable existence of human kind (Širec et al. 2018).

However concentrating on environmental perspective solely in one part of the supply chain (SC) is not sufficient for efficient improvements because environmental impacts are caused through the whole SC, from raw material extraction, production of materials and components, manufacturing final product, its distribution, use and end-of-life. Literature review suggests that environmental goals e.g. 20/20/20 objectives set by the EU can not be achieved only through inter-organizational activities and measures but in cooperation along entire value chain taking advantage of chain synergies among supply chain participants (Szegedi et al. 2017). Therefore also environmental management schemes (e.g. ISO 14001 or EMAS) include cooperation of different actors within the whole SC. Complexity of sustainable SC, circular economy as well as ecodesign require cooperation of different stakeholders on different levels, therefore systematic approach is inevitable. Companies' managers must become aware that economic and environmental goals are not contradictory to each other but can be achieved simultaneously (Preston 2012; Lieder and Rashid 2016; Ghisellini et al. 2016).

The idea of environmentally conscious (green) supply chain management (SCM) first began to take root in technical literature in the early 1970s. The integration of the disciplines of both the green operation and the complex SC (including purchasing, production and logistics) came into focus in the 1990s especially in automotive sector (Szegedi et al. 2017). Many organizations still have a very narrow perception of their environmental impact, which is mostly limited to site-specific production activities (Ammenberg and Sundin 2005). Contrary to this, one of the major trends of sustainability programs in industrialized countries is so called life cycle thinking, which expands the focus from the production site to incorporate various environmental and social aspects associated with a product over its whole life cycle (UNEP 2006). Life cycle thinking is based on the principles of pollution prevention, in which environmental impacts are reduced at the source, and on closing the loop of materials and energy (European Commission 2014). Namely, all products and services have some impact on the environment which may occur at any or all phases of the product's life cycle including raw materials extraction, manufacture, distribution, use and waste disposal (Denac et al. 2018). Companies with a higher developed traditional SC have also a more developed green supply chain management (GSCM) system (Szegedi et al. 2017).

Strong indications also confirmed that commitment to ecodesign and sustainable development within organization seems to be the most important issue for improvements and environmental labels a strong tool for communication with customers (especially green oriented). Because company managers are always interested to achieve business benefits simultaneously with environmental improvements, environmental labels are a potent tool for achieving just that—on one hand improving company's image, gaining new green oriented consumers, being able to compete on green public tenders, differentiate on a highly competitive market, decrease fees for waste or use of hazardous materials etc. and on the other hand bring benefits from environmental improvement also directly within the company walls—e.g. decreased material or energy use, decreased amount of wastes, increased efficiency, smaller water consumption.

The goal of the chapter is therefore to get better insight on greening of supply chains, raise importance of life cycle thinking for supply chain managers and to study and discuss the use of different methodologies, principles and tools like life cycle impact assessment, eco-design and environmental labels within supply chain management. Case studies of best practices on life cycle assessment and eco-design are therefore presented to consolidate the knowledge on environmental issues and to integrate it within supply chain management. Comprehensive collection of such tools, principles and methods as well as cases of solving actual problems is crucial for supply chain managers to enable them better understanding and importance of environmentally sound business models and emphasises sustainable development also for companies.

2 Greening the Supply Chains

Kyoto Protocol pioneered the concept of mandatory reduction and compliance of emission reduction norms for major polluting economies in the world by 2012 and beyond. This prompted the corporate world to go in for introduction and commercialisation of cleaner green technologies to mass produce and market wide-ranging green products. Consequently the concept of Green Marketing also developed. The concept GSCM has been a parallel development to push the green products in an ever-expanding market with huge future potential—given growing customer consciousness towards eco-friendly green products and changing lifestyles (Sarkar 2012a) and to include environmental thinking in conventional SCM.

Nowadays in most of the leading green SC's the concept of life cycle thinking can also be detected. However, the main focus is still on efficient resource use and reduction of waste (Denac et al. 2018). GSCM promotes product's and process' innovations in the area of stronger customer relations also in case of communication with customers about activities, standards and measures related with greening the SC's to impact the green consumers purchasing decisions (Sarkar 2012a, b).

Differences between conventional and green SCM are presented on Table 1.

Main goal of GSCM is integration of environmental issues in SC strategy and taking care of the environment also beyond the boundaries of single organization and can be seen as a competitive advantage. Consequently environmental labels dealing with the whole SC perspective are also a part of GSCM. GSCM also reduces risk of environmental accidents and pollution, increases flexibility in supply chains, promotes continuous improvements within SC and is focused on cooperation of all SC members and coordinates suppliers' and customers' activities related with GSCM, designing green distribution, cleaner production and reverse logistics.

SC managers can many times be convinced that greening the SC is related with higher costs. This is not necessary true anymore (Denac et al. 2018). Environmental management schemes such as ISO 14001 certification is becoming the industry standard in determining marketing channel partners in the risk of losing customers and increasing costs. Therefore adapting to stricter forthcoming environmental legislation, increasing customers' environmental awareness and share of green oriented consumers as well as requirements of business partners can only bring benefits on a long run. General framework of green SCM is presented on Fig. 1.

Sarkar (2012a) believe that introducing green SC must be based on four steps:

- Evaluating costs of transition into GSCM;
- Identify and set possibilities;
- Calculate potential benefits;
- Take decision on acting in green direction and start with changes.

However, organisation must always consider its organizational structure, needs and corporate culture since every organization is unique and specific therefore completely unique and versatile solution do not exist. Nonetheless Makover et al. (1993)

Table 1 Differences between conventional SCM and green SCM

	Factor	Conventional SCM	Green SCM
1	Orientation	Economic	Ecological + economic
2	Focus of optimisation	Integrated approach	High ecological impacts
3	Supplier selection	Price (short term relation)	Ecological aspects (long term relation)
4	Short term costs	Low	High
5	Long term costs	Higher	Lower
6	Risks	Higher	Lower
7	Environmental knowledge	Low	High

Source Adapted from Sarkar (2012a), Ho et al. (2009), Bratina et al. (2017)

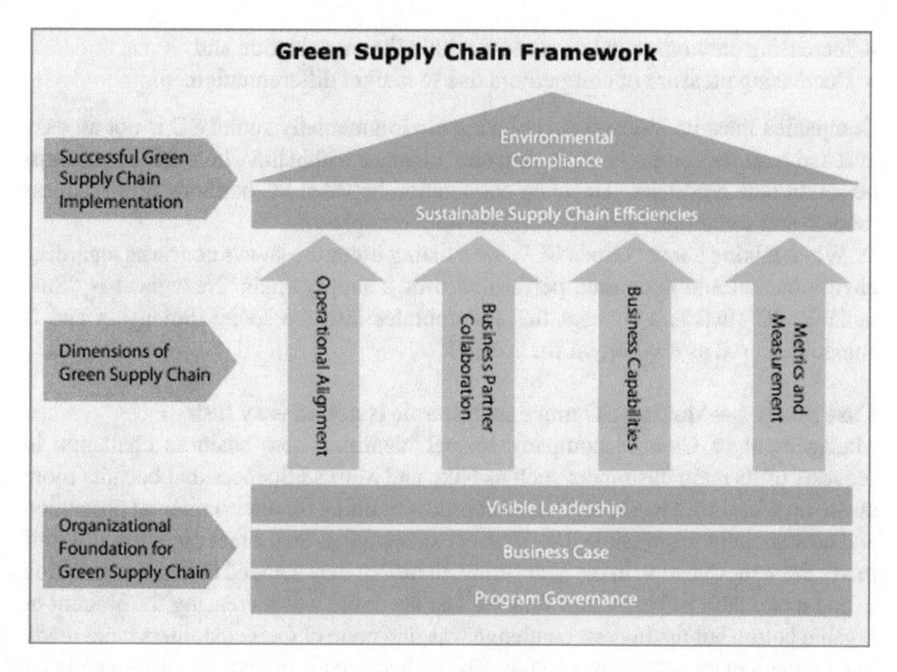

Fig. 1 Green supply chain framework. *Source* Adapted from Diamond Management and Technology Consultant

set some basic and generally applicable directions on characteristics that green SC or green product can not have such as:

- It should not endanger human (or animal) health;
- It should not harm the environment in any phase of its life cycle;
- Energy and resource use should not be inconsistent between different life cycle stages;
- It should not cause unnecessary wastes regarding additional or inappropriate packaging, inability of maintenance, etc.;
- It should not include the use or torturing of animals;
- It should not be related with materials or substances related with endangered species or restricted and environmentally protected areas;
- It should not cost the Earth.

When thinking on starting with GSCM organizations must be focused on:

- Target marketing;
- Sustainable resources;
- Reducing costs and increasing efficiency;
- Dividing products from competitive advantage;
- Adapting to legislation and reducing (current or forthcoming) environmental risks;
- (Green) branding;
- Return of investment;

- Increasing environmental awareness within the organization and
- Decreasing pressure of competitors due to market differentiation.

Companies must be aware that achieving environmentally sound SC is not an easy task and requires comprehensive structural changes within the whole SC. Focus must be set on new processes, deepened cooperation between SC partners, new business models and sometimes also cooperation with competitors.

When talking about "Green SC", the existing literature shows concerns regarding environmental and economic performance of a supply chain. Next level is "Sustainable SC" that is a concept that incorporates also the social dimension and is sometimes used as a synonym for green SC.

Case Study 1—Making SC more sustainable is not an easy task
Management of Chinese company Esquel identified new business challenge in requests of its main customers such as Nike and Marks&Spencer that became more aware on what is the world biggest shirt producer doing for environmental protection and how socially responsible it is. Esquel expected that their other buyers could start to ask the same question in the near future, therefore they decided to make a transition to and make their SC more sustainable. The key issue was increasing the amount of organic cotton but the biggest challenge was that none of their customers were ready to pay much extra for organic cotton. Esquel became aware that the challenge is far more complicated that first predicted. They could not just request from their suppliers—farmers, mostly located in north-western China, to use less water, renovate their watering systems, use less and fewer fertilizers and pesticides as well as toxic chemicals etc. because this could have catastrophic consequences for farmers because of decreased yield for 50% in case of organic farming and consequently decrease of their income. Producing organic cotton requires less chemicals but its further processing require more of it due to weaker fibres and different physical properties of organic cotton. Paradox is well known and companies' changed business model and activities carried out to achieve more environmentally sound SC can trigger a whole series of unpredictable consequences that at least partly eliminate the environmental benefits for which they strive.

Esquel therefore tried to achieve balance between sustainable oriented development, improving social corporate responsibility and or course achieving good financial results. They started with comprehensive structural changes of its SC focused on new processes and different type of cooperation between SC partners. They helped farmers with renovating water systems, educating them about the benefits of organic farming, natural ways of pest control etc. In addition, Esquel introduced other methods of harvesting cotton. Previously, farmers used chemicals so that harvesting machines carried out their work easier. Instead, farmers started to pick cotton manually to avoid excessive use of hazardous chemicals. This is on the one hand more expensive and slower, but on the other hand the cotton that is being processed is much cleaner and the subsequent phases of cleaning (where additional chemicals were used) are no longer needed. There is also significantly less waste. They also developed ecological colours for colouring their products. Crucial change was also that farmers were not just suppliers any more but became partners with guaranteed long term quantity and purchasing prices for organic cotton that were app. 30% higher than for regular cotton. This also enabled farmers to make a transition to organic farming and actually achieve higher standard of living simultaneously and in accordance with product's added value.

3 Integrating Life Cycle Thinking in Supply Chain Management

3.1 Why Is Life Cycle Thinking Important for Supply Chain Managers?

Organisations are becoming more and more aware of their environmental impacts and are taking measures on minimizing their impacts with integration of cleaner production within the organisation, increase energy efficiency to decrease energy use of final consumers, optimize transportation and distribution or dematerialize production in order to reduce costs. Due to increasing energy scarcity seen especially in the EU, Cerovac et al. (2014) noted that not just the amount of energy used in production, an important factor is also to consider the mix of energy sources used within the SC. However all of these measures are partial measures that do not concisely cover all environmental impacts related with company's SC. Increasing material prices related with material depletion, stricter environmental legislation especially in the EU and increasing customers' environmental awareness force companies to take comprehensive measures. When talking about sustainable or green SC, SC managers should consider all phases of products life cycle that do not cover only individual SC members but the whole SC. When considering only production, logistics or use of certain product only partial environmental burdens can be identified and such analyses can be misleading and might not address the most important environmental impacts and consequently most appropriate environmental improvements can not be implemented. This idea is the basic principle of Life cycle thinking, meaning that environmental impacts in all phases of the life cycle must be considered, including raw material supply, production, distribution, use and end-of-life phase that is in SCM frequently related with reverse logistics. The focus is to include comprehensive environmental burdens and to address them in accordance with their importance within the whole SC. The tricky part is that life cycle thinking requires cooperation of all parts/members of the SC and can be problematic especially for small and medium sized enterprises that do not have sufficient negotiating power in relation with larger and stronger suppliers. Nonetheless it must be clear that sustainable production and consumption can be achieved by bottom-up and a top-down approaches or by implementing new business models (Kovačič Lukman et al. 2017) meaning that this is not just a task of the top management but a commitment of a whole organization.

3.2 Life Cycle Phases

To design environmentally friendly products or services, their environmental impact must first be assessed within the whole life cycle. Life cycle assessment (LCA) has frequently been identified as an appropriate method for the comprehensive assessment of the environmental impacts of a certain product, because it evaluates environmental impacts through all phases of the life cycle, and it gives good overview of a numerous environmental impacts that are not immediately apparent. However, due to the high amount of data needed and included in LCA, it is an extremely complex and time intensive method for evaluating environmental impacts (Obrecht and Knez 2016).

Life cycle stages, impact categories, and system boundaries required in LCA are presented in Fig. 2.

Figure 2 presents the phases of products life cycle and the system boundaries of LCA, focusing on all main life cycle phases. Only after identifying and assessing environmental impacts through the whole life cycle, companies can see which impacts in their supply chain are the most critical and can therefore start to work on environmental improvements or even on avoiding them at all. Usually (but not necessary), the most common solution is to start optimising phases with the greatest

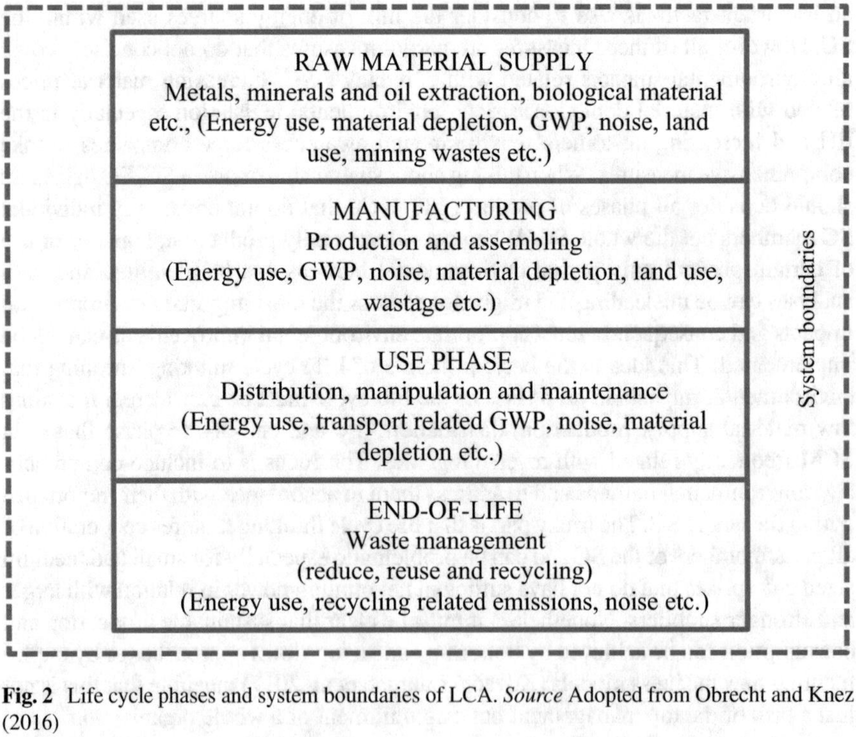

RAW MATERIAL SUPPLY
Metals, minerals and oil extraction, biological material etc., (Energy use, material depletion, GWP, noise, land use, mining wastes etc.)

MANUFACTURING
Production and assembling
(Energy use, GWP, noise, material depletion, land use, wastage etc.)

USE PHASE
Distribution, manipulation and maintenance
(Energy use, transport related GWP, noise, material depletion etc.)

END-OF-LIFE
Waste management
(reduce, reuse and recycling)
(Energy use, recycling related emissions, noise etc.)

System boundaries

Fig. 2 Life cycle phases and system boundaries of LCA. *Source* Adopted from Obrecht and Knez (2016)

environmental impact and those that seems to have the greatest opportunities for savings.

LCA is currently the only standardized method (within the ISO 14000 series) to assess environmental impacts through the whole life cycle however; LCA itself is only the first step towards more environmentally sound SC since it only reveals environmental impacts but does not actually minimise them. The next step is e.g. the use of ecodesign tools, which enable the minimisation of environmental impacts identified with LCA (Obrecht 2010). The main point of life cycle perspective for the most of manufacturers is that their obligations expanded and their (environmental as well as legal) liability does not end at the factory doors.

Case Study 2—IBM's and Apple's efforts to start with the life cycle thinking
IBM is an example of a company that pushes the envelope even more. IBM proposed initiative based on the Electronics Industry Code of Conduct—EICC to mandate their marketing channel partners to adopt environmental measures meaning that they were not focused just on their own organization but tried to push the whole supply chain with all phases of their products live cycle stages to make some improvement and to become also transparent and presented their environmental impacts to the public stakeholders. They proposed four things requested to be realized by their suppliers:

- Define and deploy an environmental management systems (EMS);
- Measure existing environmental impacts and establish goals to improve performance;
- Publicly disclose their metrics and also their results;
- "Cascade" these requirements to any suppliers that are material supplier to their (IBM's) products.

This culminates to the costly venture of life cycle analyses of their products and tracking the footprint of every step of a product and by measuring as well as being able to measurably improve environmental impact. Top managers are becoming aware that the environmentally sound path of business development is becoming more and more profitable. Consequently organizations are becoming more interested in lean, smart and green supply chains focused also on identifying and decreasing environmental burdens in their whole SC with life cycle thinking.

Apple has already done this and reported on their results in a consumer friendly and simple way as presented on a Fig. 3.

4 Ecodesign Integration

4.1 Ecodesign Principles and Ideas

Even though main environmental impacts are caused within material extraction, production, use of even after the end-of-life, most of the environmental burdens of a

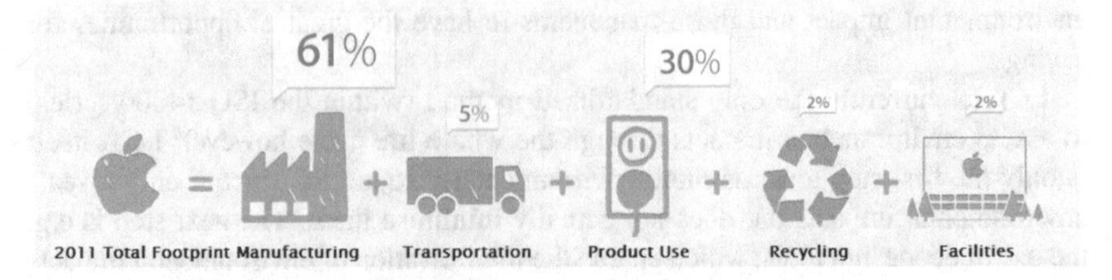

Fig. 3 Apple's communication with stakeholders about its environmental impacts with included life cycle perspective chain/life cycle. *Source* Adapted from Apple (2012)

product are determined already at the design phase. Therefore this phase represents a crucial step when improving a product's environmental performance (Obrecht and Knez 2016; Prendeville and Bocken 2017). Therefore, when talking about sustainable supply chains all phases of products life cycle must be considered and if possible optimized already when designing supply chains. If environmental considerations are taken into account preventively during the earliest phases of product or supply chain development, then it is more likely that the product's overall environmental impact through the whole supply chain can be significantly decreased.

The ecodesign is based on the integration of environmental aspects into product design and development, with the aim of reducing adverse environmental impacts throughout a product's life cycle (Denac et al. 2018). Literature review revealed that ecodesign is based on cleaner production, sustainable development and life cycle perspective. The central aims of ecodesign are to reduce the consumption of (especially scarce and primary) resources, use more renewables, use fewer hazardous materials, increase the use of recycled materials, optimize production and distribution, make production cleaner, prolong the lifespan of the product, and make end-of-life management easier and more efficient both environmentally and economically (Brezet and van Hemel 1997). This means that the potential of the economic and environmental advantages of ecodesign goes beyond the reach of the manufacturer and connects the design of a product to a wider network of supply chain members including raw materials extraction, production, transport and distribution, use and disposal, with efforts to minimize impacts in all these phases.

However, ecodesign or green product development is not that easy to implement (Albino et al. 2009) since it includes life cycle thinking, sustainable development and cleaner production simultaneously (Brezet and van Hemel 1997). This is especially true for small and medium enterprises (SMEs) (van Hemel and Cramer 2002). Although many ecodesign methods and tools are currently available, there is a gap in their integration into the design process as well as into the daily practice of designers especially if the top management is not committed to greening the company's supply chain. Existing ecodesign methods are not always suitable for all organizations or business sectors (Andriankaja et al. 2015). Consequently, ecodesign activities must be planned carefully and systematically, especially in SMEs where human and financial capital is often limited (Eco-Innovation Observatory 2016; van Hemel and Cramer 2002). This needs the support and commitment of top management including

SCM, regardless of the size of the company (Annunziata et al. 2016; Dekoninck et al. 2016).

4.2 Ecodesign Framework and Tools

When starting with ecodesign, the first step is to assess environmental impacts and burdens through the whole products or service life cycle. This can be done in different ways such as with LCA or with some simplified measures such as by using the Product Lifecycle Impact Tool (LIT) or even with ecodesign specific questionnaires. LIT can assist companies in understanding the impacts associated with their product or service environmental impacts (MRA and EEN 2013; Denac et al. 2018).

Some of the issues included in LIT on the Fig. 4 may not be relevant to a specific product however the core idea is to force product designers to start to think about environmental impacts caused outside the company's walls. For example no energy will be consumed by a lavatory in use and the issue of water consumption will be largely irrelevant in the products distribution phase, however supply chain managers must be aware of it and should minimize it when designing green SC. LIT therefore allow companies to eliminate some impacts as well as lifecycle stages and highlight areas where the major impacts arise. The matrix is useful because once it has been completed products designers as well as (SC) managers can easily see which issues and at what lifecycle stage need to be in the focus of eco-design. They can easily identify the hotspots (MRA and EEN 2013; Obrecht 2010) when they start to consider which impacts to reduce (if not all of them due to limited resources and production capabilities).

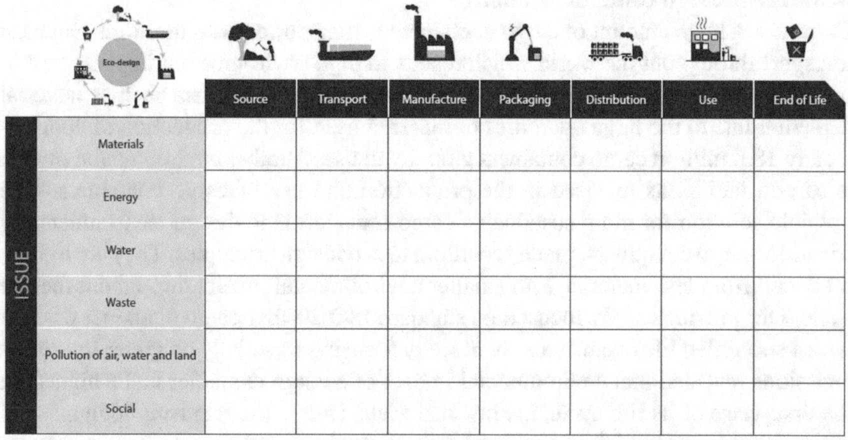

Eco-design

ISSUE	Source	Transport	Manufacture	Packaging	Distribution	Use	End of Life
Materials							
Energy							
Water							
Waste							
Pollution of air, water and land							
Social							

Fig. 4 Lifecycle impact tool (LIT). *Source* Adapted from MRA and EEN (2013), Obrecht (2010)

Having used the LIT to identify the most significant environmental impacts in the product lifecycle product designers and managers (especially technical directors and supply chain managers) will have to consider the potential design improvements which provide the greatest opportunities to reduce these impacts.

In Table 2 ecodesign questionnaire is presented with different design focus areas in accordance with ecodesign strategies which are at some extent applicable to all types of products or services. It must also be considered that due to relations within products SC and life cycle activities to improve product's environmental impact, additional costs or benefits can arise to organizations representing other supply chain members therefore comprehensive analysis is crucial to achieve the best result from the supply chain perspective.

Regarding the transport sector which is most commonly related with logistics as well as supply chains lightweight construction is identified to be a possible pathway towards future sustainability. Although many ecodesign methods and tools are currently available, there is a gap in their integration into the design process in the industry, as well as in the daily practice of designers. According to Andriankaja et al. (2015), existing ecodesign methods are not always tailored to lightweight structures. Gerrard and Kandlikar (2007) foresee that the most substantial change in the design within the transport sectors is the design of new products, involving a change in the material composition: promoting the use of lightweight materials, extending the value of end-of-life (reuse and remanufacturing) and improving the environmental communication about products. Simplifications of these methodologies are crucial the minimisation of environmental impacts because their outputs are easier to obtain and cheaper for the producers to carry out. Despite simplifications these simplified methods can still be identified as a comprehensive impact assessment.

Case Study 3—Simplified eco-design approach for carbon and resource savings in different cargo container designs

Currently, a huge amount of cargo containers is transported with maritime and road transport throughout the world which results in high environmental impacts caused by the transport as well as by the manufacturing of the containers such as material depletion due to the large quantities of material used for the production of approximately 18.6 million cargo containers globally in use. Another environmental impact is carbon emissions released in the production and use of cargo containers. One possible solution for more sustainable cargo transport is to design environmentally friendlier cargo containers, made according to ecodesign principles. They are lighter, produced from less material with smaller environmental impact throughout the life cycle. Our previous study focused on standard ISO 20-foot cargo container designs with a simplified life cycle assessment study focusing especially on green-house gas emissions revealed that environmental impact of a cargo container is the highest in the first phase of its life cycle, i.e. raw materials. Due to the relatively high mass of standard 20-foot aluminium and steel cargo containers (1877 and 2250 kg respectively), and nature of the materials manufacturing phase (raw material processing, welding, assembling, etc.) this accounts for 67% of all impacts. A solution for environmentally friendlier cargo containers is seen in an ecodesign dematerialisation

Table 2 Ecodesign questionnaire

Design focus areas	Key questions for designers	Environmental benefits	Business benefits
Design for material sourcing	When you specify materials and components, do you consider the impact on the environment related to weight, volume, use of recyclates, embedded energy and water and impacts on biodiversity?	Reduced resource depletion Reduced embodied energy/water Reduced transport burden Reduced CO_2 emiss. Reduced impact on biodiversity	Reduced transport costs Improved Image/access to markets
Design for manufacture	Have you considered changing manufacturing processes to reduce energy and water use, waste and recycling of waste?	Reduced CO_2 emissions and depletion of water resources Reduced resource depletion	Reduced energy costs Less waste—Reduced material cost
Design for transport and distribution	Have you considered size, shape and volume of your products from a packaging and transport viewpoint? When specifying packaging do you consider embodied energy and water, production of VOCs or hazardous substances?	Reduced CO_2 emissions and depletion of water resources Reduced air pollution Reduced transport use—less emission and wear and tear on infrastructure Reduced potential for proliferation of hazardous substances	Reduced transport costs Reduced packaging costs
Design for use (Including installation and maintenance)	When you design your products, do you think about their energy and/or water consumption when they are used? Do you consider the amount of consumables and any hazardous materials that may be released during use? Do you consider their longevity and ease of maintenance?	Reduced demand on new material resources Reduced CO_2 emissions Reduced depletion of water resources Reduced potential for proliferation of hazardous substances in the Environment	Lower lifecycle costs for customer—increased profits from increased prices Reduced maintenance costs Good product image
Design for end of life	When you design your products, do you think about how they could be reused/dismantled/recycled Do you consider any hazardous substances that might be released during dismantling or recycling?	Reduced use of land for landfill Reduced demand on new material resources Reduced CO_2 emissions Reduced depletion of water resources	Compliance with regulation Reduced end of life costs

Source Adapted from MRA and EEN (2013), Obrecht (2010)

strategy with particular emphasis on the use of material and production phase but without compromising its performance. Three different designs of cargo container walls presented on Fig. 5 were assessed from an environmental perspective.

Comparative analysis has shown a difference of approximately 15% (315 kg of primary material per single container) in material use when comparing cargo container wall type with the highest and the lowest impacts and significant differences also within environmental assessment as seen on Fig. 6.

Possible reductions of material used for cargo containers means that 20.97 m^2 of aluminium or steel is used just for one side wall of a standard 20-foot container and of course twice as much for a standard 40-foot ISO container when using the Wall Type 1 design. A significant reduction can be achieved when replacing Wall Type 1 containers with Wall Type 2 or Wall Type 3 containers. The amount of material used for one side wall of a standard 20-foot container can be reduced with the implementation of Wall Type 2 or Wall Type 3 design by 6.13 and 4.86 m^2, respectively.

Additional environmental improvement and cost reduction can be seen on mega container ships that can load more than 18,000 twenty-foot equivalent units (TEU). That means that loaded mass can be reduced by 4734 tons when comparing aluminium containers and for 5670 tons when comparing steel containers just by adapting container designs, *ceteris paribus*. Significant improvements can consequently be expected also in container ship fuel economy. Due to the large amount of cargo containers throughout the world and container ships on the sea, the change of Wall Types could bring a massive impact to the reduction of material use and fuel economy as well as on maritime greenhouse gas emissions.

The term "environmentally sound design" refers to the measures taken to develop the products as environmentally friendly as possible. This way it reduces the environmental impact of products in the whole product's lifetime without compromising other product features, such as functionality, price, and quality (Johansson 2002).

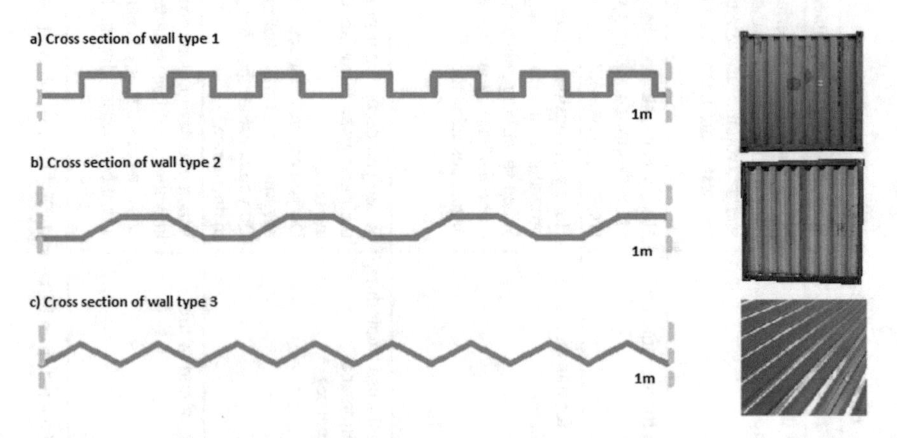

Fig. 5 Cross sections and pictures of three examined container wall types. *Source* Adopted from the study of Obrecht and Knez (2016)

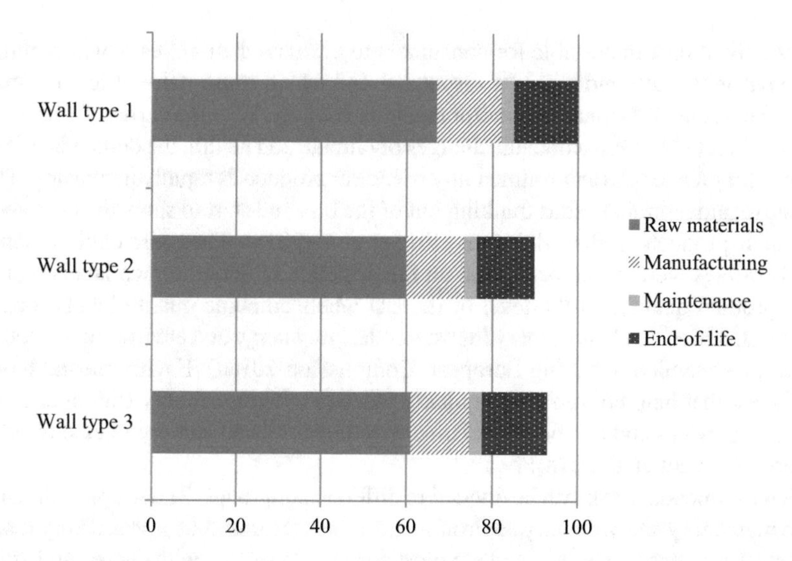

Fig. 6 Comparison of relative GWP of different studied container wall types. *Source* Adopted from the study of Obrecht and Knez (2016)

Sustainable product design, also used in some cases instead of ecodesign is the philosophy and practice of design in which products contribute social and economic prosperity and have a negligible impact on the environment and can be produced from a sustainable resource base (Niinimäki 2006; Lewis 2012).

Companies that are taking actions related with environmental protection within the whole SC such as designing their products to be environmentally friendlier usually tend to gain financial benefit out of such activities that can be in the first phase quite cost intensive. Therefore environmental improvements and efforts should be awarded with different awards as well as with labels informing potential consumers about products environmental impact to promote sustainable production and consumption. In the next part focus will be on environmental labels and certificates.

5 Environmental Labels

5.1 *Defining and Dividing Environmental Labels*

In the flood of products on the shelves of shops average consumers can get confused when thinking on environmental performance of certain product or service. When environmental awareness is on a rise and the share of green consumers is increasing, many companies and consumer organizations identified the need to introduce product's declarations on products environmental performance to enable potential consumers to get more information on environmental impacts of a certain product. It is actually almost impossible for consumers to evaluate themselves which products are environmentally more or less disputable and which cause more or less environmental burdens and which are environmentally friendly. With the current focus being mostly directed at environmental changes of climate and health, the demands of customers for greener and environmentally friendlier products is rapidly increasing. This forces manufacturers to start thinking out of the box and start to show the customers that their products fit their demands (Korent et al. 2018). Therefore environmental labels or tags were developed based on the principle of better known labels related with product quality. At the level of the EU labels combine quality labels for EU products, agricultural and fishery logos, Ecodesign, energy and environmental labels as well as chemical labelling European Commission 2018a). Environmental labels are labels that help consumers to quickly identify environmentally friendlier products and services and can be related on environment related, energy related or other criteria (Obrecht et al. 2018).

Environmental labels can be divided in different subgroups. The simplest division is on mandatory and voluntary environmental labels (Horne 2009). Mandatory labels are usually prescribed by law and are most commonly related with energy and water use. On the other hand voluntary labels can be divided on different ways. Here division of ISO is used that divides them into three groups (ISO 14024: 2018; ISO 14025: 2006; Obrecht et al. 2018; Bratina et al. 2017):

- **ISO I** (environmental labels type one based on ISO 14024 standard)—includes most important and well known labels granted and monitored/revised by independent organizations (e.g. EU Ecolabel);
- **ISO II** (environmental labels type two based on ISO 14021 standard)—more for informational purposes, voluntary declarations, used by producers, distributors, retailers etc. These are labels without independent monitoring and have the lowest scale of credibility. However due to marketing purposes and relatively easy acquisition they are very common;
- **ISO III** (environmental labels type three based on ISO 14025 standard)—informs potential consumers about impacts of certain product on environment and includes also life cycle perspective. It is most commonly used for construction materials—e.g. Environmental product declaration (EPD).

Organizations noted the need for labelling their products with environmental information due to complex customer relations related with environmental concerns, stricter environmental legislation and differentiation of their products. Even though many environmental labels exist nowadays companies are still poorly aware of the wide choice of even mandatory but especially voluntary environmental labels and tags (Obrecht et al. 2018). Even organizations that are aware of them lack the knowledge on their pros and cons as well as on their requirements related with companies supply chains. Even though they are based on different standards and different legal requirements and they might be focused on different environmental areas and differently complicated activities are required for their incorporation, all of them are somehow related with environmental information important for potential users or buyers (Yenipazarli 2015).

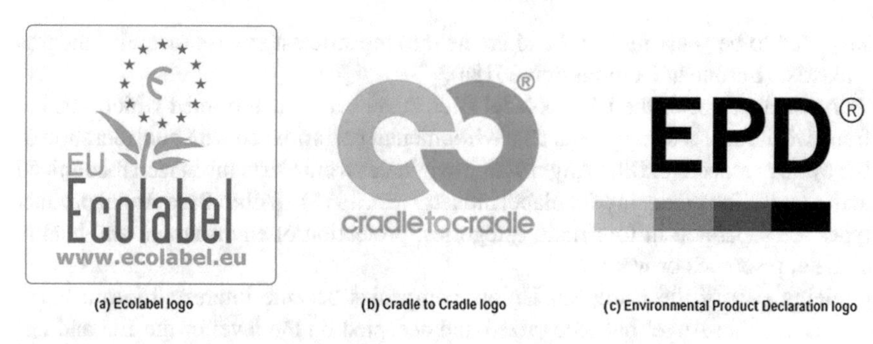

Fig. 7 Logos of **a** ecolabel, **b** C2C and **c** EPD certificates. *Source* European Commission (2018a), Cradle to Cradle Products Innovation Institute (2018), EPD International (2018)

Environmental labels can be described also as an information tool to help (especially green oriented) consumers to choose environmentally better products that need to fulfil high standards of maintaining and preserving users health and/or environmental sustainability. The most important from the perspective of comprehensive environmental knowledge are labels that incorporate environmental burdens related with the whole life cycle of a certain product from cradle to grave and can significantly increase product's added value (Chakravarthy et al. 2016) within its whole SC.

Companies that taking care of the environment and tend to contribute to a more sustainable future are most of the times interested in achieving economic benefits simultaneously with environmental ones. Improving company's image, differentiation strategy, niche products or improved communication with green consumers can be such activities that enable business benefits. Especially for communication with potential customers environmental labels are identified as an appropriate and potent tool. In this part the most common, innovative and comprehensive environmental labels that consider life cycle perspective are examined as well as their impact on SCM. Information flows enabling the certification of different ecolabels is also discussed.

In the next section the EU Ecolabel, Cradle to Cradle certificate and Environmental Product Declaration (EPD) as the most prosperous environmental labels with included life cycle perspective are examined. Their logos are presented on Fig. 7.

5.2 *Ecolabel*

An EU Ecolabel is a label or mark that can be found on everyday products such as detergent, paint and paper products, and indicates that an independent third party has evaluated the product against multiple environmental criteria to make sure that it passes the most stringent health and environmental tests. EU Ecolabel products are supposed to be "best-in-class" and are held to the strictest environmental standards that exist (European Commission 2018a).

As an ancestor of the EU Ecolabel Blue angel can be determined which sets the framework for assessing product's environmental performance with consideration of life cycle perspective. Blue Angel which originates from Germany is well recognized and is one of the pioneers of ecolabel industry (est. 1978). Within Blue Angel product types are separated in four main categories: protection of environment and health, climate, resources or water.

Being part of the European Union, companies became interested into a label, similar to Blue Angel but recognized and accepted on the level of the EU and not just focused on the German market. This was the beginning of the EU Ecolabel, also called a "Flower" launched in 1992 by the European Commission (European Commission 2018a) and first one awarded in 1996. As of March 2018, 2091 EU Ecolabels were awarded to 70,099 products and services available on the EU market with Spain as a leading country with app. 30,000 of them (European Commission 2018b). In order to get endorsement from independent professionals of EU Ecolabel organization, it is necessary to prove that products or services have reduced their impact on the environment and that they are "green" through their entire life cycle or SC. EU Ecolabel is identified as a perfect environmental label when company's target markets are the EU Member States because it is well known among the consumers across the EU (Bratina et al. 2017) and is also certified as an standardized ISO type 1 environmental label.

The advantage of Ecolabel "flower" from the B2C (business-to-customer) perspective is recognition and popularity among consumers all over the EU. From the B2B (business-to-business) perspective, the pressure on suppliers to cooperate with manufacturer that have this label is very important. It requests that improvements are done and documented through the whole SC therefore even 2nd or 3rd tier suppliers must take some environmental protection activities. The standards for implementing the Ecolabel are designed in the way that only 10–20% of products on the market correspond to the criteria to acquire the label (Obrecht et al. 2018). Products that have it give clear signal to green consumers that they are considering to buy environmentally top rated products.

Criteria of Ecolabel refer to the important environmental impacts related with products life cycle perspective (Blengini and Shields 2010) and are also in accordance with SC perspective:

- raw materials selection;
- production;
- usage and waste product.

The EU Ecolabel could be awarded for a wide range of product groups. There are currently twenty-seven different product groups covering a wide range of categories, from cleaning products to cleaning services, from home and garden to clothing and paper products, and from rinse-off cosmetics to tourist accommodation services. However it is not suitable and available for all product types and the EU actually already stopped awarding it for some special product types (e.g. heat pumps). Companies interested in the EU Ecolabel must therefore carefully examine the availability and the procedure for its acquisition since these features can change in time just like the detergent products criteria that were revised in order to guarantee consumers the best environmental performance. This means that if company would be considered to achieve criteria for gaining EU Ecolabel it is not necessary that their environmental improvements within their SC would allow them to keep it or to get it in one year when criteria could be stricken. The criteria for extension to keep the EU Ecolabel are revised every four years which means that products with the Ecolabel stand for the highest environmental performance which must also be improved (European Commission 2017, 2018b).

Case Study 4—Ecolabel based green public procurement in city of Kolding
City of Kolding (Denmark) incorporated the EU Ecolabel along with other environmental labels into its green procurement processes. The EU Ecolabel is used directly by the City of Kolding whenever they are tendering for goods or services covered by the Ecolabel. The main focus is to avoid only financial perspective and start to consider environmental perspective as a part of public procurement as well. Whenever a call for tender for a product group where ecolabelled products exist the criteria relating the product are simply copied as either technical specifications and/or as award criteria. They state that the bidders on request must document compliance with the specifications and the criteria they say they can comply to. In the last three years, the EU Ecolabel has been included in tenders for cleaning agents, copy paper, work clothes, laundry services (for the detergent used), printing services (for the paper used) tissue paper, and fleet management (for the lubricants used). For every call of tender, except the one for fleet management, City of Kolding reported that a sufficient number of bids were received (in average between 3 and 7). For several product groups the bidders offered a wide range of ecolabelled products such as for cleaning agents where approximately 45–60% of the products offered has had at least one ecolabel with the winning bidder offering 58% of ecolabelled products. In terms of costs, there does not appear to be a cost disadvantage in asking for products with ecolabel. If anything, the bidder with the highest environmental score tends to be also the one with the lowest price. That means that products that do not have Ecolabel automatically drop out at such tenders. Because trend of green public procurement is increasing, more producers will be forced to implement ecolabels and adapt to stricter environmental requirements set by their customers if they want to remain in business (adopted from Herman et al. 2018).

5.3 Cradle to Cradle Certificate

Environmental labels can also lead to a confusion when consumers are faced with same products having different labels on it. This can lead to quite a confusion on environmental labels credibility, but in short most of them are quite recognised and as mentioned by ISO 14000 series, can help boost a company's reputation in the world market by a lot. As such they are getting progressively vital for a successful or failed "market penetration". Because of that, more and more companies are in progress of obtaining them, or in the process of implementation into their processes.

One of these eco-labels or in our case more like "eco-certificate", is also the "Cradle to cradle" certificate or in short "C2C". Unlike the other more commonly know eco-labels, this certificate is at the moment much less known, but as such it is not less developed or complicated as the others from the supply chain or life cycle perspective. Even if it is globally less recognized it focuses on the whole life cycle and has very strict environmental requirements for awarding it as well as special consideration on end-of-life phase. The C2C concept foresees an own certification (Cradle to Cradle Products Innovation Institute 2018). What makes this certificate more specific or different from others, is the fact that it wants to implement a enclosed life-cycle, meaning that that the product is not entirely discarded at the end of its lifetime, but is instead again incorporated into another product.

To understand the whole meaning and purpose of the certificate, we have to explain firstly the concept or the idea on which the certificate is based on. Cradle to cradle concept was first mentioned in a research report to the European Commission in Brussels in the year 1976, by Walter Stahel and Genevieve Reday-Mulvey. The research "The potential for substituting manpower for energy" presented a vision of an economy in loops or later called circular economy. It was primarily focused on how such economy could have an impact on job creation, economic competitiveness, etc. The research was later published in a book "Jobs for Tomorrow: The potential for substituting manpower for energy", and the factors mentioned in the book, are the three pillars of sustainable development: ecologic, economic and social compatibility. Stahel then continued with the ideas and wrote a paper "The Product-Life Factor", where he discovered or better yet identified that selling utilization instead of goods is much more profitable and can lead to a better business model of a loop economy; in short selling utilization creates sustainable profits and removes the costs and risks that were connected with waste (Product-Life Institute, n.d.).

The next step was the denial of the newly developed concept "cradle to grave" that is also in accordance with life cycle thinking. Although the concept was widely recognised and already started to be implemented, one of the main reasons being rather compatible with the existing linear economic model, Stahel refused the idea, because it was still only an upgrade for gravediggers, who wanted to get rich with waste, and the whole concept was more or less just an advanced upgrade which scope opposed that it relies on end-of-pipe solutions. Rather then the concept presented by other experts, he promoted another concept, namely "cradle back to cradle" which would use durable goods in a loop and as such be much more sustainable. At the same time Braugart (2015) promoted the same concept mostly as a response to the rise of the "cradle to grave" concept, which was according to him, still relying to much on the end of the pipe solutions. The two pioneers soon started to cowork on the newly established concept (Product-Life Institute, n.d.).

Following the collaboration, they published a manifesto in 2002, called "Cradle to Cradle: Remaking the Way We Make Things" in which the whole design was described in detail. Following this published manifesto, a number of companies decided to implement it although soon criticism fall on both authors, mostly for the lack of granting the certificate to other companies, other then those in the inner circle (Sacks 2009). Following this pressure, they decided to establish a special institute, called "Cradle to Cradle Products Innovation Institute" with which they enabled for the certificate to be publicly obtainable for every company, that is prepared to implement the certificate and its demands.

One of the most important innovations related with C2C certificate is that it is free to get and works mostly on a donation system, meaning that the companies donate how much they want rather then being forced to pay for obtaining the certificate as it is practice within other recognizable environmental labels. This in a way is also much more appealing to the companies, who want to improve their processes.

The main idea of the certificate itself lies in a continual improvement process that looks at a product in five quality categories (Cradle to Cradle Products Innovation Institute 2018);

- Material health (chemical ingredients of every material in a product),
- Material reutilization (designing products with materials that can safely be returned to nature or industry),
- Renewable energy and carbon management (manufacturing powered mostly by renewable energy),
- Water stewardship (manage clean water),
- Social fairness (design operations to honor all people and natural systems which are affected by the product through is life cycle).

The product which the company certified gets achievement evaluation in each of the previously mentioned categories. This achievements are Basic, Bronze, Silver, Gold and Platinum. Unlike others, the product gets it‘s overall scoring based on the lowest achievement level he achieved (ex. if the Product has 4 Gold marks and 1 Silver mark, the overall certification level of the product is Silver). Of course this can be always improved and perfected with continues improvement of the processes connected with the product, as well as guideline for the company, what is needed to be done or which category need‘s further improvement for the product to become better.

The C2C paradigm is a new approach to business development and is higly interesting for SC managers as well. This approach is also known as circular economy (CE). CE is different from conventional business models by focusing on the SC economy and the possible business opportunities that arise when broadening the horison from gate-to-gate to cradle-to-cradle. The reason for this new horison is the fact that current and future businesses need to ensure a closed or semi-closed loop in their material cycles due to ever less resource availability. Reverse logistics is first step towards it however still having numerous opportunities for improvements and full implementation of the idea of CE. Another benefit of the C2C product certification is the identification of substitute materials in the material SC in order to promote elimination of dangerous parts from the scarce or hazardous raw materials used in the production. As a result of their change of horisons businesses are now changing their business models towards models with the entire material loop in their product. This also influences the product's design. In other words, the Cradle to Cradle paradigm teaches us to think in a whole new way and teaches us to design recyclable and adaptive solutions corresponding to new business models related with stricter environmental measures integrated within the whole supply chain (Hansen 2015).

Case Study 5—Implementing C2C on Shoes
If we conclude all what was described above in an simple example of a product such as a shoe; Shoes are being made in a factory, where the sole is made from «biodegradable materials», while the upper parts are made out of «industrially made materials», with which they compliment the material health category. The factory uses for example off-cuts of rubber soles for the development of new soles with which it drastically lowers its waste and also compliments material reutilization category. After the finished production, the shoes are sent to retailers, where the buyer pays for the shoe significantly less then for a comparable shoe. That is because the buyer pays only for the usage of the materials, for the period of time he will be using the shoe with which they compliment the social fairness category. After the shoe is for certain reason unusable, it is returned to the factory, where the soles and upper parts are divided. The soles are then returned into nature, where they biodegrade, while upper parts are again used in factory for creation of new shoes.

Although this is a simplified example, it clearly shows the difference between other concepts, where the whole cycle of a product is in many ways incomplete, open, and is not transformed into a new goods after end of life (adapted from Korent et al. 2018).

5.4 Environmental Product Declaration

Environmental product declaration (EPD) could actually be introduced as the environmental equivalent of a technical datasheet. It is an independent and registered document that allow transparent and comparable information on products environmental life cycle impacts and present comprehensive information about the environmental impacts such as waste, energy use, water use and other resource use associated with a product (or service) in a standardised form. It is applicable for different companies and products since its main idea is to focus on life cycle assessment, meaning that the whole supply chain is again included in such evaluations. EPD clearly defines product composition, production processes, environmental efficiency, sources of raw material supply and relations with waste management. It is a voluntary ISO type 3 (according to ISO 14025: 2006 standard) declaration that does not directly means that one product is environmentally better that others but it means that it has evaluated and assessed information flow what is actually happening with this product through its life cycle phases (ISO 14025: 2006; ZAG, n.d.; EPD International 2018).

According to ISO 14025: 2006 and EPD International (2018) acquiring EPD is done in four steps as presented on Fig. 8.

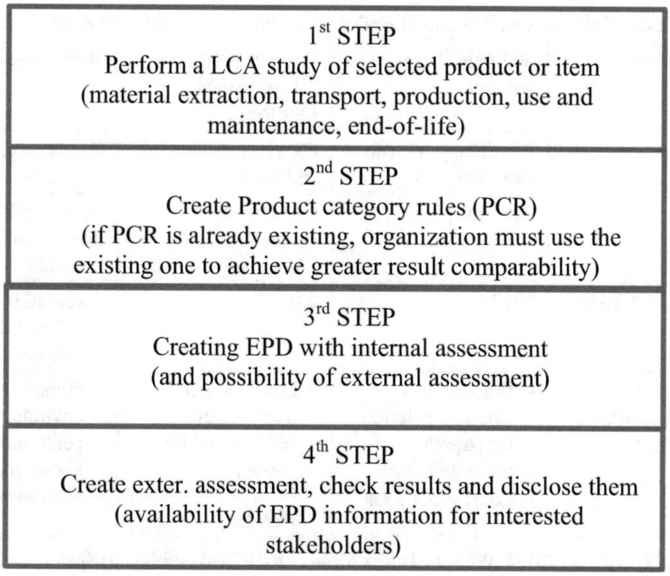

1st STEP
Perform a LCA study of selected product or item (material extraction, transport, production, use and maintenance, end-of-life)

2nd STEP
Create Product category rules (PCR) (if PCR is already existing, organization must use the existing one to achieve greater result comparability)

3rd STEP
Creating EPD with internal assessment (and possibility of external assessment)

4th STEP
Create exter. assessment, check results and disclose them (availability of EPD information for interested stakeholders)

Fig. 8 Acquiring environmental product declaration (EPD). *Source* ISO 14025: 2006 and EPD International (2018)

The term Environmental Product Declaration and its acronym as well as logo are registered within the EU and is valid for all EU Member States. It is mandatory also for some European structural project funding (Mikuš Marzidovšek 2009) and can be registered within the programmes such as International EPD System. It can be used for several applications including green public procurement and schemes for assessing building's sustainability. Due to lack of comparable information especially in construction sector (e.g. construction materials, buildings) the need for comparable information tool was identified in the EU to enable comparison of different materials or building designs with life cycle perspective taken into consideration. The basic idea was to develop an information tool for communication between companies (B2B) however it can be used also for communication with final consumers (B2C) about products environmental performance. Since it is based on LCA with evaluation of impacts through the whole supply chain it is highly complex but also needed and reliable basement especially for construction sector where sustainable use of natural resources will have to be proven in the future. EPD is issued for the period of 5 years with the possibility of prolonging it if there were no changes impacting its environmental performance (ZAG, n.d.).

Case Study 6: EPD as a tool for achieving environmental and economic benefits
On Table 3 three companies are compared—two (Squiggle glass (2017) and Wiesner-Hager Möbel (2018) that already acquired EPD and one that could do get in the future (Nobis 2018).

Table 3 Cases of business with acquired EPD and business potential for its acquisition

Organization	Squiggle (Great Britain)	Wiesner-Hager Möbel GmbH (Austria)	Nobis (Slovenia)
Sector	Furniture production and sales	Furniture production and sales	Furniture production and sales
Product	Squiggle glass	Office desks, cabinets, drawers, other office furniture	Office desks, drawers, other office furniture
Acquisition of EPD	2017	2012	Potentially interested, not yet acquired
Environmental protection standards and activities	– Product idea (lifetime solution for paperless presentations) – Local production	– Environmental management (EMAS and ISO 14001) – LCA in accordance with ISO 14040	– High environmental perception – Focus on supply of local materials

Source Squiggle glass (2017), Wiesner-Hager Möbel (2018) and Nobis (2018)

Nobis is Slovenian office furniture producer that is focused in local material supply and has all the basics to start implementing ISO 14025 standard. They should first study their organization structure, business processes and supply chain. When all required data are gathered an EPD can be obtained and issued by different independent evaluators. Due to local raw material supply, very important issue—source of raw materials, is well documented and can bring advantage in obtaining EPD more easily than in case of imported raw materials without any information about its extraction, processing or even origin. At the end of life, their products could be re-collected and reused or recycled by the company. With the acquisition of the EPD company could gain additional economic benefits. Especially differentiation on a highly competitive market and achieving additional added value, just like Squiggle or Wiesner-Hager Möbel, are the most important two benefits.

An EPD could actually be introduced as the environmental equivalent of a technical datasheet and present standardized form of comprehensive information about the environmental impacts—waste, energy use, water use and other resource use associated with a product. It is applicable for different companies and products since its main idea is to focus on life cycle assessment, meaning that the whole supply chain is again included in such evaluations.

6 Discussion on Information Flow Within Environmental Assessments and Improvements Related to the Supply Chains

To perform environmental performance analysis or to assess environmental impacts, concise data acquisition is a preconditioned. When considering life cycle perspective, data must be acquired on the completely different level that when assessing only environmental performance within the company's walls. Therefore cooperation of supply chain partners in additional environmental information flow is crucial. Many times these data include also information of a sensitive nature about the business of a company (e.g. supplier) which the company has a legitimate interest in keeping confidential.

When SC manager starts with new business model to form closer cooperation between partners within the supply chain and manage its "greening", agreement on data disclosure within their partnership is urgent for efficient implementation of best possible measures. On Table 4 basic issues regarding environmental information within different studied concepts of entrepreneurial environmental protection are cross-compared and related with SC, consideration of life cycle perspective and marketing effect.

6.1 *Life Cycle Assessment*

LCA as the standardized assessment of environmental impacts through the whole life cycle has well determined required information on scope and margins of the study, inventory data analysis, assessing data on environmental impacts of inventory and interpretation of the results to create useful information for managers. However LCA itself only focuses on environmental assessment and do not actually require any activities related with environmental improvements, nor from the company performing LCA nor from their supply chain partners. Nonetheless companies that decide to perform LCA usually perform it to get sufficient information on their product's or service's environmental impacts and to focus on improvements in most potential focus areas that are many times hidden if life cycle perspective is not included and might be related with environmental performance of primary raw material, energy used in the manufacturing processes, decomposition etc. Consequently it can inspire the whole supply chain or at least a group of supply chain partners to make environmental improvements which are becoming more and more often also profitable (at least on a long run). It can be performed due to business requirements (e.g. if public tender demands EPD for materials included in procurement process) or voluntarily (e.g. if company wants to make environmental improvements and focuses on comprehensive environmental assessment first) and can also be used for marketing purposes however interpretation of results must be done according to the ISO standard.

Table 4 Comparison of information and performance factors of different approaches and labeling schemes

	LCA	Ecodesign	EU-Ecolabel	C2C	EPD
Life-cycle perspective	Included	Included	Included	Included	Included
Environ. improvem.	Focus on environmental assessment	Basic idea is to achieve environmental improvements	Needed for acquisition and sometimes extension	Needed for acquisition and extension	Just certificate on environ. performance
Supply chain effect	Just assessment but gives insight about SC partners environmental performance (which to keep and which to replace)	Focuses also on SC partners' environmental performance (which to avoid or replace and which to keep)	Can impact suppliers	Can impact suppliers	Can impact suppliers
Marketing effect	Indirectly it can be used in marketing and communication with customers	Improvements can be used in marketing and communication with customers	Possible direct implement. in marketing	Possible direct implement. in marketing	Possible direct implement. in marketing
B2B or B2C approach	B2B and B2C	B2B and B2C	Especially B2C, also B2B	B2B and B2C	B2B and B2C

Source Authors

6.2 Ecodesign

Ecodesign's basic idea is to minimize product's environmental impacts already in the design phase therefore information of products life cycle environmental performance are fundamental for effective implementation. LCA can be a starting point for ecodesign; however it is far from being the only possibility. Ecodesign questionnaires are more frequently used because they are simpler and most of the time less time and cost intensive than LCA. It can be successfully distributed among supply chain partners to gather required information however even eco-design questionnaires can sometimes ask to much from SC partners in terms of revealing information that one partner could consider as a business secret (e.g. energy used in production process). Despite simplifications in comparison with standardized LCA, ecodesign questionnaires also enable good insight into life cycle related environmental impacts within the whole SC. Since its structure is divided into strategy for material sourcing, for production, for distribution, for use and for end-of-life phase it covers environmental impacts and possible improvements within the whole SC comprehensively. It can also include information on pre-ecodesign and post-ecodesign phase therefore it can easily be implemented in business marketing strategy to gain green oriented customers and to give signal to environmentally sound business on company's improving environmental performance.

6.3 Environmental Labelling Schemes

The whole process of gaining one of the environmental labels or certificate begins with the company reviewing its products and its compatibility with the demands of gaining the certificate or label. If the product is appropriate for the certification, the company needs to select an accredited assessment organization, which then conducts a careful analysis. The data, which is then collected with the help of assessor is send to the accredited assessment organization, which decides if everything is according to the rules. If everything is alright, the company receives the certificate, after which it works with the institute in promoting the product, as well as improving its processes. After that, on certain period of time re-certification is needed (2 years for C2C and 5 years for EPD) with intention to review the reported progress and to keep awarded certificates or labels only to environmental performance market leaders. (Cradle to Cradle Products Innovation Institute 2018; European Commission 2018a; EPD International 2018).

Requested information in all three cases are combined from data on environmental performance through the products life cycle. Some environmental labels also require exact information of products (or companies) state before and after acquisition of the certificate or label. The last is required also for the self-declarative labels, that are not standardized and do not require independent assessment and accreditation, to prove that environmental activities and improvements were really implemented.

Environmental labels can also be seen as a strong marketing tool to promote sustainable production and consumption, gain environmentally aware customers and enable priority in green public procurement as in case of City of Kolding.

All of these information can have positive or negative effects on SC partners. If partners (e.g. suppliers or manufacturers) are assessed to be of poor environmental performance they should be contacted and communicated about possible environmental improvements that are sometimes a precondition for acquisition of certain environmental label or certificate (e.g. minimization and substitution of hazardous materials at suppliers side). If they are not interested to perform improvements, SC manager must assess cost and benefits of avoiding or replacing such e.g. supplier or manufacturer with environmentally more sound partners which enable that the whole SC becomes more environmentally undisputable and allow them to enter into the ecolabeling scheme to improve companies' images within the whole SC, make environmental improvements also within the whole SC, increase products added value across its value chain and convince more green oriented consumers to buy or use their product or service.

7 Conclusion

Above described concepts, certificated, labels and ideas of greening the SC as well as cases and real business show that "greening" a SC is not necessary that complicated if well planned and organized. It is just about squeezing more economic and simultaneously also environmental benefits out of current operations. SCM is nowadays confronted with new challenges regarding just in time production, increased products variations, batch size one production series, shortening of products and services life cycles, fast changing environment as well as increased environmental pressure. Recently the last is becoming priority among SC managers and innovative ways of greening the SC are being investigated. In this chapter life cycle perspective as a prosperous concept for assessment of well known and hidden environmental burdens, ecodesign as a tool for environmentally friendly design of products and services enabling environmentally sound SC already in the products and SC design phase and environmental labelling schemes including life cycle thinking as potential tool for making SC more environmentally sound as well as for communication with customers are being examined. All of examined principles, ideas and labelling standards regarding environmental sustainability are at least partly in accordance with the concept of circular economy, promoting circular instead of linear flow of materials. Due to scarce and limited raw materials and awareness that the future wellbeing of societies as well as companies is related with nowadays environmental protection and performance these ideas are becoming more relevant than ever before. All of these principles are in favour of the idea that economic growth and environmental sustainability are not two contradictory but complementary goals that connects more and more stakeholders within the SC. It can be concluded that environmental protection is an interdisciplinary issue and it must be carried out on with tide cooperation of partners on different SC levels comprehensively to achieve the best results.

References

Albino V, Balice A, Dangelico RM (2009) Environmental strategies and green product development: an overview on sustainability-driven companies. Bus Strategy Environ 18:83–96. https://doi.org/10.1002/bse.638

Alexandratos, N Bruinsma, J (2012) World agriculture towards 2030/2050: the 2012 revision. ESA working paper 3. Food and Agricultural Organisation [FAO] of the United Nations. Rim

Ammenberg J, Sundin E (2005) Products in environmental management systems: drivers, barriers and experiences. J Clean Prod 13:405–415. https://doi.org/10.1016/j.jclepro.2003.12.005

Andriankaja H, Vallet F, Le Duigou J, Eynard B (2015) A method to ecodesign structural parts in the transport sector based on product life cycle management. J Clean Prod 94:165–176. https://doi.org/10.1016/j.jclepro.2015.02.026
Annunziata E, Testa F, Iraldo F, Frey M (2016) Environmental responsibility in building design: an Italian regional study. J Clean Prod 112:639–648. https://doi.org/10.1016/j.jclepro.2015.07.137
Apple (2012) Environmental progress report. https://www.apple.com/environment/reports/. Retrieved 2 Jun 2018
Bešter J (2017) Economically efficient circular economy. Institute for Economic Research, Ljubljana
Blengini GA, Shields DJ (2010) Green labels and sustainability reporting: overview of the building products supply chain in Italy. Manage Environ Qual Int J 21(4):477–493
Bratina T, Šinko S, Šlajmer V, Obrecht M (2017) Ecolabels and Ecodesign potential for greening companies supply chains. In: 7th international student symposium on logistics and international business. Plzen
Braugart M (2015) A cradle to cradle economy is Europes only future. Friends of Europe. https://www.friendsofeurope.org/greener-europe/cradle-cradle-economy-europes-future-certified. Retrieved 14 May 2018
Brezet H, van Hemel C (1997) Ecodesign: a promising approach to sustainable production and consumption. United Nations Environmental Programme, Paris
Cerovac T, Ćosić B, Pukšec T, Duić N (2014) Wind energy integration into future energy systems based on conventional plants—the case study of Croatia. Appl Energy 135:643–655
Chakravarthy Y, Potdar A, Singh A, Unnikrishnan S, Naik N (2016) Role of ecolabeling in reducing ecotoxicology. Ecotoxicol Environ Saf 134:383–389
Cradle to Cradle Products Innovation Institute (2018) Get certified. https://www.c2ccertified.org/get-certified. Accessed May 2018
Dangelico RM, Pujari D, Pontrandolfo P (2017) Green product innovation in manufacturing firms: a sustainability-oriented dynamic capability perspective. Bus Strategy Environ 26(4):490–506. https://doi.org/10.1002/bse.1932
Dekoninck EA, Domingo L, O'Hare JA, Pigosso DCA, Reyes T, Nadège Troussier N (2016) Defining the challenges for ecodesign implementation in companies: development and consolidation of a framework. J Clean Prod 135:410–425. https://doi.org/10.1016/j.jclepro.2016.06.045
Denac M, Obrecht M, Radonjič G (2018) Current and potential ecodesign integration in small and medium enterprises: construction and related industries. Bus Strategy Environ, 1–13. https://doi.org/10.1002/bse.2034
Diamond Management and Technology Consultant. N.d. Green supply chain framework. http://www.diamondmc.com/. Retrieved 20 June 2017
Eco-Innovation Observatory (2016) Eco-innovate. In: Miedzinski M, Charter M, O'Brien M (eds) A guide to eco-innovation for SMEs and business coaches. Eco-Innovation Observatory, Brussels
EPD International AB (2018) The international EPD system. https://www.environdec.com/The-International-EPD-System/. Retrieved 16 Sept 2018
European Commission (2014) Towards a circular economy: a zero waste programme for Europe. COM (2014) 398 final. European Commission, Brussels. http://www.ipex.eu/IPEXL-WEB/dossier/files/download/082dbcc54653729e014700aed53e6209.do. Retrieved 20 June 2017
European Commission (2017) The EU ecolabel product catalogue. http://ec.europa.eu/ecat/. European Commission. EU labels. https://ec.europa.eu/info/business-economy-euro/product-safety-and-requirements/eu-labels_en#eu-quality-standards. Retrieved 3 Jun 2018
European Commission (2018a) Environment—ecolabel. http://ec.europa.eu/environment/ecolabel/. Retrieved 26 Oct 2018
European Commission (2018b) Fact and figures. http://ec.europa.eu/environment/ecolabel/facts-and-figures.html. Retrieved 20 Oct 2018
Gerrard J, Kandlikar M (2007) Is European end-of-life vehicle legislation living up to expectations? Assessing the impact of the ELV Directive on 'green' innovation and vehicle recovery. J Clean Prod 15(1):17–27. https://doi.org/10.1016/j.jclepro.2005.06.004

Gerstlberger W, Præst Knudsen M, Stampe I (2014) Environmental requirements, knowledge sharing and green innovation: empirical evidence from the electronics industry China. Bus Strategy Environ 23(2):131–144. https://doi.org/10.1002/bse.1746

Ghisellini P, Cialani C, Ulgiati S (2016) A review on circular economy: the expected transition to a balanced interplay of environmental and economic systems. J Clean Prod 114:11–32. https://doi.org/10.1016/j.jclepro.2015.09.007

Hansen H (2015) What cradle to cradle teaches us about sustainable development. Linked In. Retrieved June 2, 2018 from https://www.linkedin.com/pulse/what-cradle-teaches-us-sustainable-development-hanne-tine-ring-hansen

Herman L, Gračner T, Močnik A, Obrecht M (2018) Analysing best practices of making supply chains more environmentally sound. In: 8th international student symposium on logistics and international business, 5th–9th November. Yasar University, Izmir. Turkey (In press)

Ho JC, Shalishali MK, Tseng TL, Ang DS (2009) Opportunities in green supply chain management. Coastal Bus J 8(1):18–31

Horne RE (2009) Limits to labels: the role of eco-labels in the assessment of product sustainability and routes to sustainable consumption. Int J Consum Stud 33:175–182. https://doi.org/10.1111/j.1470-6431.2009.00752.x

International Energy Agency (2017) World energy outlook 2017—executive summary. Paris, France. https://www.iea.org/publications/freepublications/publication/WorldEnergyOutlook2016ExecutiveSummaryEnglish.pdf. Retrieved 20 Sep 2018

International Organization for Standardization (ISO) (2012) Environmental labels and declarations—how ISO standards help (Standard No. 14000) https://www.iso.org/files/live/sites/isoorg/files/archive/pdf/en/environmental-labelling.pdf. Accessed 20 Sep 2018

International Organization for Standardization (ISO) ISO 14025:2006. 2006–2007. Environmental labels and declarations—type III environmental declarations—Principles and procedures. ISO

International Organization for Standardization (ISO) ISO 14024:2018. (2018) Environmental labels and declarations—type I environmental labelling—principles and procedures. ISO

Johansson G (2002) Success factors for integration of ecodesign in product development. A review of state of the art. Environ Manage Health 13:98–107. https://doi.org/10.1108/09566160210417868

Korent Z, Berglez K, Obrecht M (2018) Ecolabels and cradle to cradle certificate. In: 8th international student symposium on logistics and international business. Yasar University Izmir, Turkey, 5th–8th Nov 2018 (in press)

Kovačič Lukman R, Glavič P, Carpenter A, Virtič P (2016) Sustainable consumption and production: research, experience, and development: the Europe we want. J Clean Prod 138:139–147. https://doi.org/10.1016/j.jclepro.2016.08.049

Lewis H (2012) Designing for sustainability. In: Verghese K, Lewis H Fitzpatrick L (eds) Packaging for sustainability. Springer, London, pp 41–107

Lieder M, Rashid A (2016) Towards circular economy implementation: a comprehensive review in context of manufacturing industry. J Clean Prod 115:36–51. https://doi.org/10.1016/j.jclepro.2015.12.042

Makover J, Elkingtov J, Hailes J (1993) The green consumer. Tilden Press

Maribor Development Agency (MRA) and Enterprise Europe Network (EEN) (2013) Ecodesign—environmentally friendly design in construction industry. MRA, Maribor

Mikuš Marzidovšek M (2009) Analysis of EU structural funds from European regional development fund. Master thesis. University of Ljubljana, Ljubljana

Niinimäki K (2006) Ecodesign and Textiles. Res J Text Apparel 10:67–75

Nobis (2018) Office furniture Nobis Maribor. http://www.nobis.si/. Retrieved Jun 2018 (in Slovenian language only)

Obrecht M (2010) Ecodesign of buildings. Faculty of Economics and Business University of Mariboru, Maribor (in Slovenian language only)

Obrecht M, Denac M, Bratina T, Gračner T, Mohorko K, Šinko S, Šipek G, Šlajmer V (2018) Greening of the supply chain and possibilities for implementation of ecolabels in the company

Orca energija d. o. o. Final report of a research project. Creative wat to the knowledge. Faculty of Logistics University of Mariboru, Celje
Obrecht M, Knez M (2016) Carbon and resource savings of different cargo container designs. J Clean Prod 155:151–156. https://doi.org/10.1016/j.jclepro.2016.11.076
Plouffe S, Lanoie P, Berneman C, Vernier MF (2011) Economic benefits tied to ecodesign. J Clean Prod 19:573–579. https://doi.org/10.1016/j.jclepro.2010.12.003
Prendeville S, Bocken N (2017) Design for remanufacturing and circular business models. In: Matsumoto M, Masui K, Fukushige S, Kondoh S (eds) Sustainability through innovation in product life cycle design. EcoProduction (Environmental issues in logistics and manufacturing). Springer, Singapore
Preston F (2012) A global redesign? Shaping the circular economy. Briefing Paper. Chatham House, London
Product-Life Institute (n.d.) Cradle to cradle. http://www.product-life.org/en/cradle-to-cradle. Accessed 30 May 2018
Sacks D (2009) Green Guru William McDonough must change, demand his biggest fans. Fast Company. https://www.fastcompany.com/1186727/green-guru-william-mcdonough-must-change-demand-his-biggest-fans. Retrieved 30 May 2018
Sarkar AN (2012a) Green supply chain management: a potent tool for sustainable green marketing. Asia-Pac J Manag 8(4):491–507. https://doi.org/10.1177/2319510X13481911
Sarkar AN (2012b) Green branding and eco-innovations for evolving a sustainable green marketing strategy. Asia-Pac J Manage Res Innov 8(1):39–58. https://doi.org/10.1177/2319510X1200800106
Slovenian National Building and Civil Engineering Institute (ZAG) No date. Rules for preparation, issuing and prolonging Environmental product declaration. ZAG, Ljubljana
Squiggle Glass (2017) Green squiggle? https://squiggleglass.com/squiggle-glass-environment/. Accessed 1 Jun 2018
Szegedi Z, Gabriel M, Papp I (2017) Green supply chain awareness in the Hungarian automotive industry. Pol J Manage Stud 16(1):259–268
Širec K, Bradač Hojnik B, Denac M, Močnik D, Rebernik M (2018) Slovenian companies and circular economy: Slovenian business observatory 2017. University of Maribor publishing house, Maribor (in slovenian language only)
The 2030 Water Resource Group (2009) Charting our water future economic frameworks to inform decision-making. https://www.mckinsey.com/~/media/McKinsey/Business%20Functions/Sustainability%20and%20Resource%20Productivity/Our%20Insights/Charting%20our%20water%20future/Charting%20our%20water%20future%20Full%20Report.ashx. Retrieved 2 Jan 2018
UNEP—United Nations Environmental Programme (2006) UNEP guide to life cycle management. In: Jensen AA, Remmen A (eds) Report. UNEP, Paris
van Hemel C, Cramer J (2002) Barriers and stimuli for ecodesign in SMEs. J Clean Prod 10:439–453. https://doi.org/10.1016/S0959-6526(02)00013-6
Yenipazarli A (2015) The economics of eco-labeling: standards, costs and prices. Int J Prod Econ 170(A):275–286
Wiesner-Hager (2018) Wiesner-Hager sustainability. https://www.wiesner-hager.com/en/about-us/about-us/company-profile/. Retrieved 15 May 2018

Modelling and Simulation of Logistics Processes as a Decision-Making Tool

Value of ICT Integration Model of e-Booking System and Intelligent Truck Traffic Management System in the Sea Port of TEN-T Corridor

Boguslaw Sliwczynski

Abstract Systematic growth of cargo flow in the TEN-T corridor passing through the all Sea Ports served both by the existing terminals and those planned in the Port's development strategy. The forecast growth rate of turnover in the Port results from the strategy adopted by the Port Authority to increase competitiveness and attractiveness of the Port as a business partner, in response to the implementation of the strategy of economic development of Europe, the strategy of development of the New Silk Belt and Route China-Europe and the dynamic development of European import/export. Currently, traffic in the European Sea Ports (e.g. Rotterdam, Hamburg, Antwerp, Dover, Felixstowe, Aalborg, Le Havre, Lisbon) periodically reaches the saturation level in the existing road network. This causes, periodically and in the perspective of the development of goods turnover, the reasons for this:

- lorry queues before entering terminals and on access roads, creating traffic jams and blocking roads in the Port and City and the main access routes,
- extension of unloading/loading time of ships,
- reducing the efficiency and effectiveness of terminals' work,
- inefficient use of parking lots in the port, as well as in the period of traffic jams when there are no parking spaces for lorries in the port's parking lots and its surroundings,
- reduction of traffic safety and greater susceptibility to collisions with high or critical saturation of traffic in the road network and reduction of transport comfort for the society of the port agglomeration.

The problems identified have a significant impact on the efficiency and duration (and delays) of cross-border road freight transport (road hauliers from all over Europe), in global maritime supply chains served by global shipping lines, as well as in international rail freight transport.

Keywords Supply chain efficiency · Value analysis · Intelligent transport system · Integration of information flow · IT solutions · e-booking

---

B. Sliwczynski (✉)
Poznan School of Logistics, Estkowskiego 6, 61-755 Poznan, Poland
e-mail: boguslaw.sliwczynski@wsl.com.pl

A. Kolinski et al. (eds.), *Integration of Information Flow for Greening Supply Chain Management*, Environmental Issues in Logistics and Manufacturing, https://doi.org/10.1007/978-3-030-24355-5_14

1 Introduction

The increase in the traffic of truck servicing the flow of goods in the network of the Port and access roads takes place in the dynamically developing maritime traffic and the growing of road traffic of coaches and cars. A significant share of sea terminals, carriers, forwarders, etc. that deal with transport and supply chain management in Europe recognise interoperability and information exchange as a relevant challenge (Kolinski and Jaskolska 2018). Interoperability is expensive and complex. Information exchange implementations in this respect have so far often triggered by traffic management needs (e.g., road, rail, maritime and ship-to-shore communications) and public ITS systems for route, journey and parking management. ITS components that are integrated can result in synergistic effects when considered as an entire system.

To achieve higher efficiency levels, large companies have significant costs in making sure that their transport/supply chains are connected. Many SMEs lack the capability and resources to interoperate and collaborate with other companies.

In response to these needs, the European Commission has launched ITS Action Plan (Action Plan for the Deployment of Intelligent Transport Systems in Europe COM (2008) 886 final) and Directive 2010/40/EU on the framework for the deployment of Intelligent Transport Systems in the field of road transport and for interfaces with other modes of transport.

Intelligent Transport System is only a means for intelligent traffic control, while the needs for careful coordination in the sea port include transhipment operations in sea terminals and warehouses correlated with maritime/rail/road transport, road traffic management and parking area management (Zhang et al. 2011; Meneguette et al. 2018; Greer et al. 2018). The critical role of sea port as multi-modal hub require intelligent, integrated and efficient time booking system for tracks that will integrate all transhipments operations in sea terminals with harmonised road transport in port and surrounding area.

The objectives and assumptions presented in the article—systemic assurance of better and integrated traffic management, effective use of transport infrastructure and road capacity as well as tracking of traffic and optimization of cargo flow—are completely consistent with the assumptions of many EU direction documents (including the Plan for the creation of The Single European Transport Area—the 'White Book' and the EU Guidelines for the development of the Trans-European Transport Network—Regulation of the European Parliament and of the Council No. 1315/2013). According to the guidelines presented in Regulation 1315/2013 and the Plan for the creation of a Single European Transport Area (White Paper), the intelligent traffic management system in the Port should:

- integrate the use of several systems:
  - ITS system,
  - cargo information management system,
  - parking area management system and provide commercial users with an adequate number of parking spaces,

in order to alleviate the impact of traffic congestion on roads,

- be designed, developed and operated in a resource-efficient manner—i.e. by implementing new technologies and financially justified telematics applications,
- use modern information and communication technologies to solve bottlenecks and increase road capacity,
- meet the requirements of the EP and Council Directive 2010/40/EU on the framework for the deployment of intelligent transport systems in the field of road transport and interfaces with other modes of transport,
- provide information flow between the port, sea terminals in the port, logistics platforms and individual transport modes in the logistics chain to enable real-time data access on available road capacity infrastructure, traffic flows and traffic tracking,
- provide reservation services for secure and protected parking spaces for trucks,
- ensure the functioning of the system and availability of digital services defined in the system on a continuous basis (24 h/day/365 days),
- take into account the principles of evaluating socio-economic costs and benefits and European added value.

An important horizontal guideline for the central and integrated traffic management system in the Port is the EC position presented in Communication COM (2016) 766 Final concerning the European strategy for cooperating intelligent transport systems—digital technologies and implemented digital services in transport should allow new forms of value creation, including the data sharing economy and their reusability. In the case of the analysed system, the EC guidelines concern the commonality, smooth exchange and data transfer, sharing and re-use of data among partners of intelligent traffic management system (e.g. terminals, logistics operators, carriers, forwarders, port, National System of Traffic Management as well as National Access Point to traffic data). The EC guidelines resulting from Regulation 1316/2013 establishing the Connecting Europe Facility specify requirements for increasing the capacity of road transport infrastructure in the core TEN-T network and in the TEN-T core network corridors—including ITS and other traffic management systems identified for improving the efficiency of transport operations in the area of a seaport as a key element of the last mile sections of the TEN-T network.

2 Operational Analysis of Logistic Processes in Sea Ports

The port's ability to reduce delays and meet customers' requirements enhances supply chain efficiency. Exchanging timely, accurate information about terminal operations, road/maritime/rail traffic and parking fulfilling/capacity with suppliers, sea terminals, carriers, forwarders, etc. benefits everyone. Operational coordination of transports and transhipments in the existing and developing network of feeder-district roads requires intelligent and multimodal synchronisation of the goods flow (Wagener 2017). The proposed project of a multidimensional integrated traffic management system will have the task of linking the planned and effective operation of terminals in the port with sea, road and rail transport, ensuring smooth delivery and transport of goods. Multidimensional integration of intelligent traffic management will include:

- all types of cargo—containers, general cargo, bulk cargo—solid and liquid, oversized and special-purpose cargo,
- synchronization of road transport traffic with priority maritime transport,
- synchronization of road transport with reloading operations of all terminals in the port,
- coordination and management of traffic on the internal roads of the sea port with traffic on access roads using ITS and IoT systems,
- integrated management of parameters for a common parking space (central parking or several parking lots) and by access roads,
- an integrated digital platform for e-service upgrades by hourly time slots (slot booking system) for customers and partners around the world, at all port's terminals,
- an integrated platform for communication and data exchange between the port traffic management centre, terminals, forwarders, carriers, drivers, customs and border services.

The objective of the intelligent truck traffic management system in the Port is:

- balancing traffic saturation in the road network,
- reduction of congestion and elimination or reduction of queues before entry to terminals and access roads,
- improving the efficiency and effectiveness of terminals and shortening the time of ship service by synchronising the delivery/departure of cargo in road transport with the operations in the terminals,
- optimal allocation of trucks in the parking lots of the Port and its surroundings and integrated management of the parking space,
- improvement of safety and comfort of transport for the society of City agglomeration.

Transshipment operations in the sea port include container, general cargo, bulk cargo (loose, liquid) and oversized cargo (including ferry terminals). The bulk cargoes are dominated by coal, coke, ore, cement, cereals and fodder, biomass, middlings, sugar and flour, aggregate minerals, fertilizers, wood, chemical industry products, oil and petroleum products, diesel and heating oil, waste.

The volume and route of cargo flow and lorry traffic flows in a seaport are fundamentally influenced by factors:

- the volume of cargo turnover in individual terminals,
- Location of terminals in the port area—e.g. container, ferry, mass specialized and universal terminals,
- Car arrival and import procedures—e.g. including examination of the goods before unloading and internal journeys of the terminal—parking—terminal,

- journeys in the Port between terminals and wharfs of the Port and effective use of car, access and return routes from the Port, i.e. arrival—unloading—internal passage—loading—departure,
- Location of parking lots and mandatory car registration/weighing procedures in parking lots before entering the terminal,
- cargo handling tactics in ship handling processes (direct loading/unloading of lorries or indirect, including warehouses and depots in port terminals).

The volumes of transshipments in the sea port by product groups (general cargo and bulk—loose and liquid) and container transshipments are presented in Table 1.

The volume of transshipments allocated to particular terminals according to their turnover and cargo specialization is the basic premise for the size of truck streams on access roads. Table 2 presents vehicle streams including damming up and maximum truck streams of entry/exit to/from individual terminals, based on data from cargo turnover records in IT systems, managers' estimates and port operational data. Moreover, the access/output traffic of lorries on the port roads also generates logistic services and operations of terminal operators, logistic operators and other companies operating in the Port area (Table 2).

Truck traffic is the result of transshipments in port terminals, and the flow of goods and traffic flows are the result of the location of terminals and the specialization of the handled goods.

Road traffic on the access roads and internal roads of the Port is also generated by passenger cars and coaches. The needs of travel/ departure to/from work of employees in particular terminals and enterprises, travel/ departure of passengers in ferry traffic and coach transport in passenger traffic service are usually accumulated at specific times of start/end of work and dates of ferry trips (Table 3).

Individual terminals provide in their development strategies for an increase in reloading (often at the level of 10–15%/annually), expansion of the warehouse and warehouse base both in the Port area and in logistics centres in the vicinity of the Port, in the extended integrated formula Sea Port—Dry Port. The plans also include

Table 1 Transhipment volumes in seaport by product group

Transshipment by assortment groups	Volume
Coal and coke	2134.5
Redhead	15.1
Other bulk	1079.3
Cereals	3482.7
Wood	234.0
Mineral matter	12,460.4
Fuels	1819.1
Total	21,225.2
Container handling/in TEU/	710,698

Source Own study based on the Port development and terminals data

Table 2 Volume of transhipments in a seaport

Terminals	Number of trucks per day	Maximum number of trucks/hour	
		Entry	Exit
Cement terminal	40	5	5
Mass terminal	30	4	4
Grain terminal	300	15	15
Container terminal	1300	110	110
Tank terminal	45	5	5
Bulk terminal	300	25	25
Gas terminal	40	5	5
Universal cargo terminal	430	45	45
Ferry terminal	1200	80	80
Other cargo transhipment terminal	500	25	25
Total	4185	319	319

Source Own study based on the Port development and terminals data

Table 3 Road traffic on access roads and internal roads of the Port

Traffic of cars and coaches	Number of vehicles per day	Maximum number of vehicles/hour	
		Entry	Exit
Terminals, enterprises, institutions in the Port	2200	550	550
Cars and coaches (city)	1200	100	100
Total	3400	650	650

Source Own study based on the Port development and terminals data

greater product diversification, expansion into new foreign markets and acquisition of new contractors. The resulting increase in transport streams may increase both road and rail transport.

A systematic increase of cargo flow is expected in the TEN-T corridor passing through European ports, serviced both by existing terminals and those planned in port development strategies. On the basis of the operational analysis, the forecasts presented in Table 4 were adopted.

According to the adopted assumptions for the development of the Port and its terminals, the forecast traffic of trucks resulting from the forecast container turnover is presented in Table 5.

In addition, the development of universal terminals, LNG and ferry terminals is also planned, as shown in Table 6.

Table 4 Basic assumptions of the operational analysis

Forecasted operational data	Value
Transhipment of containers/TEU/	1000,000
The structure of container transhipment/%/	
−20′ containers 20′	40%
−40′ containers	60%
Coefficient of load flow unevenness	1.3
Empty tractors arriving for containers and departing after unloading containers	60%
Number of transhipment work days per year	300
Number of containers transhipped daily in/TEU/	3340
The number of containers per day according to the structure 40%/60%	2338
The number of IN/OUT trucks per day (including uneven flow and empty commuting/ departures)	**4900**
Effective time of IN/OUT operational traffic in the terminal/man-hour/	20

Source Own study based on the Port development and terminals data

Table 5 Forecasted truck traffic resulting from the forecasted container turnover

Traffic of cargo trucks	Number of trucks per day	Maximum number of trucks/hour	
		Entry	Exit
Development and extension of container terminal	9800	244	244

Source Own study based on the Port development and terminals data

Table 6 Analysis of the development of terminals—universal, LNG and ferry terminals

Traffic of cargo trucks	Number of trucks per day	Maximum number of trucks/hour	
		Entry	Exit
Development and extension of general cargo terminal	600	45	45
Development and extension of ferry terminal	1500	100	100
Total	2100	145	145

Source Own study based on the Port development and terminals data

Table 7 Car traffic growth forecast

Traffic of cargo trucks	Number of trucks per day	Maximum number of trucks/hour	
		Entry	Exit
Development and extension of container terminal	9800	244	244
Development and extension of general cargo terminal	600	45	45
Development and extension of ferry terminal	1500	100	100
Total	11,900	389	389

Source Own study based on the Port development and terminals data

Table 8 Forecast of growth in passenger car and coach traffic

Traffic of cars and coaches	Number of vehicles per day	Maximum number of vehicles/hour	
		Entry	Exit
Development and extension of container terminal	1600	400	400
Development and extension of general cargo terminal	100	40	40
Development and extension of ferry terminal	200	30	30
Total	1900	470	470

Source Own study based on the Port development and terminals data

In total, the increase in truck traffic resulting from transshipment operations in the Port, forecast for the future, is presented in Table 7.

The forecast for the future growth of passenger car and coach traffic, resulting from the growth of goods turnover in the Port, is presented in Table 8.

The road accessibility of the Port depends on the capacity and throughput of the system of access roads[1] to individual port terminals. Within the analysis, both feeder/carriage traffic of lorries as well as road traffic of cars and coaches, commuting to terminals, enterprises and institutions, and urban traffic of cars and coaches in the road network of the Port and its surroundings have been taken into account (Table 9).

The capacity assessment was made for roads forming the framework of the road transport system for the Port. The capacity of access roads to the Port conditioning the availability of particular terminals in the transport processes of goods flow in the

[1] Road capacity—the largest number of motor vehicles that can pass through a given cross-section of a road or lane in one direction (for two-way roads in both directions) per hour in favourable weather conditions. The throughput is expressed in real vehicles per hour [P/h].

Table 9 Analysis of Port road accessibility

	Number of vehicles per day	Maximum number of vehicles/hour	
		Entry	Exit
Traffic of cars and coaches – Resulting from the flow of cargo in the Port	2200	550	550
City passenger cars and coaches	1200	100	100
Total	3400	650	650

Source Own study based on the Port development and terminals data

base network of TEN-T was estimated on the basis of the applied in practice road traffic engineering method HCM-85 (Highway Capacity Manual) and subsequent editions of the method HCM-2010 and HCM-6 (HCM 2010).

In the estimation analyses, the required parameters and engineering coefficients were taken into account, both for road infrastructure and vehicle driving behaviours (affecting the road throughput), including, among others, design speed, lane width and side width, type of side roads, number of lanes, traffic directional and generic structure, traffic organization and signalling, visibility, average response time of drivers, blocking elements, side obstacles and other accompanying elements. An adequate Free-Road Level (FRL)[2] is also taken into account:

- Level A—free traffic, low traffic and relatively free choice of speed and manoeuvres,
- Level B—uniform traffic, slightly limited driving speed and freedom of manoeuvring,
- Level C—uniform traffic, considerably limited driving speed, manoeuvring requires much attention due to the presence of other vehicles,
- Level D—uneven traffic, very limited choice of speed and manoeuvres, poor comfort of driving, temporary increases in traffic cause traffic disturbances,
- Level E—uneven traffic, traffic intensity reflects road capacity. Poorly stabilised speed, manoeuvring based on forced movements, slight increase in traffic or temporary traffic halts lead to serious disturbances.

The capacity of road sections is calculated according to the real traffic speed (currently the permissible speed on the main access road to the Port is 70 km/h, in the area of 50 km/h) and the estimated traffic intensity, taking into account the above road parameters, traffic coefficients and the level of freedom of traffic.

[2]Free-Road Level (FRL) is a qualitative measure of traffic conditions, taking into account the feelings of drivers and other road users. The whole range of variability of traffic conditions is divided into 6 classes—marked with the letters A–F. The level of freedom of movement A corresponds to the best, and FRL F to the worst traffic conditions. The level of freedom E determines the traffic intensity corresponding to the road throughput.

The throughput analysis was performed for FRL level D and E—assuming real conditions of momentary increases in traffic intensity, causing congestion and slow-down of traffic.

The capacity on the access road to the Port was estimated in the range (750 P/h/FRL—E/; 1650 P/h/FRL—D/), while on the access roads in the range (125 P/h/FRL—E/; 1400 P/h/FRL—D/).

The main balance of the current freight flow created by cargo turnover at the terminals of the Port and the traffic load induced—in relation to the capacity of access roads (Fig. 1) calculated according to the real traffic model is presented in Table 10.

The forecasted increase in the volume of freight flow in the Port, presented in Fig. 1, will result in a significant increase in the traffic load on access roads in the

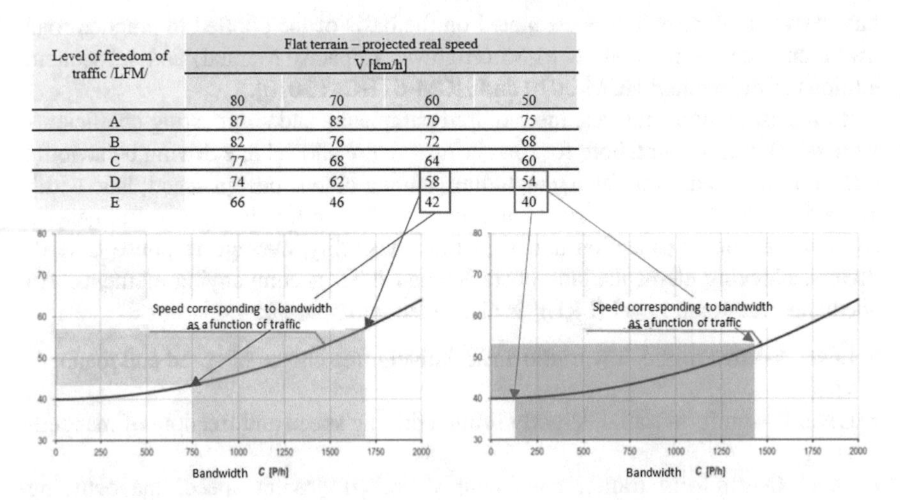

Level of freedom of traffic /LFM/	Flat terrain – projected real speed V [km/h]			
	80	70	60	50
A	87	83	79	75
B	82	76	72	68
C	77	68	64	60
D	74	62	58	54
E	66	46	42	40

Fig. 1 Current main access roads to terminals in the Port, own study based on the Port development and terminals data

Table 10 Balance of truck traffic resulting from the goods turnover in the Port in relation to the capacity of access road

Aggregated traffic / day (entry + exit)	The average top access traffic / h	The average top exit traffic / h
Port		
14 670	1 014	1 049
Capacity of the access road to the Port /real traffic model/ 1500 - 1600 truck & cars / h		
Internal roads in the Port		
11 200	535	570
Capacity of the internal roads in the Port /real traffic model/ 900 - 1000 truck & cars / h		

Table 11 Forecasted load on access roads resulting from the Port development strategy

Aggregated forecast traffic / day (entry + exit)	Forecasted average peak access traffic / h	Forecasted average peak exit traffic / h
Port		
27 870	**1 780**	**1 880**
Capacity of the access road to the Port /real traffic model/ **1500 – 1600 truck & cars / h**		
Internal roads in the Port		
16 670	**1 245**	**1 310**
Capacity of the internal roads in the Port /real traffic model/ **900 – 1000 truck & cars / h**		

Source Own study based on the Port development and terminals data
The bold numbers mean that the road capacity limit is exceeded by forecasted traffic according to the real traffic model

perspective of several years of the Port's development (Table 11). The increase of goods and truck traffic is forecast mainly in the eastern part of the Port.

The results of the analyses and operational forecasts made so far, as well as the capacity of the access and internal roads of the Port, point to prospective difficulties in balancing road traffic and road capacity especially in non-investment scenario, which may cause bottlenecks in cross-border freight transport, in the European Transport Corridor of the TEN-T network Baltic Sea—Adriatic Sea and in the global maritime supply chains Europe—Asia—Europe. The analysis of the target traffic saturation in the access and exit direction from the Port indicates the possibility of the situation—lack of the required capacity of the access road network to the Port for the planned traffic load and thus the lack of the required Port accessibility.

The results of the research indicate that the existing road system may be a limitation (in the absence of road development investments) for the development of the Port and an increase in the volume of transhipments. The basic limitation will be the road capacity analysed for the optimistic Level of Freedom of Movement—D. However, for the level 'E' of FRL, the restrictions are significantly higher. The current estimated capacity of the main access road to the Port—1650 vehicles per hour, compared to the forecasted road traffic (1828 vehicles per hour), may be a limitation.

The results of the analyses and operational forecasts made so far, as well as the capacity of the access and internal roads of the Port, point to prospective difficulties in balancing road traffic and road capacity especially in non-investment scenario, which may cause bottlenecks in cross-border freight transport, in the European Transport Corridor of the TEN-T network and in the global maritime supply chains Europe—Asia—Europe. The analysis of the target traffic saturation in the access and exit direction from the Port indicates the possibility of the situation—lack of the required capacity of the access road network to the Port for the planned traffic load and thus the lack of the required Port accessibility.

The problems identified have a significant impact on the efficiency and duration (and delays) of cross-border road freight transport (road hauliers from all over Europe are served in the Port), in global maritime supply chains served by global shipping lines, as well as in international rail freight transport.

3 The Concept of Intelligent Truck Traffic Management System in the Sea Port

To achieve higher efficiency levels, large companies have significant costs in making sure that their transport/supply chains are connected. A significant share of sea terminals, carriers, forwarders, etc. that deal with transport and supply chain management in Europe recognise interoperability and information exchange as a relevant challenge. Interoperability is expensive and complex. Information exchange implementations in this respect have so far often triggered by traffic management needs (e.g., road, rail, maritime and ship-to-shore communications) and public ITS systems for route, journey and parking management. ITS components that are integrated can result in synergistic effects when considered as an entire system. Many SMEs lack the capability and resources to interoperate and collaborate with other companies.

In response to these needs, the European Commission has launched ITS Action Plan (Action Plan for the Deployment of Intelligent Transport Systems in Europe COM (2008) 886 final) and Directive 2010/40/EU on the framework for the deployment of Intelligent Transport Systems in the field of road transport and for interfaces with other modes of transport. But Intelligent Transport System is only a means for intelligent traffic control, while the needs for careful coordination in the Port include transhipment operations in sea terminals and warehouses correlated with maritime/rail/road transport, road traffic management and parking area management. Main aims of the proposed concept comes to issues:

- integrated and connected information management in maritime inland and core network port operations,
- facilitate access of cargo through the port area,
- optimise handling of cargo,
- improve connectivity and reduce impact of port operations with the city.

The competitiveness of Port will depend on their ability to innovate in terms of ICT platform, ITS and IoT technology, organisation and management. Their critical roles of port as multi-modal hub require intelligent, integrated and efficient time booking system for tracks that will integrate all transhipments operations in sea terminals with harmonised road transport in Port City and surrounding area.

The main objective of the project is to develop technical documentation of intelligent truck traffic management system in the Port—but the main future impact effect is eliminating the bottleneck in the Port as well as undisturbed operations of port terminals and smooth flow of cargo delivery/departure.

To this end, it is fundamental to create systemic support for the dynamic development of the Port and the increase in the flow of cargo and transhipments, as well as further development of port terminals. The basis for achieving the objective is to balance the growing flow of goods in the Port with limited capacity of internal and access roads of the Port.

The concept of integration on the Port IT platform includes three IT systems and at the same time mutually resulting and logically related data areas:

- integrated digital truck traffic management platform with two-level digital e-availability process for the Customers of individual terminals (1st level) and for the Port (2nd level)—a source of aggregated data on the total planned feeder/departure traffic of the Port terminals,
- intelligent traffic management system for coach/carrier traffic, supplied with data from port notifications, to control intelligent infrastructure of roads leading to terminals and traffic on roads—integrated with ITS system in port and on roads leading to the port,
- a management system for a common and IT-integrated parking space, which is a virtual waiting area—ensuring dosing of cars' exit to terminals according to the planned announcements and the needs of optimal traffic control and its distribution.

The functional and process logic of the traffic management system and digital booking in the seaport—from the planning of the ship's call to the port, to the entry of the car into the terminal to the collection of cargo—is shown in Fig. 2.

The main goal of the Port's digital promotion process is a comprehensive service in the IT system of operational activities, from booking cargo handling in the terminal, to coordinating the entry of a car into the terminal in accordance with the time, date and place of the promotion. The process of digital booking of vehicle service in the Port is shown in Fig. 3.

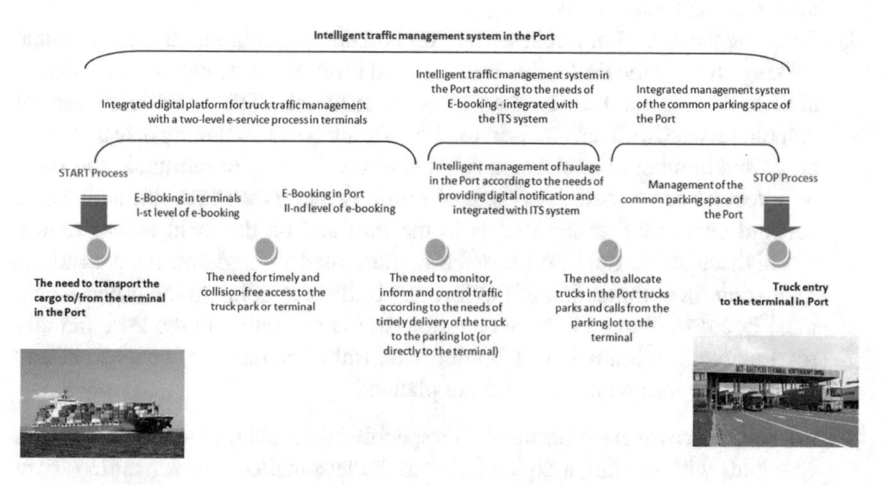

Fig. 2 Integrated process of the Port traffic management system with the Central e-booking System, own study and results of researches

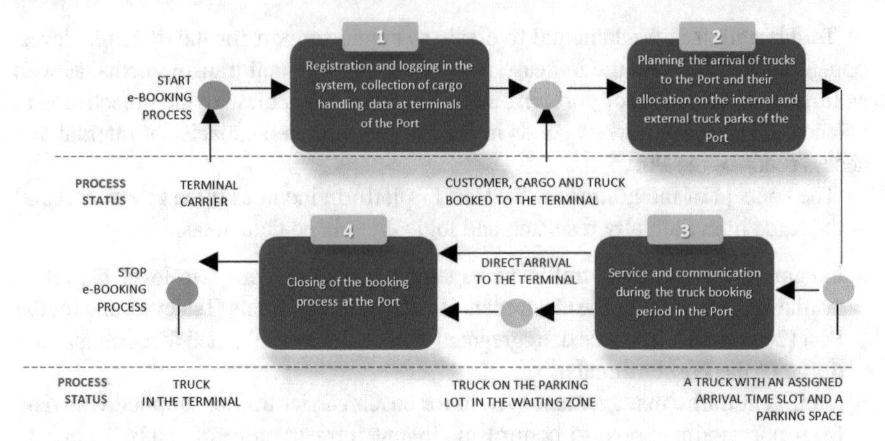

Fig. 3 e-Booking process in Sea Port, own study and results of researches

The process shown in Fig. 3 is divided into four sub-processes:

(1) Registration and logging in the system, collection of cargo handling data in the Port terminals—this is the first level of booking (in the terminal)—booking by the Customers of particular cargo handling terminals and the arrival of cars, according to their mutual agreements, schedule of ship arrivals, arrivals and departures of trains, preferences of shippers/consignees/carriers/forwarders and terminal service capabilities. The preferences of the driver of an earlier arrival at the Port parking lot and/or procedural needs of the terminal and cargo handling are taken into account. This stage indicates the priority for the whole traffic management system in the Port—i.e. the planned and smooth course of the transhipment process in the Port terminals and the timely control of the car's access to the terminal gate.

(2) Planning the arrival of trucks at the Port and allocation in internal and external parking areas of the Port—this is the second level of notification (in the Port)—after collecting notification data from all terminals. The operational data of vehicle service in all terminals in the Port are analysed, including date and time, order and number of vehicles in relation to the location of terminals and route on internal access roads. An optimal traffic variant is worked out, taking into account temporary traffic models in the Port and on the main access routes, the highest traffic flow rate and the highest road throughput. As a result of multi-criteria optimization of the forecast traffic according to the criteria—the most liquid traffic and the least possible traffic saturation in the Port and city road network, elimination of bottlenecks, traffic congestion, congestion and queues—the following scenarios are planned:

    (a) direct arrival at the terminal on a specified date and time—the sub-process ends with sending a digital ticket to the terminal/carrier/ forwarder/driver including data—date and time of arrival of the car at the terminal,

(b) access on a specified date and time to a designated parking lot within the common parking space of the Port Waiting Zone. Time requirements for vehicle service in terminals are mapped into time windows for the use of internal and external parking lots of the Port. As a result, a forecast time schedule of car park distribution is obtained, a dynamic schedule of car park filling and a forecast dosage of exit from car parks to individual terminals. The sub-process ends with the allocation of a place and time window for a car in the waiting area of the Port and sending a digital ticket to the terminal/transporter/forwarder/driver containing the data—date and time of arrival of the car at the Port and to the car park in the waiting area.

(3) Service in the digital period of trucks' promotion in the Port—subprocess includes many areas of activity—communication with the carrier and driver in the digital period of promotion, time of arrival reminder, possibility of dynamic changes and cancellations of the promotion, traffic estimation on access roads and planning of the passage road to the parking lot of the Port. In the process (3) it is planned to supply data from the planned route to ITS and other telematics systems on access roads to the Port, recognition of a car on access roads, registration of arrival at the parking lot in the waiting area of the Port, monitoring of the parking lot filling level in the waiting area and continuous estimation of the parking lot load. The final phase will include communication with the driver and calling for departure to the terminal, planning the route to the terminal and ETA calculation. The communication system provides the driver with a lot of additional information about traffic, road conditions, driving conditions and recommended detours, parking lots, weather conditions, etc. The communication system provides the driver with a lot of additional information about the traffic, road conditions, driving conditions and recommended detours, parking lots, weather conditions, etc.

(4) Closing the digital notification in the Port—confirmation of arrival or entry of the notified lorry to the terminal on the basis of data from gate systems and ANPR cameras (existing or optionally from cameras installed within the Port Traffic Management System) and closing the notification in the Port Traffic Management System.

Aggregate e-availability data from all terminals in the Port enable multi-sectional analysis of the freight road traffic resulting from the truck flow in 1-hour time windows on a daily and monthly basis, located in the eastern and western part of the Port. An example of the daily e-availability schedule for the Port terminals is presented in Table 12.

The time schedule of arrival at individual port terminals, for each day of the forecasting horizon, requires advancement in the process of analysis:

- the locational transposition into individual parking lots in the common waiting area of the Port, with an appropriate allocation in the Port,
- time of arrival at individual parking lots in the waiting area of the Port—including advance notice of arrival at the parking lot according to the earlier arrival time

Table 12 Example of daily distribution of cargo e-booking at Port terminals

Terminals	e-Booking time slots/1 h slots—start time/												Total
	0^{00}	1^{00}	2^{00}	3^{00}	4^{00}	5^{00}	6^{00}	7^{00}	8^{00}	9^{00}	10^{00}	11^{00}	
	12^{00}	13^{00}	14^{00}	15^{00}	16^{00}	17^{00}	18^{00}	19^{00}	20^{00}	21^{00}	22^{00}	23^{00}	
Terminal 1	0	0	0	0	0	0	2	5	5	6	4	5	63
	8	9	4	6	4	3	2	0	0	0	0	0	
Terminal 2	0	0	0	0	0	0	0	5	5	4	7	5	55
	5	6	4	5	4	3	2	0	0	0	0	0	
Terminal 3	0	0	0	0	5	10	12	18	25	26	17	25	325
	25	28	34	19	21	17	19	10	7	4	3	0	
Terminal 4	19	16	13	8	9	10	13	23	29	47	57	45	852
	63	61	63	65	59	57	37	43	40	30	26	19	
Terminal 5	11	13	23	18	14	12	23	21	39	42	77	55	827
	53	62	63	68	49	35	31	3	40	30	26	19	
Terminal 6	0	0	0	0	0	1	1	3	5	6	7	5	63
	9	7	5	5	4	3	2	0	0	0	0	0	
Terminal 7	4	8	3	9	5	10	12	18	25	36	27	25	409
	45	38	34	29	21	17	19	10	7	4	3	0	
Terminal 8	0	0	0	0	0	2	0	6	4	6	3	5	59
	5	7	7	4	5	3	2	0	0	0	0	0	
Terminal 9	12	8	9	10	15	12	12	18	28	26	27	35	414
	25	28	34	19	21	17	19	10	11	8	7	3	
Terminal 10	0	0	0	0	0	1	1	3	5	6	7	5	63

(continued)

Table 12 (continued)

Terminals	e-Booking time slots/1 h slots—start time/												Total
	0^{00}	1^{00}	2^{00}	3^{00}	4^{00}	5^{00}	6^{00}	7^{00}	8^{00}	9^{00}	10^{00}	11^{00}	
	12^{00}	13^{00}	14^{00}	15^{00}	16^{00}	17^{00}	18^{00}	19^{00}	20^{00}	21^{00}	22^{00}	23^{00}	
	9	7	5	5	4	3	2	0	0	0	0	0	
Terminal 11	2	5	17	19	23	19	20	27	21	29	14	4	365
	0	1	0	2	10	23	28	30	26	28	12	5	
Port in total	48	50	65	64	71	77	96	147	191	234	247	214	3495
	247	254	253	227	202	181	163	106	131	104	77	46	

Source Own study and results of researches

notified in the notification form or standard 1-hour advance notice of arrival at the parking lot in the waiting area of the Port,

- adoption of assumptions for direct access to the terminal and indirect access through the parking lot in the waiting area of the Port (for the 24-h schedule of the notification within the planning horizon of the next 30 days):
  - time windows to reach the terminal, in which there may be a direct arrival,
  - time windows, in which it is necessary to plan the arrival of the lorry at the parking lot and then control the exit to the terminal, according to the appropriate traffic dosing on the access roads and internal roads of the Port.

Operations and transhipments affect road traffic on the main access road, as well as traffic on the access and internal roads of the Port. The analyses should consider the assumption that the traffic of lorries in the Port road network consists of two traffic streams:

- 80%—truck traffic stream resulting from e-availability data—planned and controllable stream,
- 20%—free flow of internal traffic in the Port, which will be monitored by means of telematics devices in the Port's road network, but will not be planned and controlled.

The division of streams into planned and free streams should be specified in detail at the stage of detailed measurements and design estimations of the traffic management system in the Port.

Free traffic results from many needs of the current operational situation in the terminals:

- truck rides under the associated transport order (import and collection of cargo)—after unloading, the cars have to drive on the Port's internal roads to another quay for loading. Often, in order to load, the cars have to pass through the parking lot before entering the terminal, weigh the empty vehicle and only drive to the indicated place of loading. This requires several car passes on the roads of the Port, generating additional traffic on internal roads,
- internal procedures resulting from the specificity of the goods handled—e.g. examination of goods before unloading, followed by internal passage of terminal—parking (waiting for the results of the sampling) and subsequent passage of parking—terminal,
- truck journeys to border and customs control posts,
- access/ departure traffic of lorries servicing the logistic services and operations of terminal operators, logistic operators and other companies operating in the Port area.

The model of traffic on access roads and internal roads of the Port should take into account the traffic of passenger cars and coaches. The needs of the employees in particular terminals and enterprises to get to/from work are usually accumulated in two periods of peak traffic load:

- morning traffic peak: 6.45–9.00—including 6.45–7.15—average peak; 7.15–8.20—high peak; 8.20–9.00—low peak.

The needs of passengers in ferry traffic to/from and coach transport in passenger traffic service are most often accumulated at specific times of commencement/end of work and dates of ferry cruises.

Traffic estimation will allow to determine in which time windows the hourly stream of trucks may exceed the threshold of the assumed Level of Freedom of Movement or cause traffic congestion and queues on roads. In such cases, it is necessary to agree with the terminal on the possibility of postponing the time of notification in order to avoid unplanned delays and untimely arrivals of cars and waiting for the terminal to handle the cargo.

4 ICT Integration Platform of e-Booking System and Intelligent Truck Traffic Management System in the Sea Port of TEN-T Corridor

This objective is in line with the sectoral objectives of EU transport system development (Priority—Deployment of innovation and new technology actions, including a focus on safety and objective—Support, through digitalisation, for maritime and inland port operations) to eliminate bottlenecks in transport systems and to ensure sustainable and efficient transport systems, as well as to optimise the integration (Speier et al. 2008; Prajogo and Olhager 2012; Sliwczynski et al. 2012; Wong et al. 2015) and interconnection of the different modes of transport.

Intelligent road transport management of all types of cargo—containers, general cargo, bulk cargo—solid and liquid, oversized cargo—requires taking into account the processes and operations of reloading, warehousing, identification and control of cargo characteristic for their handling at individual Port Terminals. On the basis of the analyses performed so far, the Traffic Management System and central reservation of cargo handling in the port terminals will contain components:

- digital platform (integrated IT system) of traffic management,
- equipment of ICT and telematics systems on the road and lane networks—including: cameras, devices for measuring traffic parameters, variable message signs, communication networks (Pedersen 2012),
- IoT (Internet of Things) devices for detection, identification, measurement and monitoring of vehicle traffic,
- organisational structure of traffic management in the area of the Port—including: Traffic Management Centre in the Port as an organisational unit of the Port and the operator of the traffic management system as well as digital waiting zone virtually defined in a common parking space.

Digital platform (integrated IT system) of traffic management including:

- central digital advancement system for the Port, including:

- module of processing and analysis of operational data from the digital advancement of cargo and vehicle handling in Port terminals (analysis of time, sequence and number of vehicles/transport units—data from the first level of advancement),
- module of intelligent allocation of lorries on internal and external parking lots of the Port according to the time requirements of handling vehicles and loads in all terminals of the Port
- digital truck advancement module of the Port—with functionality:
  allocation of space and time window—digital ticket,
  digital identification and registration at the designated car park (in the waiting zone) and communication and sequential control of telematics devices at the car park,
  exchange of confirmation data with terminal systems,
  communication and service in the waiting zone and calls to leave for the Terminal,
  digital route planning to the terminal,
  monitoring the passageway to the terminal and entry to the terminal,

- e-Booking portal for terminals (mainly in the group of SMEs without their own digitalisation systems),
- Internet and mobile communication module with carriers, forwarders and drivers,
- API communication interfaces with terminal IT systems (ERP class systems),
- a layer of digital communication services with ITS systems in the region and on national roads,
- central system of intelligent traffic control and parking space, communicated with telematics devices in the Port's road network,
- system of data exchange with maritime and customs and fiscal services,
- central software for system management and data administration (Fig. 4).

All IT components of the digital traffic management system platform and communication with external systems will be performed in the digital services model. IT applications will be available through a web browser, using the data processing technology in the cloud computing of the Port, as well as selected services and data (e.g. weather information services, navigation services) in the public from the public cloud computing.

The action will take up several results from previous e-Booking and ITS implementation experiences, such as:

- implementation of the e-booking system: Port Aalborg, Port of Hamburg—Slot Booking, Port of Lisbon—Gate Operational System, Port La Havre—Truck Appointment System, Port Dover—Traffic Access Protocol and Traffic Management Improvement, Tallinn—Smart Port and other,
- implementation and functionality of ITS system in the area of Port City and access roads to the Port,
- implementation and functionality of the system and ITS services and access points to traffic data.

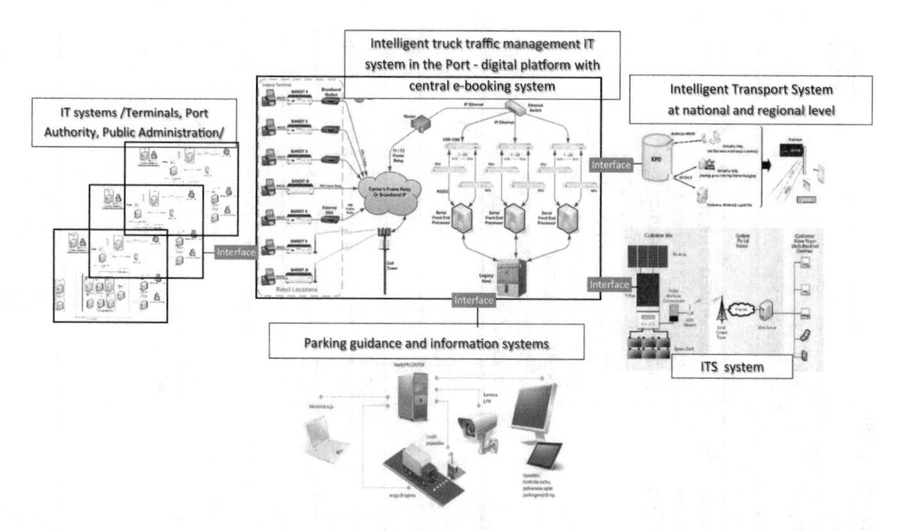

Fig. 4 Integrated environment of the Port traffic management system with the Central e-booking System, own study and results of researches

In conclusion, the presented traffic management systems in sea ports are characterized by different range of activities and functionality (including functions: e-availability, parking management, intelligent traffic control). Table 13 presents a comparison of functionality and range of operation of the analysed systems, taking into account the criteria of valuable comparison, for four distinguished features and functions of the systems.

5 Value Conclusions and Further Research

The functional goal and useful purpose of traffic management at the Port is the use of e-booking planning in advance of entry/exit of trucks, which will avoid bottlenecks in the transshipments service process. The mechanism of 1-h time slots will enable a convenient choosing the time of arrival and service for terminal customer as well as defining the 30 days/24 h trans-shipment capacity for the terminal. The mechanism will have the functionality of limiting or increasing capabilities of booking by the terminal in particular commodity groups or for individual contractors. For this purpose, they will be designed e-booking portals for terminals (or APIs for sharing data from terminal booking systems/truck appointment systems/), which will allow aggregation of booking data on the Port's digital platform (e.g. time, order and quantity of planned trucks in successive time slots). The list of truck service times in all Port terminals is aimed at intelligent allocation of trucks on both the internal and external truck parks of the Port.

The basic functionality of the truck e-booking system at Port terminals includes:

Table 13 Comparison of functionality and range of operation of the analysed intelligent traffic management systems

No	System	Features and functions of the system				Total rating
		Range of the system	e-booking system	Parking management system	Intelligent transport system (ITS)	
		Comparison criteria				
		4—country/region 3—city 2—port 1—terminal	3—hour time slots 2—daily 1—booking without specifying the time	2—integrated with the e-booking system 1—not integrated	3—integrated with the e-booking system and parking management 2—integrated with the e-booking system or parking management 1—traffic control without integration with external systems	
1	National intelligent transport system	4	–	–	1	5
2	Port Tallinn	2	3	2	–	7
3	Port Dover	3	–	1	2	6
4	Port Aalborg	3	–	–	1	4
5	Port Felixstowe	2	3	–	–	5
6	Port Le Havre	2	3	–	–	5
7	Port Lisbon	2	3	2	–	7
8	Port Cork	2	3	–	–	5
9	Port Hamburg	2	3	–	–	5
10	The concept of the traffic management system of the seaport	3	3	2	3	11

Source Own study and results of researches

- defining and managing the booking calendar by each of terminals (including defining the operating capacity for each time slot),
- sharing the calendar for e-customers—carriers, shippers, forwarders, drivers—with the ability to track the booking and make changes,
- e-booking and communication of any change or delay—both from the terminal and the carrier side,
- registration the arrival of the truck with e-booking in the parking lot of the waiting zone and in the entry of terminal.

The e-booking and intelligent traffic management system requires a comprehensive coverage of data for each stage of booking process implementation—from its 'opening' to 'closing'. It should be designed and installed comprehensively because can't be interrupted an information chain and flow.

The Sea Port digital platform will be designed in the Service Oriented Architecture (SOA) and the Service Component Architecture (SCA). It will be an open and technologically independent ICT environment. The flexible, standardized and open approach to logical and IT architecture results from the dynamic development of the Port and the open needs of integration/communication of many future ICT systems. All IT components of the digital platform of the traffic management system and communication with external systems will be made in an open digital service model.

The projected benefits of implementing the presented system concept result from the direct impact on traffic in the Port's road network and include, inter alia, eliminating bottlenecks in truck and stacking traffic in truck traffic, as well as reducing travel time, increasing freedom of movement and reducing traffic saturation in the road network. Numerous measurable benefits of increased reliability of deliveries and punctuality in delivering cargo are key reasons for the growth of competitiveness of the Port, among other seaports in the TEN-T network corridors.

The estimated analysis of the measurable benefits of implementing an intelligent truck traffic management system at the Port includes the potential of:

- improvement of terminal capacity and increase in transhipment by 18%, with the current available infrastructure and work organization—as a result of elimination of road congestion, ensuring timely commuting of trucks to terminals in accordance with planned time slots, and thus stable, planned, efficient and effective work terminals in the Port,
- increase the capacity of the road system in the current road network by 12%, including a reduction of travel times in Port's road traffic by 5.5% and an increase in the speed of travel by 16–20%,
- shortening the service time of the truck at the Port by 20%—including shortening by more than 25% of waiting time for trucks in the terminal, approx. 20% of the average time of entry and exit processes to/from the terminal and a reduction of over 10 min of administrative service time as a result computerization of the process of e-booking integrated with service automation at the entrance gates of terminals,
- financial benefits for all stakeholders of the Port, including:

- reducing the operational costs of terminals by approx. 20%—by improving the efficiency of operations and reducing staff costs, better use of machines and devices and storage and infrastructure,
- more courses and higher revenues for drivers, as well as an increase in revenues for the carrier and forwarder, as a result of more transport orders and deliveries/pickups,
- lowering the carrier's costs and ordering transport of shippers, as a result of shortening the transport process and delivery cycle,
- reducing the cost of frozen capital in a shorter period of more efficiently flowing loads both for shippers and for customers,
- a reduction of fuel consumption (about 33%) and transport costs of cargo as a result of an efficient and short journey, with a much more efficient driving technique possible,
- reducing administration costs due to computerization of the e-booking process and automation of access to terminals,
- postponing (delay/postponing) the need for investment to expand and develop facilities, equipment and terminal infrastructure, as a result of better use of existing machines, equipment and better use of infrastructure and rotation in warehouses and squares,
- increase in terminal revenues due to increased transhipments by approx. 15%, resulting from the increased service capacity and terminal capacity achieved,
- increase in revenues for the Port Authority due to the greater volume of goods transhipped, transhipments handled and the number of vessel visits,
- increase in revenues for logistic and intermodal operators as well as other service providers in the back of the Port due to the greater number of services provided and the flow of cargo handled,
- reduction of financial costs (operational capital costs)—by shortening the time of operations, the cycle of cash turnover and the cycle time of operational capital,

- social and environmental benefits—including:
  - improvement of road traffic safety by over 20%—measured by a reduction in the number of road accidents and accidents involving trucks participating in the carriage of cargo to/from the Port and an increase in the level of freedom of movement,
  - increasing the living comfort of the region's inhabitants by eliminating queues of trucks in the city of road network as well as timely commuting to work and shorter time allocated for journeys,
  - to improve the protection of the natural environment and healthy living conditions of the region's inhabitants—by reducing the emissions by about 30%, as well as greenhouse gases and noise.

References

Greer L, Fraser JL, Hicks D, Mercer M, Thompson K (2018) Intelligent transportation systems benefits, costs, and lessons learned: 2018 update report (No. FHWA-JPO-18-641). United States. Department of Transportation. ITS Joint Program Office

HCM (2010) Highway capacity manual. Transportation Research Board of National Academies, Washington

Kolinski A, Jaskolska E (2018) Analysis of the information flow efficiency in the intermodal supply chain-research results. Bus Logistics Mod Manage 135–155

Meneguette RI, Robson E, Loureiro AA (2018) Intelligent transport system in smart cities. Springer International Publishing, Basel

Pedersen JT (2012) One common framework for information and communication systems in transport and logistics: facilitating interoperability. In: Go-linska P, Hajdul M (eds) Sustainable transport. Springer, Berlin, pp 165–196

Prajogo D, Olhager J (2012) Supply chain integration and performance: the effects of long-term relationships, information technology and sharing, and logistics integration. Int J Prod Econ 135(1):514–522

Sliwczynski B, Hajdul M, Golinska P (2012) Standards for transport data ex-change in the supply chain–pilot studies. In: KES international symposium on agent and multi-agent systems: technologies and applications. Springer, pp 586–594

Speier C, Mollenkopf D, Stank TP (2008) The role of information integration in facilitating 21(st) century supply chains: a theory-based perspective. Transp J 47(2):21–38

Wagener N (2017) Intermodal logistics centres and freight corridors—concepts and trends. LogForum 13(3):273–283

Wong CW, Lai KH, Cheng TCE, Lun YV (2015) The role of IT-enabled collaborative decision making in inter-organizational information integration to improve customer service performance. Int J Prod Econ 159:56–65

Zhang J, Wang FY, Wang K, Lin WH, Xu X, Chen C (2011) Data-driven intelligent transportation systems: a survey. IEEE Trans Intell Transp Syst 12(4):1624–1639

Model Setting and Segregating the Reverse Logistics Process

Slobodan Aćimović and Veljko M. Mijušković

Abstract Reverse logistics can be identified as the key dimension of the green supply chain. The necessity to segregate the regular and reverse logistics flows is expressed through the differences in aspects such as product quality, transport organization, warehouse and inventory management, the general visibility of the entire process, but also by the differences in costs basis of these activities. That is why it is important to adequately set the model for the phases of the reverse logistics process, which is the subject of preoccupation within this chapter. Although there is no unified interpretation of the reverse logistics phases, that does not mean that different approaches to this issue can not be jointly analyzed, and a relatively precisely segregated model can not be established and set. Being driven by such logic, the following phases of the reverse logistics process have been analyzed in detail: returns initiation, collection of returned goods, consideration of different modes of treating the returned goods, decision-making concerning the treatment of the returned goods, client/supplier crediting and returns analysis with performance measurement. An extremely important moment within this process is connected to the consideration and decision- making concerning the treatment of the returned goods. The potential modes are the following: direct product reuse, product reuse after slight modification, using the recycled goods and waste management, when products can not be used any longer. The aim of this analysis is to indicate the complexity of the return logistics process, as well as the importance of the adequate model setting of its particular phases.

Keywords Green supply chain · Reverse logistics · Process model setting · Process segregation

S. Aćimović (✉) · V. M. Mijušković
Faculty of Economics, Belgrade University, Belgrade, Serbia
e-mail: asloba@ekof.bg.ac.rs
URL: http://www.ekof.bg.ac.rs/slobodan-acimovic-phd/?lang=en

V. M. Mijušković
e-mail: mijuskovic@ekof.bg.ac.rs
URL: http://www.ekof.bg.ac.rs/veljko-mijuskovic/?lang=en

A. Kolinski et al. (eds.), *Integration of Information Flow for Greening Supply Chain Management*, Environmental Issues in Logistics and Manufacturing, https://doi.org/10.1007/978-3-030-24355-5_15

1 Introduction

Effective and efficient management of reverse flows represents a relevant issue in company business management. The proven strategic importance of this set of activities for the success of the supply chain participants in developed western economies is especially highlighted with the increase of returns during the last few years. For example, only within the territory of the USA during one calendar year more than 100 billion $ of goods worth are returned (Zhou and Wang 2008). The mentioned example is proof alone that the reverse flow process must be very carefully considered.

Creating the reverse logistics process includes a great number of decisions that need to be introduced on various levels, both of operative and strategic nature, as well as special issues that need to be considered. For example, some of the first things that need to be defined are the type of product return, i.e. which products can enter the reverse channel and which can not, at which stage of a product life cycle can the product enter the reverse flows, what is the residual value of a product, which return option fits best the precise circumstances and similar. The complexity of these issues is additionally stressed by the non-existence of a unique expert approach towards the formulation and formatting of the very reverse process (Genchev et al. 2011). That does not in any way mean that different approaches to this issue could not be sublimed into relatively precise categories. Being driven by this logic, the essence of this chapter is to explain in detail the segregated model setting of the reverse logistics process described by various renowned authors in the field. The aforementioned analysis follows.

2 Important Issues Preceding the Reverse Logistics Process Model Setting

The complexity of establishing the unique set of sequential steps comprising the reverse logistics process is determined by a certain number of open issues which must be dealt with before the very process differentiating. The initial dilemma is connected to the determination of the types of returns. Namely, it is possible to identify a great number of different types or return categories, where each type represents a special challenge for process formatting. According to the analysis carried out by the Global Supply Chain Forum, there are five types of returns (Global Supply Chain Forum 2018) The mentioned types, along with their explanations, are shown within Table 1.

Analysing the different types or product categories it can be concluded that their differentiation is an important precondition of the efficient reverse logistics process design. Namely, the managers must analyse every type or return category and develop an adequate procedure, i.e. process steps for it. This is important since the type of return can have a direct influence not just onto the very company, but also onto other participants within the supply chain. For example, the return which affects

Table 1 Types of returned goods

Type of returns	Explanation
Client returns	Return of goods which is generated based on the client's change of heart concerning the product buying or due to the existence of a defect onto the very product. This type of return represents the biggest category
Marketing returns	This type of returns representing the products from the direct, following link in the chain. It is most usually caused by the slow pace of sales, issues concerning the quality and the need to reorganize the inventory. It includes the following subtypes: (a) close-out returns; (b) buy-out returns or lifts, (c) job-out returns, (d) surplus returns
Value returns	Refers to the return and repositioning of certain goods which the company management wishes to return back. A good example are the package returns which can be used again
Product recalls	Assumes a type of return caused by the issues concerning quality or security. These kinds of recalls can be voluntary or compulsory
Ecological returns	Refers to the management of hazardous materials and abiding the green regulations. This type of returns is different from other types, since it can be legally abiding and it can demand a more extensive paper works

Source Created based on data from Global Supply Chain Forum (2018)

the client causes a long term effect onto the company market perception. Therefore, the company management must take into consideration the marketing effect of this return in order for the process to be formulated as simple as possible for the client. On the other hand, if a specific type of return does not have any influence onto the client, the main criterion while determining the returns process model can be to find the most cost effective solution.

Besides the type of return, an important issue is to determine the preconditions of strategic nature in order for the returns process to function without any issues. Figure 1 shows the strategic decisions connected with the process of returns logistics, as well as key activities for each of those strategic decisions.

The analysed set of decisions of strategic nature connected to product returns is set in a logical order and has the task to establish adequate infrastructure for the implementation of the process within the company, but also with the key participants within the supply chain. In such manner, determining the goals and strategies of product returns can improve the general company performance. The adequate returns policies can have a positive impact on client loyalty, profit increase and creating a positive perception of the public concerning the company image. These policies must be in accordance with the ecological and legal regulations which treat the returns process (Wang 2009).

The next important issue deals with determining the guidelines concerning the avoidance, selection and decision-making of modes of treating the returned product. This decision is narrowly connected to determining the previously analysed types of returns, in order for the policies and control mechanisms of managing the anticipated returns to be developed. The usage of effective procedures of avoidance, selection and decision-making of modes of treating the returned product significantly reduces

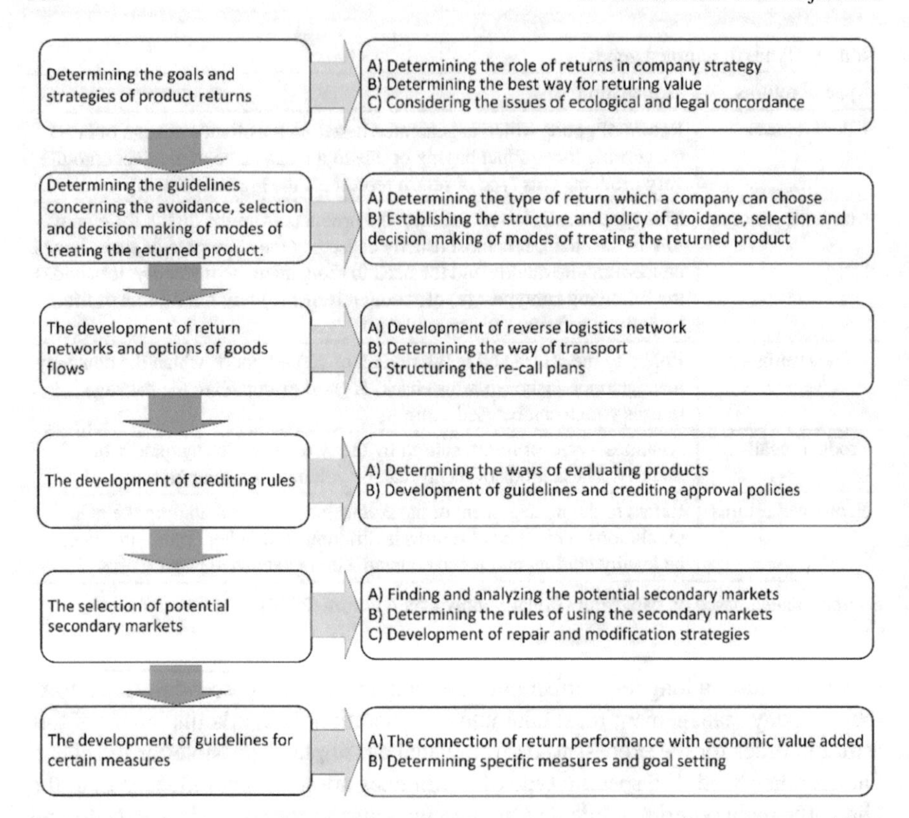

Fig. 1 Key strategic decisions and following activities connected with the reverse logistics model setting process. *Source* Revised according to Rogers et al. (2002)

the costs of the entire process. Precisely, the role of avoidance is the reduction of number of the demands for returns, while the selection is aimed to early identify the goods which shall be treated as returns (Rogers and Tibben-Lembke 2001). Finally, the decision-making concerning the modes of treating the returned product should speed up the product placement back towards the most suitable destination and in the most adequate way (Gobbi 2011).

The development of return networks and options of goods flows assumes the determination of the degree in which the reverse logistics process shall be outsourced for the operative realization by the specialized market participants (3PL and 4PL providers). Also, this decision equally affects the selection of modes of transport, as well as the determination of plans regarding product recalls (De Kostner et al. 2002). The development of crediting rules refers to the formulation of general guidelines based on the inputs gained from our suppliers and clients, which shall determine the manner of evaluating the returned goods. This decision assumes the confirmation of policy for crediting approval, as well as the determination of values of used, returned products and those which have been returned unused due to bad sales dynamics (Gobbi 2011).

Once the location for the products return within the supply chain is chosen, a decision needs to be made concerning the selection of potential secondary markets. The choice needs to be such to most adequately fit the process and the product which is being returned. The companies which decide to acquire the residual value in such a manner must also take into consideration the hazards of such an option, for example, the potential cannibalization of sales of new prime quality products, as well as the influence of the secondary market onto the brand image of that company (Chan 2008). The last decision of strategic nature which must be considered when it comes to the reverse logistics process is the development of guidelines for certain measures. The defined and used metrics can include such indicators as different rates and such financial impact of returns. As a part of this decision it is needed to create valid procedures which most adequately analyse the rates of return and determine the source of their creation. Good examples of metrics are for example the participation of used products in total sales or the return on engaged recycled materials (De Brito and Dekker 2004). Bearing in mind and appreciating all elements up to now, preceding the segregation of the logistics process, it can be concluded that it is a very complex task, where the very process can be modelled in multiple ways.

3 Segregating the Steps of the Reverse Logistics Process Model

The supply chain management experts do not have a unique opinion regarding the steps comprising the reverse logistics process model setting. Some authors are prone to identify a great number of small, individual steps (Prahniski and Kocabasoglu 2006), while others regard the process in an integral way, i.e. as a uniform whole (Russo and Cardinali 2012). The analysis which follows is based on the biggest affirmation in practice, and it takes a partially compromise solution, compared to the previous two extreme points of view (Srivastava and Srivastava 2006).

Therefore, the chosen reverse logistics process model consists of the following steps: (Rogers et al. 2002)

- Returns initiation
- Gathering of returned goods
- Defining the modes of treating the returned goods
- Decision-making about the mode of treating the returned goods
- Client/Supplier crediting
- Returns analysis and performance measurement.

3.1 Returns Initiation

Returns initiation represents the first step within the reverse logistics process model. Figure 2 shows the summed up key activities of returns initiation, as the first phase of the reverse logistics process model setting.

As shown within Fig. 2, the process begins with the receipt of the returns demand from the client who can be either a final consumer or another company upstream in the supply chain. The returns can appear either in a way that a consumer brings back the product into the retail store or based on marketing returns by the retailer or distributor due to slow sales, clearing of credit lines or inventory rotation. For companies which have catalogue sales such as the German seller of fast moving consumer goods-Neckermann, the client may submit a return initiation by internet, phone or by simple returning the goods via post. On the other hand, the American producer and distributor of computer hardware and software Hewlett-Packard commences the returns process by using toner cartridges for printers once the client makes a phone call to the local service for gathering used cartridges. The service of gathering is often handed over to a specialized entity, who performs the communication with the client and carries the returned goods from its location. Hewlett-Packard is known as a company which has a global rate of return of 16% for used toner cartridges. During the last 25 years, Hewlett-Packard has recycled over 47 million of toner cartridges (HP-Product Return and Recycling 2018).

An important moment during this first phase of the reverse logistics process is the efficient initial returns selection which assumes the identification of goods which can not be the subject of return. That means when the client asks a return permission during the contact between the technical service and the client, it is decided whether this permission is possible or not. Of course, the majority of retailers does not perform this kind of technical support, thus it is more difficult for the very producer to

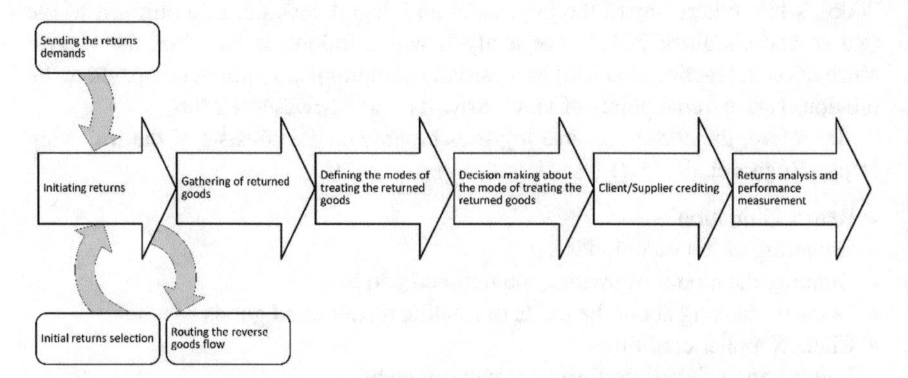

Fig. 2 The key activities of returns initiation as the first phase of the reverse logistics process model setting. *Source* Created according to analysis in Srivastava and Srivastava (2006), Rogers et al. (2002)

perform an initial efficient selection of entering the reverse channel (Klausner and Hendrickson 2000).

Finally, the last important activity of the first phase of the reverse logistics process is routing or determining the return goods flow. The function of routing is primarily of planning nature. During this activity process phase a return materials authorization is created, generated based on the demand of return. By issuing such an authorisation a signal is sent to the receipt location that the returned goods shall soon arrive. In the majority of cases this activity is left over to specialized providers. In such a manner, GENCO, a 3PL provider specialized for return logistics, organizes specific centralized centers for return for several clients-companies. As a part of the company service range both internal and external transport of return flows is organized. For some of its biggest clients, GENCO coordinates the transporters to gather the returned goods from retail objects in consolidated dispatches, and then such consolidated dispatches are delivered to the centralized center at a lower rate, then in the case when the retail would independently return the goods back (GENCO- the Reverse Logistics 2018).

3.2 Gathering of Returned Goods

After the phase of returns initiation, the following phase is the gathering of returned goods which are placed back into the reverse logistics channel. Figure 3 shows the summed up key activities of gathering the returned goods, as the second phase of the reverse logistics process model setting, based on different findings from several renown authors in the field.

As shown within Fig. 3, this is a very important phase in an operative way, where a selection of goods is performed based on type, in order to for it to be located, unified

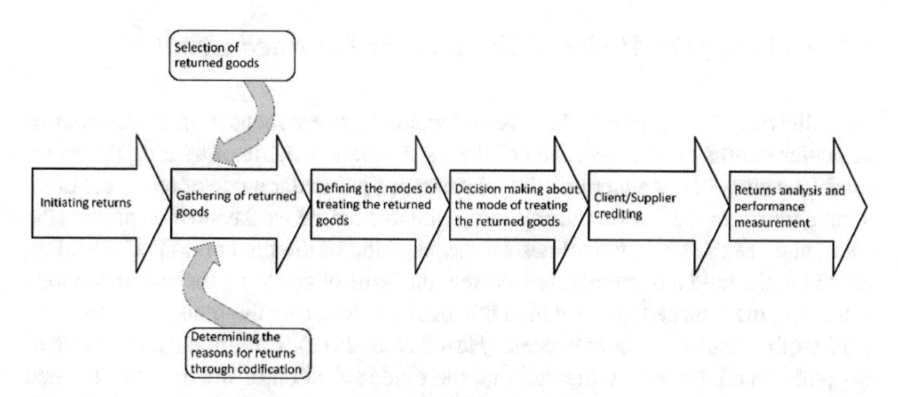

Fig. 3 The key activities of gathering the returned goods as the second phase of the reverse logistics process model setting. *Source* Created according to analysis in Zhou and Wang (2008), Cruz-Rivera and Ertel (2009)

and if needed transported further on for modification/repair. Although the used goods come from multiple sources, they are grouped within one returns location, which results in a specific convergence. Once the returned good has arrived at the warehouse or the centralized returns centre, the products must be checked and processed. Certain authors suggest classification schemes for gathered returns based on whether the initial transport is performed by the client or the entity specifically in charge of gathering the returns (Cairncross 1992). This is usually a manual type of activity which needs to be performed as soon as possible, in order to improve the flow of financial funds. Although the process of returns initial selection has been performed while the goods entered the reverse channel, a detailed return evaluation must be performed both within the warehouse and within the centralized centre for return.

An important activity of this process phase is to determine the reason for return which is coded in a certain manner. The typical return reasons, integrated into defined codes, have been developed by the Executive committee for the reverse logistics and fall under the following categories: repair/service, errors while processing orders, damaged goods and contract issues (Roggers and Tibben-Lembke 1999). Every return of goods needs to be given an adequate code in order for the appropriate performance matrix to be developed and which can obtain valuable information, both within the company and with external entities, such as suppliers and clients.

In some cases it is not cost-efficient to analyze in detail the returns reasons, since it is often very expensive to follow particular goods, but it rather needs to be done generally. However, going into the other extreme, which is a rather vague and imprecise determination of the return reasons can lead to generating wrong information and inadequate organization of the entire reverse process, which causes even greater costs (Roggers and Tibben-Lembke 1999). It can be concluded that the determination of the returns reasons must be performed as precise as possible, and the costs of being too/under precise maintained under control.

3.3 Defining the Modes of Treating the Returned Goods

The following, third phase of the reverse logistics process focuses on the analysis of particular returns and the selection of the most adequate mode of treating the return good. The guidelines concerning the selection of a particular mode of return are often contained within the information system, available for the workers which process the return and use the given guidelines for the selection of the final option. The leading authors in the field agree that it is precisely the issue of choosing the adequate mode of treating the returned goods which practically determines the model, and thus the quality of the entire logistics process (Hazen et al. 2012). Additionally, the value of this phase is confirmed by the fact that the mode selection of treating the returned good also determines the degree of value which can be abstracted from that good, but also the perceived level of green supply chain competitiveness which the precise option of reverse logistics creates. Figure 4 shows the summed up key activities of

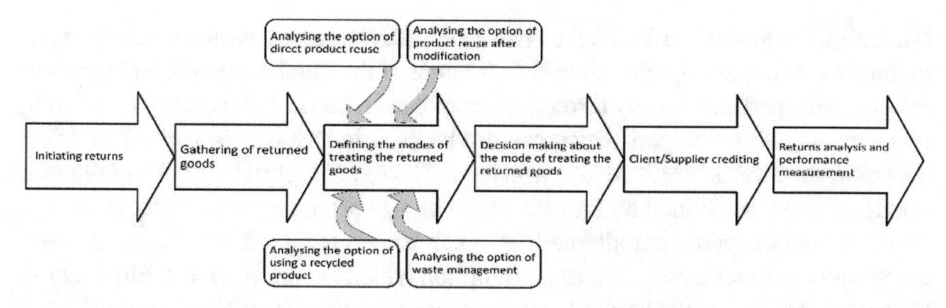

Fig. 4 The key activities of defining the modes of treating the returned goods as the third phase of the reverse logistics process model setting. *Source* Created according to analysis in Srivastava and Srivastava (2006), Rogers et al. (2002)

defining the modes of treating the returned goods, as the third phase of the reverse logistics process model.

The term mode of treating the returned good shown within Fig. 4 is a synonym for the term return option. At the beginning of the analysis of this issue, while forming the models for option returns the intention is to clearly identify them and make a distinction between them. The best model of the initial period which describes the standard process of return was offered by Thierry and associates. According to the analysis performed within their model, there are three alternative return options. First, the company can implement a direct re-usage, which assumes either an immediate re-usage or the re-selling of the returned good without modification. Second, there is the option of managing the returned good thorough the activities of repair, smaller or bigger modifications, the cannibalization of usable components, as well as recycling valuable materials. Finally, there is the alternative of waste management, which assumes the incineration of goods or their adequate waste disposal (Thierry et al. 1995).

After such a division, several similar models were conceptualized, created within other relevant research by equally famous authors within the area (Kirkke et al. 2004; Rogers et al. 2002). Although by the passing of the time, each of the stated studies emphasized to a certain degree different alternatives and definitions, today there is a relatively unified opinion according to which four return options are segregated: (Gobbi 2011)

- Reuse
- Refurbishment/Remanufacturing
- Recycling
- Waste management.

It is interested to state that, going from the first until the last alternative, the necessity to decompose the good decreases, with the simultaneous decrease of the possibility to preserve its integral structure. The four return options can additionally be classified into two groups. Namely, using reused and refurbished/remanufactured goods falls into the category of first class return, since these manners allow the return of residual value in a form of a good which can be reused, with minimum waste and ecological burdening (Simpson 2010). On the other hand, recycling and waste management are defined as secondary options since the structure of the good is wasted, and the value return performed exclusively through components, materials and energy released by incineration, while the part which can not be used in any way is sent off to waste. This classification is in accordance with the ladder of ecological hierarchy of returned goods, which was defined by Lanskin and which is presented within Fig. 5.

Analysing the particular alternatives, it can be pointed out that the direct reuse of the product without modifications is the option which occurs when the client returns the unused product to the point of sale, thus initiating its entry into the return channel and opening the possibility for its reuse. However, the moment the product demands any kind of modification (cleaning, change of spare parts, re-packing, refurbishing and similar) direct reuse is no longer an option. Generally seen, this alternative is possible only if the entity towards which the product is being placed back has the possibility to return the product into the original state. That means that the products must be either unused, or used in such a small extent that it is not needed to modify anything in order to get the status of the new product. If that condition is fulfilled, there are several options what to do further: the product can be offered for sale by the retailer, it can be sent lateral, to another retailer, it can be returned to the supplier, or sent to any other place of direct or reverse flow within the chain where the supply structure misses such a product (Stock and Mulki 2009).

The special issue from the logistics point of view is connected with the forecasting the quantity and quality of such returns. Namely, the unknown quantities of return introduce an additional element of variability into the forecasting process which is already prone to mistakes, stressing the potential bullwhip effect which results in mounting inventories over the needed measure (Vlachos and Decker 2003). If an adequate forecast is performed and the returned goods are managed without needed refurbishing, it possible to reduce the costs of procurement, transport and warehousing simultaneously improving productivity, since every good which is returned back through the value chain, eliminates the need for some other good, to be place forward thorough a regular channel (Vlachos and Decker 2003).

The following alternative is connected to refurbishment or remanufacturing. The basic aim of this return option is the expansion of life span and maximum value abstraction from the original good. This option opens up when direct reuse is no

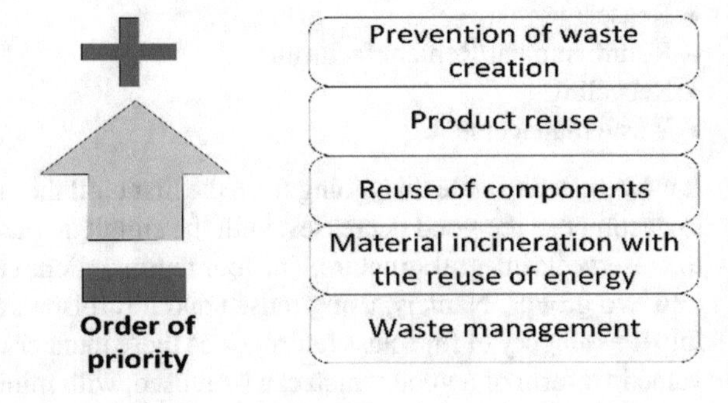

Fig. 5 The ladder of ecological hierarchy of returned goods. *Source* Revised according to Duflou et al. (2008)

longer possible or economic. If adequately carried out it can create significant business possibilities in the regard of value return which would be lost otherwise. The term refurbishment/remanufacturing refers to the process of product improvement from the stage at which it is positioned at the end of the life cycle, towards the stage at which the product is acceptable for future usage or sales (Steinhilper 1994). Therefore, the state and quality of remanufactured good significantly varies depending on the technique used in the process of refurbishing and its purpose. Usually, a difference can be made between refurbishment and remanufacturing. However, some authors suggest that for this return option the better term is remanufacturing (Steinhilper 1994). Remanufacturing is defined as "process of decomposing the used products, their control, component repair and their usage in new production structures. The returned good can be considered significantly remanufactured if its primary components come from used products" (Majumder and Groenevelt 2001).

The usage of recycled products and materials, as a third alternative, assumes the preservation of any part of the returned good which can contain some value. Within this option it is possible to cannibalize parts of the good which do not demand remanufacturing in order to be directly reused again or materials abstracted which can be recycled in order to be reused or resold (Kirkke et al. 2004). Some authors consider that precisely this alternative of treating the returned good is mostly connected with the usage of green initiatives of a company within the supply chain and improving its ecological performance (Daugherty 2011).

Finally, the fourth alternative refers to waste management, and the following can be stated: Once the returned good can not be abstracted of value through reuse, refurbishment or recycling, the good becomes waste. Waste can be defined as "the entity which is perceived of no value for the owner or for which the quantity of efforts needed to abstract value is much greater than the benefit of usage. Practically, waste represents the residual which is thrown away" (Veerakamolmal and Gupta 1999). The selection of this alternative indicates that it was not possible to abstract any additional value from the returned good. Precisely due to that reason this is the least desirable option. Regardless of that, different types of economic incentives or the area regulations impose an ecologically oriented and cost effective waste management as an important activity within the reverse logistics process model.

3.4 Decision-Making About the Mode of Treating the Returned Goods

Once the analysis of all alternatives has been performed, i.e. the modes of treating the returned goods have been analysed, the fourth phase of the logistics process model follows. This phase concerns the decision-making about the mode of treating the returned goods. We have already expressed the opinion that the selection, i.e. the decision on adequate alternative is the basis for the establishment and quality of the entire reverse flow. Consequently, the decision-making of this sort, although it may not have a strategic importance for every supply chain participant, is very important and by its character resembles to the process of strategic decision-making. Precisely, due to that reason, Hazen and associates have decided that the decision-making on the treatment of returned goods should be approximated by the process of strategic decision-making. On that basis, a specific model would be created to significantly simplify the decision process on the returns option (Hazen et al. 2012). Out of greater number of alternatives concerning the strategic decision-making, these authors have chosen a framework set by Hunger and Wheelen, presented within the following steps: (Hunger and Wheelen 2007)

- Result estimate at current performance
- Corporate management revision
- The analysis of external surroundings
- The analysis of internal surroundings
- The analysis of relevant factors
- Creating, estimate and selection of the best alternative
- The implementation of the chosen alternative
- The evaluation of the implemented alternative.

Integrating the return alternative into the mentioned process of strategic decision-making a model on decision-making concerning the returned goods is obtained. The model is shown within Fig. 6.

The first step of the suggested model of decision-making assumes the estimate of company results at current performance. That demands a review of the great number of elements, such as the company mission, business goals or any valid policies which refer to reverse logistics. The aim of this step is to give insight to decision makers about the current state of the company. The next step is connected with the revision of the acceptable way of managing the company. This usually demands the obtaining

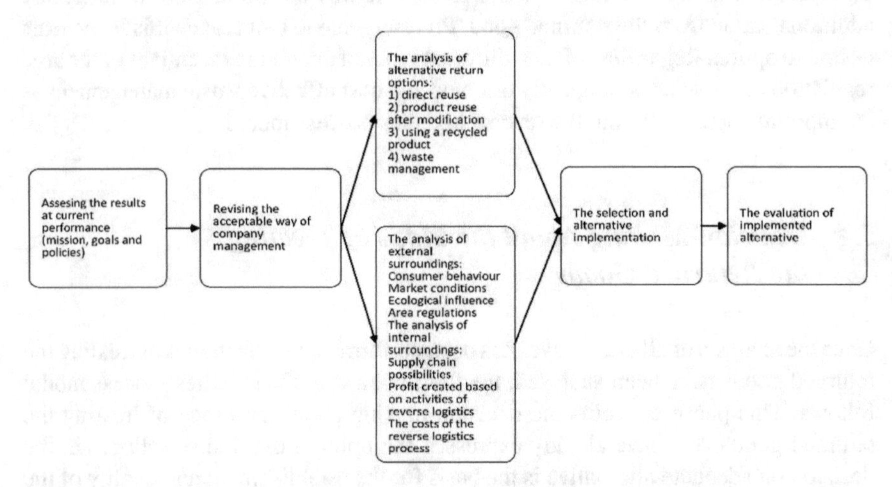

Fig. 6 The model on decision-making about the mode of treating the returned goods. *Source* revised according to Hazen et al. (2012)

of wider support of company management for testing the new return options or for determining whether such option is adequate for the company or not. After the initial preparations, carried through the first two steps of the model, within the third step the decision makers should in detail revise the modes of treating the returned goods/the options of returns which stand at company disposal. The revision assumes the creation of clear perception about what are the possibilities of every alternative and the ways in which they are in accordance with the first two steps of the process. It is interesting to notice that this step is somewhat earlier set within this model compared to the framework set by Hunger and Wheelen. That has been done so since the return options are here already familiar to the decision makers (Hazen et al. 2012).

Continuing, the process of decision-making spans into two familiar branches which assume the analysis of both internal and external surroundings. Speaking about the considerations concerning external surroundings, the decision makers should analyse client behavior, market conditions, existing regulations and the ecological impact made by each of the alternatives. Focusing on the revision of internal surroundings, it is needed to determine which are the capabilities and possibilities of the present supply chain as well as to perform a unique cost-benefit analysis i.e. to determine the costs and the potential profit generated by each option of return. The following step assumes the precise decision-making on which option to choose, which can be simplified by weighing every alternative according to the set of characteristics which every individual company marks as the most important. Finally, the last step in the process presented by the model assumes a periodical revision of the chosen alternative in order for the continuity to assure the desired results.

The conclusion is that the predicted model represents a good framework for managers which introduce decision about the selection of the precise return option, but also a solid base for future research which considers the issues of reverse logistics alternatives decision-making. That is why its usage at this stage of the reverse logistics process is multifold useful and adequate. Based on the given model, Fig. 7

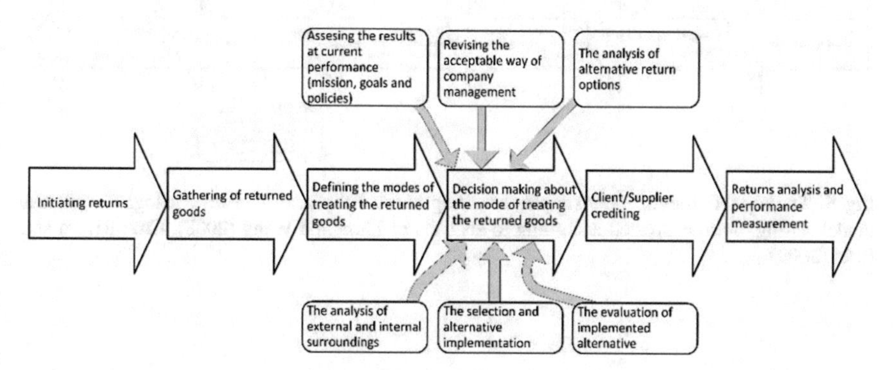

Fig. 7 The key activities of decision-making about the mode of treating the returned goods, as the fourth phase of the reverse logistics process model setting. *Source* Created according to analysis in Rogers et al. (2002)

shows the summed up key activities of decision-making about the mode of treating the returned goods, as the fourth phase of the reverse logistics process model.

3.5 Client/Supplier Crediting

Once the processing of returned good has been performed, i.e. after a precise return option has been identified and implemented, it is needed to determine in which way to approve crediting of certain clients, i.e. consumers and suppliers. Figure 8 shows the summed up key activities of client/supplier crediting, as the fifth phase of the reverse logistics process model.

As shown within Fig. 8, this phase of the reverse logistics process model can be very complex and demand negotiation between different participants of the supply chain. At this place, a significant help in the form of guidelines can be offered by the crediting rules developed as one of the preconditions of strategic nature which have been previously considered.

The financial issues concerning client/supplier crediting significantly influence the efficiency of the entire returns process. As is the usual practice with retailers to tend to intensify the sales volume at the end of the business quarter in order to have more successful results, the ownership concerning the returned goods and supplies of slow turn over can be transferred back to the supplier (Vlachos and Dekker 2003).

Also, some downward participants in the supply chain can use specific price discounts in order to be faster credited by the supplier. These discounts refer to price reduction of the delivered goods for the amount of goods which is delivered back to the supplier or for which the return is yet being planned (Vlachos and Dekker 2003).

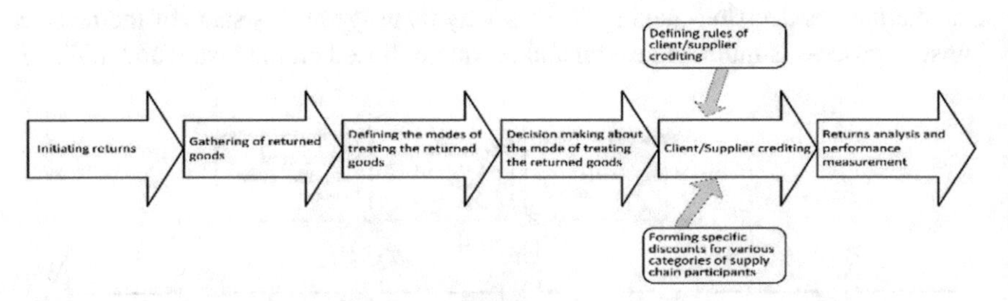

Fig. 8 The key activities of client/supplier crediting, as the fifth phase of the reverse logistics process model setting. *Source* Created according to analysis in Zhou and Wang (2008), Cruz-Rivera and Ertel (2009)

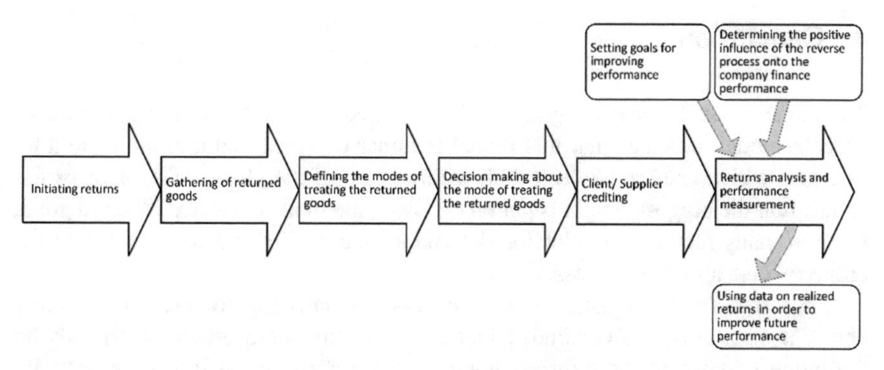

Fig. 9 The key activities of returns analysis and performance measurement, as the sixth phase of the reverse logistics process model setting. *Source* Created according to analysis in Rubio and Jimenez-Parra (2018), Rogers et al. (2002)

3.6 Returns Analysis and Performance Measurement

The last phase of the reverse logistics process model is connected to returns analysis and performance measurement. Figure 9 shows the summed up key activities of returns analysis and performance measurement, as the sixth phase of the reverse logistics process model.

As shown within Fig. 9, a very important activity of this process phase is using the data on realized returns in order to identify the possibilities for their avoidance or to improve the product or the very process. The critical measures, i.e. the metrics used at this process phase is connected with the determination of the degree of defect products, the rate of returns, as well as establishing the needed time for the realization of the cycle of the chosen return option (De Brito and Dekker 2004). It is needed to invest all possible efforts to document the positive influence of the reverse logistics process onto the company financial performance expressed by economic value added. Determining the effect of this process onto sales is surely more difficult than to determine their effects onto cost reduction, but it should not be forgotten that the sales volume is the one which mostly determines profitability. However, since the cost reduction is much more easy to be measured, that is one of the basic reasons why in the biggest number of practical cases this category is exclusively followed (Asif et al. 2014).

The final activity of this phase of the reverse logistics process model is goal setting for performance improvement. Following a detailed analysis of the realized company performance, goals that need to be improved are set up, and then they are communicated among all important entities within the company, as well as the supply chain. It is important to emphasize that the scalability of performance improvement of the reverse logistics process is very difficult to determine, so one needs to be realistic when setting goals, as well as regarding expectations invested within this phase.

4 Conclusion

The carried out detailed analysis of all identified phases of the reverse logistics process model indicates that this model is rather complex and that there are a lot of issues which need to be dealt with within every of the phases. The analysis has shown that the biggest weight is put on choosing the mode of treated returned good, since in many factors this selection determines the quality and composition of the entire reverse logistics process.

Since the reverse logistics is the key operationalisation of the green supply chain, it can finally be concluded that its competitiveness precisely depends on the mode of treating the returned good i.e. the return option (reuse, remanufacturing/refurbishment, recycling, waste management), as the key phase of the reverse logistics process model. This opens up a lot of new issues to be the subject of thorough future research papers and scientific investigation efforts. For example, is the identified model universal for all global economies which want to implement it? Or if national peculiarities exist, what drives those specifics? Since this is a globally relatively new topic without much empirical testing, our presented analysis has mostly focused on extensive literary review and that is its main shortcoming. Since we have not carried out an empirical research, at this particular moment we do not wish to suggest any modifications to the model defined by renowned authors in the field. Therefore, in order to define a model of our own, or to test the presented, theoretical one, in the research to come we plan to carry out a survey or a series of in-depth interviews with middle and top management of national and international companies within the region of Western Balkans and to see what the main phases of the process model of reverse logistics look like in practice. We hope that such effort will contribute much more to the already existent, but relatively scarce practical scientific findings dealing with this topic.

References

Asif F, Bianchi C, Rashid A, Nicolescu C (2014) Performance analysis of the closed-loop supply chain. J Remanufact 2(4):1–21

Cairncross F (1992) Costing the Earth. Harvard Business School Press, Boston

Chan J (2008) Product end-of-life options selection: grey relational analysis approach. Int J Prod Res 46(3):2889–2912

Cruz-Rivera R, Ertel J (2009) Reverse logistics network design for the collection of end-of-life vehicles in Mexico. Eur J Oper Res 196:930–939

Daugherty P (2011) Review of logistics and supply chain relationship literature and suggested research agenda. Int J Phys Distrib Logistics Manage 41(1):16–31

De Brito M, Dekker R (2004) A framework for reverse logistics. In: Dekker R, Inderfurth K, van Wassenhove L, Fleischmann M (eds) Reverse logistics: quantitative models for closed-loop supply chains, Chapter 1. Springer, Berlin

De Koster R, De Brito M, van de Vendel M (2002) Return handling: an exploratory study with nine retailer warehouses. Int J Retail Distrib Manage 30(8):407–421

Duflou J, Selinger G, Kara S, Umeda Y, Ommeto A, Willems B (2008) Efficiency and feasibility of product disassembly: a case-based study. CIRP Ann Manuf Technol 57(2):583–600
Genchev S, Glenn-Richey R, Gabler C (2011) Evaluating reverse logistics programs: a suggested process formalization. Int J Logistics Manage 22(2):242–263
GENCO-Reverse logistics, available at: http://www.genco.com/Reverse-Logistics/reverse-logistics.php. Accessed on 03/08/2018
Gobbi C (2011) Designing the reverse supply chain: the impact of the product residual value. Int J Phys Distrib Logistics Manage 41(8):768–796
Hazen B, Cegielski C, Hanna J (2012) Diffusion of green supply chain management. Int J Logistics Manage 22(3):379–389
HP-Product return and recycling, available at: http://www8.hp.com/us/en/hp-information/environment/product-recycling.html#.VgHJW9-qpBc. Accessed on 03/08/2018
Hunger J, Wheelen T (2007) Essentials of strategic management, 4th edn. Pearson Prentice-Hall, Upper Saddle River, New Jersey
Krikke H, le Blanc I, van de Velde S (2004) Product modularity and the design of closed–loop supply chains. Calif Manag Rev 46(2):23–39
Klausner M, Hendrickson C (2000) Reverse logistics strategy for product take-back. Interfaces 30:156–165
Majumder P, Groenevelt H (2001) Competition in remanufacturing. Prod Oper Manage 10(2):125–141
Prahniski C, Kocabasoglu C (2006) Empirical research opportunities in reverse supply chains. Omega 34(6):519–532
Rogers D, Lambert D, Croxton K, Garcia-Dastugue S (2002) The returns management process. Int J Logistics Manage 13(2):1–18
Roggers D, Tibben-Lembke R (1999) Going backwards: reverse logistics trends and practices. Reverse Logistics Executive Council Press, Pittsburg
Roggers D, Tibben-Lembke R (2001) An examination of reverse logistic practices. J Bus Logistics 22(2):129–148
Rubio S, Jimenez-Parra B (2018) Reverse logistics: overview and challenges for supply chain management. Int J Eng Bus Manage 1–7
Russo I, Cardinali S (2012) Product returns and customer value: a footware industry case. In: Jodlbauer H, Olhager J, Schonberger R (eds) Modelling value: contribution to management science, part 2. Springer, Berlin
Simpson D (2010) Use of supply relationships to recycle secondary materials. Int J Prod Res 48(1):227–249
Srivastava S, Srivastava R (2006) Managing product returns for reverse logistics. Int J Phys Distrib Logistics Manage 36(7):524–546
Steinhilper R (1994) Design for recycling and remanufacturing of mechatronic and electronic products: challenges, solutions and practical examples from the European Viewpoint. In: Proceedings of the 1994 ASME national design engineering conference DE- vol 67, pp 65–67
Stock J, Mulki J (2009) Products return processing: an examination of practices of manufacturers, wholesalers/distributors and retailers. J Bus Logistics 20(1):33–61
Thierry M, Salomon M, Van Nunen J, Van Wassenhove L (1995) Strategic issues in product recovery management. Calif Manag Rev 37(2):114–135
Types of returns- Global supply chain forum, available at: http://fisher.osu.edu/centers/scm/media/. Accessed on 03/08/2018
Veerakamolmal P, Gupta S (1999) Analysis of design efficiency for the disassembly of modular electronic products. J Electron Manufact 9(3):79–95
Vlachos D, Dekker R (2003) Return handling options and order quantities for single period products. Eur J Oper Res 151(1):38–52
Wang H (2009) Web-based green products life cycle management systems: reverse supply chain utilization. Information Science Reference, New York
Zhou J, Wang Sh (2008) Generic model of reverse logistics network design. J Transp Syst Eng Inf Technol 8(3):71–78

Information Flow in the Context of the Green Concept, Industry 4.0, and Supply Chain Integration

Brigita Gajšek and Marjan Sternad

Abstract Supply Chain Integration describes how employees in a specific company along a supply chain and its trading partners work together to achieve common business objectives via integrated business processes and information sharing. In this way, material and information flows are established. Due to the competitiveness between supply chains and the trend to be more efficient and sustainable, we distinguish several phases of supply chain integration, from the oral negotiation and repetition of the same activities to automation and reduction in the number of activities. Recently, much attention has been paid to the concept of Industry 4.0, which is based on computerisation and connectivity. Digitalisation is recognised as a springboard for visibility, transparency, predictive capacity and adaptability or, in other words, for Industry 4.0, which inherently requires technological development. Information systems, as an example of technology, are an enabler for increased sustainability and greener supply chains. Currently, companies are at different stages of technology development, and the establishment of external integration depends on the digital development of each element of the supply chain. The current situation in companies reveals different degrees of maturity, both from the aspect of supply chain integration and from the aspect of Industry 4.0. In this paper, we reveal possible overlapping areas between different maturity phases of supply chain integration and the Industry 4.0 concept, as well as how different combinations of maturity levels characterise information flow.

Keywords Information flow · Green · Industry 4.0 · Supply chain integration · Maturity models

B. Gajšek (✉)
Department for Technical Logistics, Faculty of Logistics,
University of Maribor, Maribor, Slovenia
e-mail: brigita.gajsek@um.si

M. Sternad
Department of Sustainable Logistics and Mobility, Faculty of Logistics,
University of Maribor, Maribor, Slovenia
e-mail: marjan.sternad@um.si

A. Kolinski et al. (eds.), *Integration of Information Flow for Greening Supply Chain Management*, Environmental Issues in Logistics and Manufacturing, https://doi.org/10.1007/978-3-030-24355-5_16

1 Introduction and Motivation

The Industry 4.0 concept has recently gained much attention due to the awareness that its initiatives can influence whole business systems via transforming the means by which products are designed, produced, delivered and discarded, and that it is significantly transforming the behaviour of supply chain management (SCM) (Luthra and Mangla 2018). The dependence of Industry 4.0 and SCM is indicated in research works of several authors, for example Brettel and Friederichsen (2014). Industry 4.0 (also known as the fourth industrial revolution) is shaping a future heavily reliant on data acquisition and sharing throughout the supply chain (Brettel and Friederichsen 2014; European Commission 2016). This indicates that information flow is in some way also changing with the advancement of this breakthrough concept. Researchers have been identifying developments in Industry 4.0, such as virtualising the process- and supply-chain, ensuring supply chain flexibility, raising supply chain visibility, establishing collaborative networks, achieving end-to-end digital integration, and the use of individualised traced data, while leveraging technologies and concepts less common in less mature supply chains, such as the Internet of Things (IoT) and services, and cloud computing (Brettel and Friederichsen 2014). This factor and many similar additional ones prove our assumption of possible essential relations between Industry 4.0 and SCM.

Although the general impression is that the introduction of new technologies under the umbrella of Industry 4.0 can mostly have a positive effect on supply chain performance, there are also concerns. Dallasega et al. (2018) observed that Industry 4.0 concepts are applied without much reflection on context. This includes the degree of proximity that is desirable for an organisation. Proximity comprises more than the geographic distance between two actors. Some authors differentiate between three to five dimensions of proximity, depending on the scope of their research. If proximity is understood as a multi-faceted concept, it can be seen as a construct involving geographic, organisational, cognitive, social, cultural, institutional, and technological dimensions (Capaldo and Petruzzelli 2014). Dallasega et al. (2018) prove that the use of more technology would increase proximity at the expense of the efficiency of cooperation. In this situation, the movement towards increased distance, not only geographic, would be a better solution, because it would lead to increased efficiency. Efficiency affects the company's performance as it aims to assist connecting supply to demand (Dragan et al. 2018). For each company and supply chain, there is an optimal degree of proximity beyond which greater closeness might be harmful. Greater proximity would result in reduced efficiency. The concept of proximity is close to the concept of integration.

Our study promotes the advisability of exploring relationships between Industry 4.0, supply chain integration, information flow, and the green concept. We chose not to focus on the green concept because Erol (2016) has already proven an unambiguous link between it and Industry 4.0 in his paper 'Where is the green in Industry 4.0?', in which he argues for a fourth industrial revolution that is not only targeted at leveraging competitiveness but is also built upon the concept of sustainability as a basis for long-term economic prosperity and welfare and that information systems are a significant enabler for this vision.

Following these findings, the present paper will answer the following research questions.

RQ1: Is there an overlapping area(s) between the concepts of Industry 4.0 and supply chain integration?

RQ2: What are the characteristics of the information flow in the areas of overlap?

2 Theoretical Background

2.1 Information Flow

The supply chain practice focuses on material movement (Chopra and Meindl 2001), while information sharing focuses on information flow (Premkumar and William 1994; Zhou and Benton 2007). Information, as the basic constituent of the information flow, is useable data, inferences from data, or data descriptions (Ackoff 1989; Checkland 1988; Durugbo et al. 2013). It is crucial to the existence of organisations: so much so, that it is likened to oxygen for human life (Al-Hakim 2008). In organisations and beyond, information flows in verbal, written or electronic form (Yazici 2002), from a sender to a receiver (Westrum 2004) and is dependent on access to information resources (Atani and Kabore 2007). Information moves between: (i) individuals in an organisation or organisations, (ii) organisational departments, (iii) multiple organisations, and (iv) an organisation and its environment (Henczel 2001). Information flow is the movement of information between people and systems (Spacey 2017) and an essential part of workflows (Al-Hakim 2008; Mentzas et al. 2001) that requires a synergy between humans and computer systems in modern organisations (Burstein and Diller 2004; Hinton 2002). Efficient and secure information flows are a central factor in the performance of decision making, processes, and communications (Spacey 2017). Publish/Subscribe, Push, Pull, Choreography, Orchestration, Event Handling, Communication, and Knowledge are common types of information flow (Spacey 2017). The information flow can be visualised using an information flow diagram, which shows how information is communicated (or 'flows') from a source to a receiver or target, through a medium (Wintraecken 1990). The medium acts as a bridge, a means of transmitting the information. Examples of media include word of mouth, radio, email, and others.

The increasing complexity of systems and supply chains places the information flow at the foreground of research. The management of the supply chain has become dependent on the information flow (Madenas et al. 2014). Information links various stakeholders along a supply chain and allows shared supply chain management on different decision-making levels. An efficient supply chain must establish short links (Hearnshaw and Wilson 2013) or direct communication with as few intermediaries as possible, as this decisively contributes to the efficient operation of the network.

The establishment of an information flow requires reflection on information sharing from at least three aspects: information sharing support technology (supply chain management (SCM) IT applications), information content (supplier information, manufacturer information, customer information, distribution information, and retailer information), and information quality (accuracy; availability; timeliness; internal connectivity; external connectivity; completeness; relevance; accessibility; and frequently updated information (Zhou and Benton 2007).

Information sharing among two or more organisations from different tiers of the supply chains can be categorised based on the intended decision time horizon (Kembro et al. 2017):

- operational level: companies share sales and order information to facilitate customer orders, reduce information distortion, and lower stock levels;
- tactical level: involves monthly and quarterly forecasts and plans to help partners reserve adequate capacities for production and logistics activities;
- strategic level: organisations share annual demand forecasts and promotion plans as well as marketing strategies to enable planning of future purchases and growth within the alliance.

One of the most critical information flow enablers is an information system (IS), defined as an organised system for the collection, organisation, storage, and communication of information, consisting of hardware, software, data, business processes, and functions which can be used to increase the efficiency and management of an organisation. Information systems interrelate with both data systems and activity systems (Beynon-Davies 2009). Yousefi and Alibabaei (2015) find that the supply chain information system is useful in synchronising all the supply chain actors and helps planning and predict demand. Its use can increase operational efficiency, flexibility, and quality of service while reducing costs. For example, Lee et al. (2000) state that the exchange of information reduces the costs of inventory management.

Vickery et al. (2002) define as the primary objective of the integrated supply chain the coordination of the requirements of the final customer with the flow of goods and information along the supply chain in order to strike a balance between the high service for users and the costs that arise thereon. The integration of the supply chain is linked to integrated information technologies, which increase and accelerate the flow of relevant information among participants within the supply chain. Prajogo and Olhager (2012) have similar conclusions, noting that the integration of information relates to the exchange of crucial information in the supply chain through information technology.

Integration is developing in parallel with the development of information communication technology (ICT) and is focused on direct communication in real time between sources and users of data and information without intermediaries through the most suitable medium. Traditional IT systems are declining as modern digital technologies, such as artificial intelligence (AI) and virtual reality (VR) are proving to be strong business-driving forces. A Forbes magazine (Columbus 2015) report recently revealed, 'By 2018, more than 60% of enterprises will have at least half of their infrastructure on cloud-based platforms.' Leveraging the hybrid cloud, an integrated computing environment that blends elements of the public and private cloud to perform various functions in an enterprise, helps a firm achieve several organisational goals, such as making real-time business decisions, improving customer experience, and enabling digital business transformation. Modern information flow consequently also increasingly includes data from the environment, from publicly available databases. The latter is of utmost importance for the realistic planning of transport and logistics activities.

Integrated information systems and integrated electronic data exchange are more important for suppliers than previously and have become of great importance for customers. They like to compare products, want to fulfil their wishes immediately and have the experience under control, are always connected and ready to search/compare/buy, and expect personal interactions. Consequently, we are witnessing an accelerated development of customer information in addition to all others.

2.2 Green Supply Chain Integration

The concept of the supply chain has been known since the end of the twentieth century and represents a new way of understanding integration. The role of the environment in the management of supply chains becomes increasingly important. Chin et al. (2015) define the concept of the green supply chain as the integration of environmental thinking into supply chain management and represents an environmental innovation. A green supply chain is initially defined as a set of three or more entities (organisations or individuals) directly involved in the upstream and downstream flows of products, services, finances, and/or information from a source to a customer (Mentzer et al. 2001) and later as a network of relationships (Stevens and Johnson 2016). The green supply chain is important in influencing the total environmental impact of any organisations involved in supply chain activities (Chen et al. 2009). Wu (2013) also defines green supply chain integration as the collaboration between a firm and its supply chain partners to direct both intra- and inter-organisational environmental practices.

However, independently of perceived structure, the mere formal positioning of the entity in the supply chain is not a guarantee of achieving any positive effects. Supply chains' primary goal is to connect entities and establish basic formation for the introduction of supply chain management (SCM), as a driver and enabler of business performance.

Supply chain management takes a system approach to viewing the supply chain as a single entity, rather than as a set of fragmented parts, each performing its function (Houlihan 1988; Ellram and Cooper 1990; Tyndall et al. 1998), and manages the supply chain as an integrated whole. Stevens (1989) states that the goal of SCM is to synchronise the customer's requirements with the flow of materials from suppliers in order to strike a balance between high customer service, optimal stock, and low unit costs. The next phase in the evolution of SCM happened when Bowersox and Closs (1996) argued that to be fully effective in a competitive environment, firms must expand their integrated behaviour to incorporate customers and suppliers, and referred to that as SCM. Related to integrated behaviour, mutually sharing information among supply chain members is required to implement an SCM philosophy, especially for planning and monitoring processes (Mentzer et al. 2001). Cooper et al. (1997) emphasise frequent information updating among the chain members for effective supply chain management. The Global Logistics Research Team at Michigan State University (1995) defines information sharing as the willingness to make strategic and tactical data available to other members of the supply chain. Open sharing of information such as inventory levels, forecasts, sales promotion strategies, and marketing strategies reduces the uncertainty between supply partners and results in enhanced performance (Mentzer et al. 2001).

In addition, effective SCM requires mutually sharing risks and rewards, cooperation, the same goal and the same focus on serving customers, integration of processes, and partners to build and maintain long-term relationships (Mentzer et al. 2001). Integration of processes could be understood as the previous step or a prerequisite for information sharing and supply chain management. There are different integration models in practice. The most important models are (De Olivera et al. 2011) the Supply Chain Operations Reference Model (SCOR), the Computer Sciences Corporation framework (CSC), and the Supply Chain Process Management Maturity model (SCPM3). The SCOR model is appropriate for evaluating and comparing supply chain activities and performance. The purpose of the model is to describe process architecture (SCOR 2012). The primary objective of the CSC model is to identify the logistics functions development stage in companies (Poirier and Quinn 2003). The SCPM3 model is the first process model to use statistical analysis to define maturity levels; it has five levels: Foundation, Structure, Vision, Integration and Dynamics (De Olivera et al. 2011). In the present research, we focus on Stevens's model.

According to Stevens (1989), the implementation of SCM necessitates the integration of processes from sourcing, to manufacturing, and distribution across the supply chain. Achieving a state of 'integration' requires a firm to progress through a number of defined stages of development (Stevens 1989; Stevens and Johnson 2016):

- Stage 1: The supply chain is a function of fragmented operations within the individual company and is characterised by staged inventories, independent and incompatible control systems and procedures, and functional segregation;
- Stage 2: Begins to focus on internal integration, characterised by an emphasis on cost reduction rather than performance improvement, buffer inventory, initial evaluations of internal trade-offs, and reactive customer service;
- Stage 3: Reaches toward internal corporate integration and characterised by full visibility of purchasing through distribution, medium-term planning, tactical rather than strategic focus, emphasis on efficiency, extended use of electronics support for linkages, and a continued reactive approach to customers;
- Stage 4: Achieves supply chain integration by extending the scope of integration outside the company to embrace suppliers and customers;

- Stage 5: The supply chain is a non-linear goal-directed network with connections between firms;
- Stage 6: The future global integrated supply chain model will be devolved, collaborative, supply chain clusters. With devoted, collaborative supply chain clusters, supply chain integration moves away from being a monolithic approach to one that enables the 'modular' connecting of the focal firm to different clusters (Fig. 1).

In recent years, numerous studies have emphasised the importance of information technology-based information sharing within the supply chain to improve its integration (Mason-Jones and Towill 1997; Barrat 2004; Trkman et al. 2007). Hanfield and Nichols (1999) present three main drivers of integration: the information revolution, increased levels of global competition, and the emergence of new types of inter-organisational relationships. However, surprisingly, adoption of the Internet by itself demonstrates no benefits in terms of reduced transaction costs or improved supply chain efficiency in Scottish SMEs (Wagner et al. 2003) and has not led to a

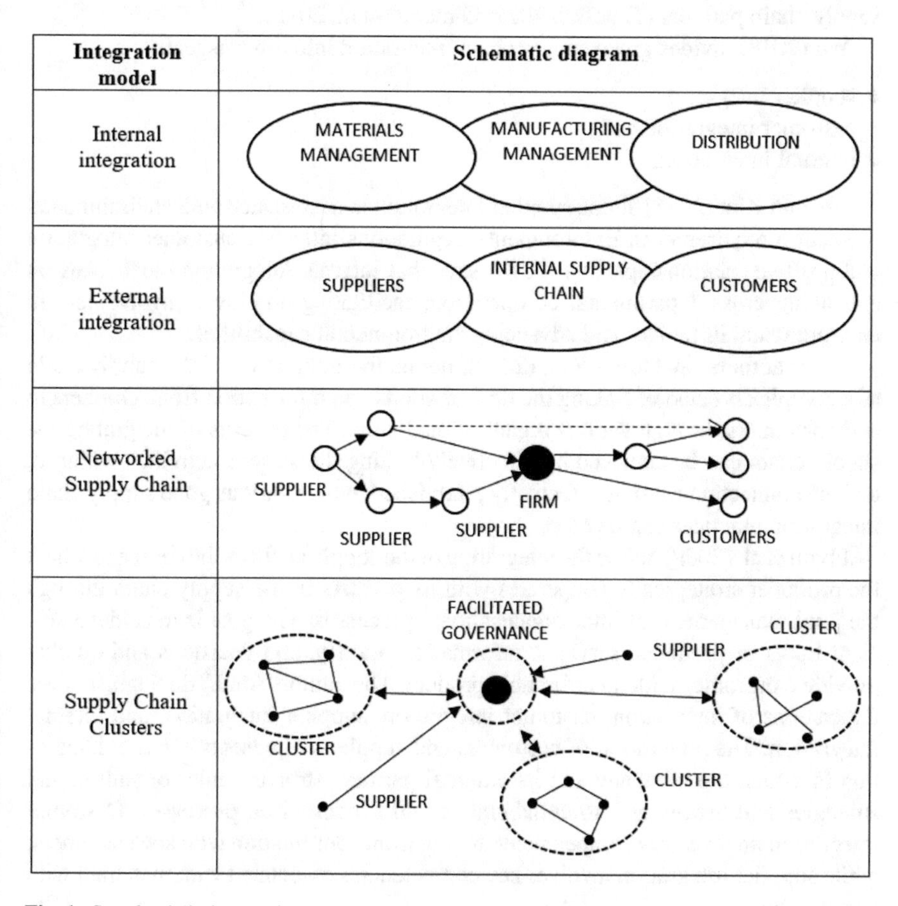

Fig. 1 Supply chain integration stages. *Source* Adapted from Stevens and Johnson (2016)

decrease in the inventory level in Slovenian enterprises (Trkman 2000). Additionally, studies have shown that information transfer brings few benefits and that most benefits from IT are due to the shorter lead-times and enabling the production of smaller batches (Cachon and Fisher 2000). While there is no doubt that the use of IT can reduce costs along the supply chain, the formation of a business model and utilisation of information is crucial (Trkman et al. 2007).

Supply chain integration is a set of activities concerned with the coordination of product flows between supply chain partners, including transactions, materials movements, procedures, and optimisation processes, taking into consideration the underlying information flows (Frohlich and Westbrook 2001; Sahin and Robinson 2002). At the tactical level, the literature suggests two interrelated forms of integration, i.e., information exchange and operational integration tactics (e.g., Frohlich and Westbrook 2001). Information exchange refers to the coordination of information transfer and communication, while operational integration points to joint activity development, collaborative work processes, and coordinated decision making among supply chain partners (Topolšek 2011; Gimenez et al. 2012).

Wu (2013) divides green supply chain integration into three aspects:

- supplier integration;
- customer integration and
- internal integration.

Lee and Kim (2011) define supplier integration in accordance with environmental protection requirements. Environmental regulations influence customer integration and positive relationships. Wu (2013) says that internal integration must focus on enhancing cross-departmental co-operation, facilitating employee involvement in environmental initiatives and advancing environmental capabilities.

Other authors, including Kim (2009), define the integration of the supply chain as a complex process of linking the flow of goods and information from suppliers to end consumers through various organisational units. The benefits of integrating the supply chain can be achieved by effectively linking the various activities within it, and the connection must be effectively established, where various good supply chain integration practices can be used.

Flynn et al. (2009) define the integration of the supply chain as the degree to which the producer strategically cooperates with its partners in the supply chain through the joint management of inter-organisational processes. The goal is to achieve efficient flows of products, services, information, money, and decisions and quickly provide a customer with an affordable product. The authors (ibid.) distinguish three dimensions of integration: customer integration, supplier integration, and internal integration. The integration of customers and suppliers represents external integration in which the company and its external partners structure inter-organisational strategies and processes into collaborative and synchronised processes. Customer integration involves key competencies arising from coordination with key customers, while supplier integration involves key competencies associated with matching with key suppliers.

Chen et al. (2009) note that the implementation of the integration of supply chain processes, covering different functional areas in and between companies, requires the involvement of customers in integration processes. Only if integration is customer-oriented, integration of processes in the supply chain can create value for customers and bring the desired financial results for the company. In order to successfully integrate processes in the supply chain, it is essential to develop strategic priorities. As priorities, the authors (ibid.) indicated their orientation to the customer and cost orientation.

Frohlich and Westbrook (2001) identified two types of integration:

- the harmonisation and integration of commodity flows between suppliers, manufacturers, and customers;
- a feedback flow from customers to suppliers.

Gimenez and Ventura (2005) note that internal integration has a positive impact on external integration because coordination between internal functions facilitates coordination between different companies. External integration also has a positive impact on internal integration, since external cooperation has a positive effect on internal cooperation: if companies want to work with their supply chain members, they need to improve internal integration. Companies have realised that cooperation and integration between different functional areas increases the success of external integrated relationships. The authors (ibid.) also find that it has the most significant influence on the implementation of logistics services' external integration.

Long-term supply chain strategies provide effective external integration in line with environmental aspects. Kumar and Chandrakar (2012) emphasise one of the key aspects to green supply chain integration, which is to improve economic and environmental performance simultaneously throughout the chains by establishing a long-term buyer-supplier relationship.

2.3 Industry 4.0 Concept

Lean production is a widely recognised foundational element of production systems (Herlyn 2011; Kolberg and Zühlke 2015). It contributes to faster reaction to changing market demands, smaller batches, as well as transparent and standardised processes for mass and batch production (Womack et al. 2007; Ōno 1988). At present, it seems that lean production has reached its limit as industry is faced with strong deviations in market demands. Thus, production decoupled from market demand is needed (Erlach 2013; Ōno 1988; Dickmann 2007), since the previously successful traditional organisation of production environments has become cost-inefficient. Looking forward, the Industry 4.0 concept gives a boost to the modern development of production with increased integration of information communications technology (ICT) into production.

'Industrie 4.0' (Industry 4.0, I40) is a national strategic initiative from the German government through the Ministry of Education and Research (BMBF) and the Ministry for Economic Affairs and Energy (BMWI) (Klitou et al. 2017). It aims to drive digital manufacturing forward by increasing digitisation and the interconnection of products, value chains, and business models. It also aims to support research, the networking of industry partners and standardisation. The initiative was launched in 2011 and has become institutionalised with the Platform Industrie 4.0 (Platform I40) that now serves as a central point of contact for policy-makers. In spring 2014, VDMA, BITKOM, and ZVEI, three leading German associations of mechanical engineering, ICT, and the electrical industry, released a definition for I40 (Kolberg and Zühlke 2015). According to them, I40 aims for optimisation of value chains by implementing autonomously controlled and dynamic production. Enablers are the availability of real-time information and networked systems (Acatech-Platform Industrie 4.0 2014) that require a higher degree of automation. Instruments to reach this increased automation are Cyber-Physical Systems (CPS) equipped with microcontroller, actuators, sensors and a communication interface. CPS can work autonomously and interact with the production environment; in parallel, a factory becomes 'smart' (Broy 2010; Lee 2008; Kolberg and Zühlke 2015). The department of Innovative Factory Systems (IFS) at the German Research Center for Artificial Intelligence (DFKI) identified four enablers for the Smart Factory, named Smart Products, Smart Machines, Smart Planner, and Smart Operators (Kolberg and Zühlke 2015). Smart Products know their production process and negotiate it with Smart Machines. The Smart Planner optimises processes in nearly real time. In this environment, humans take a central position. Supported by innovative ICT, they become Smart Operators who supervise and control ongoing activities.

With regard to industry, this means that the existing and, in parts, rather inflexible processes can be revolutionised by high-performance computers, a powerful internet and intelligent products and machines via the active exchange of information. According to the Platform Industry 4.0 interest group, production processes of the future will be decentralised, which means a shift away from today's still centrally controlled factories (Bundesverband Informationswirtschaft, Telekommunikation und neue Medien e.V. 2015). Industry 4.0 describes the increasing digitisation and automation of the manufacturing environment, as well as the creation of digital value chains to enable communication between products, their environment, and business partners (Lasi et al. 2014).

The primary economic potential of Industry 4.0 lies in its ability to accelerate corporate decision-making and acceptance processes (Schuh et al. 2017); the same authors define Industry 4.0 as 'real-time, high data volume, multilateral communication and interconnectedness between CPSs and people'. The availability of vast quantities of data and information at affordable prices and in real time, if necessary, enables a better understanding of how things relate to each other and provide the basis for faster decision-making processes.

The turbulent events in practice are accompanied by a response from the scientific community, which is seen in the sharp increase in the number of scientific publications. 'Industrie 4.0' in general is still classified as highly technical. The idea is spreading rapidly to other areas. Industry 4.0, therefore, has to be considered as an overall operational strategy.

Industry 4.0 is also very much related to environmental protection. Szalavetz (2017) have determined that advanced manufacturing technologies will not only improve operational excellence and increase financial effectiveness, but will have a beneficial impact on environmental performance. Szalavetz (2017) also identified three functional areas for improved eco-efficiency:

- production control and quality management,
- process optimisation and virtual product,
- process engineering.

Erol (2016) states that only those production systems that incorporate sustainability in their concept of intelligence will be competitive in the long-term. Kiel et al. (2017) assert that Industry 4.0 provides economic, environmental and social benefits. If we want to create sustainable industrial value, manufacturers are required to pursue all three dimensions simultaneously (Sridhar and Jones 2013).

2.4 Findings from Theoretical Background

Industry 4.0 and supply chain integration are desirable areas of research with an increasing number of publications in scientific journals. In 2017, within the Web of Science (WoS), supply chain integration was a topic in 440 scientific papers and Industry 4.0 in 977 scientific papers. Industry 4.0 attracts mostly researchers from engineering research while supply chain integration primarily attracts researchers from business economics and computer science. Information flow is the leading thread of both concepts and the fundamental enabler of business activities at all levels of operations (i.e., operational, tactical, and strategic) of the individual company and the supply chain as a whole. It is becoming digitised and directed according to the requirements of digital transformation. Although originally applied to manufacturing, digital transformation in the era of Industry 4.0 is slowly but strongly changing all sectors. Decisions-makers largely agree that digitisation will affect every process in their industry. To overcome these challenges, four keys to digital transformation are seen as decisive: digital data, automation, connectivity, and digital access. All of them are and will continue to be reflected in information flow and, through, it on supply chain integration.

There are already a few studies looking for cross-connections between Industry 4.0 and supply chain management. However, all of them explore quite narrow areas: for example, cloud-based platforms and social media apps can improve collaboration (Dallasega et al. 2018) and perhaps consequently supply chain integration. Furthermore, used in combination with mobile devices or wearable computing, technologies like Augmented Reality, Virtual Reality or Mixed Reality can increase customers' understanding of the final product early in the design phase to avoid wasteful changes during project execution (Jones 2016).

One significant obstacle is a fact that the technical characteristics of Industry 4.0 are limited to technological development inside individual enterprises. In contrast to

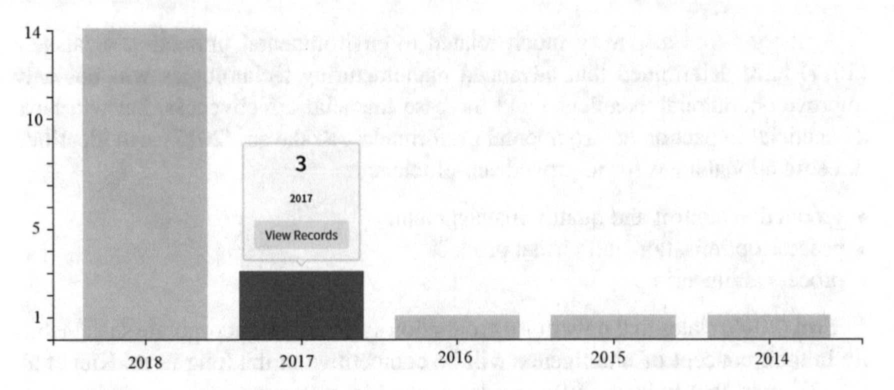

Fig. 2 Records for Industry 4.0 and SC integration combined research. *Source* Authors

this, supply chain integration deals with the supply chain as a whole, also including supply chain members as the supplier at the beginning of the chain and the end customer at the end of the chain. Our underlying assumption is that Industry 4.0 concepts may affect supply chain integration through transformational requirements regarding supply chain information flow.

Combined research of Industry 4.0 and supply chain integration topics is something new, with the first publication in 2014 (based on WoS, Fig. 2). In contrast, several authors have already demonstrated the overlap between Industry 4.0 and green concept/approach through digitisation. Implementation of the concept of Industry 4.0 increases greenness and sustainability.

3 Methodology

We answer the research question using the idea of a maturity model. Since the Capability Maturity Model (CMM) has been introduced, maturity as a measure to evaluate the capabilities of an organisation in regards to a certain sector has become popular (De Bruin et al. 2005a, b). From a theoretical perspective, the maturity model is appropriate. Such models differ in terms of their structure, complexity of the questions, and accessibility, in order to help companies and supply chains direct the transformation to Industry 4.0 and also to supply chain integration. Its basic purpose is to describe evolution stages and maturation paths. Quality models explicate characteristics for each stage and the logical relationship between successive stages. Carvalho et al. (2016) say that maturity models are based on the premise that people, organisations, functional areas, processes evolve through a process of development and growth towards a more advanced maturity stage.

We see the main obstacle in not yet knowing the final result of the transformation to Industry 4.0. That means that the existing Industry 4.0 and supply chain integration

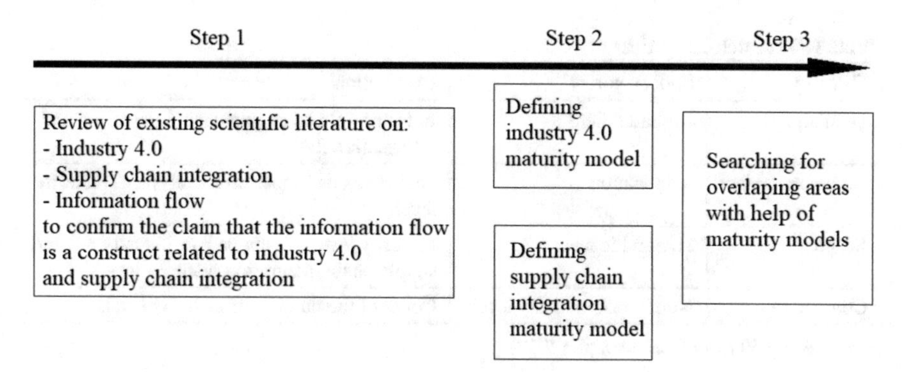

Fig. 3 Research model. *Source* Authors

maturity models are inaccurate and only describe the possible realisation of the vision, which has not yet been fully realised in the real business environment.

The three steps of the research methodology are illustrated in Fig. 3. A similar methodology was used in other research (e.g., Correia et al. 2017). Systematic literature reviews, which seek to provide answers to specific questions were used. Denyer and Tranfield (2009) propose five phases, but for the purpose of our research we combined the first and second phases (problem formulation and question identification; literature search) into Step 1, the third phase (evolution of research) is Step 2 and the fourth and five phases (research analysis and interpretation; presentation of results) are Step 3.

Review of existing scientific literature on (1) Industry 4.0, (2) Supply chain integration, and (3) Information flow was conducted in WoS by searching the topics 'Industry 4.0', 'Supply chain integration', 'Industry 4.0 AND Information flow', 'Supply chain integration AND Information flow', and 'Industry 4.0 AND Supply chain integration'. The combination of 'Supply chain integration AND Information flow' topics is present in 195 publications, the combination of 'Industry 4.0 AND Information flow' in 39 publications, and the combination of 'Industry 4.0 AND Supply chain integration' in 20 publications. However, none reveals the relationship between Industry 4.0 and supply chain integration.

For deduction, we used different categories (Table 1) based on the literature review analysis. Similar categories were also used in other research (e.g., Correia et al. 2017).

WoS database was used to search for Industry 4.0 and supply chain integration maturity models. We searched for overlapping areas with the deduction research method.

4 Maturity Models

The maturity model is not entirely new. The Software Engineering Institute launched the Capability Maturity Model (CMM) more than twenty years ago (Paulk et al.

Table 1 Research categories

Category	Subcategories	Description
Domain	Research field	Industry 4.0 and supply chain integration and information flow
Research subject	Application	Self-check or cooperative analysis of maturity level
Scope	Hierarchic level	Maturity levels on the path to Industry 4.0 and supply chain integration maturity levels
Components	Number of maturity levels	Count of the number of maturity levels

Source Adapted from Correia et al. (2017)

1993). Based on the assumption of predictable patterns of evolution and change, maturity models usually include a sequence of levels that together form an anticipated, desired, or logical path from an initial state to maturity (Pöppelbuß and Röglinger 2011). Typically, three application-specific purposes of maturity models usage are distinguished (Pöppelbuß and Röglinger 2011; Iversen et al. 1999; De Bruin et al. 2005a, b; Maier et al. 2009): descriptive, prescriptive, and comparative. A maturity model serves a descriptive purpose of use if it is applied for as-is assessments where the current capabilities of the entity under investigation are assessed with respect to given criteria (Becker et al. 2009). It serves a prescriptive purpose of use if it indicates how to identify desirable maturity levels and provides guidelines on improvement measures (Becker et al. 2009). Comparative purpose of use is achieved if it allows for internal or external benchmarking (Pöppelbuß and Röglinger 2011). From their beginning, maturity models have been subject to criticism (Pöppelbuß and Röglinger 2011):

- they have been characterised as 'step-by-step recipes' that oversimplify reality and lack empirical foundation;
- they tend to neglect the potential existence of multiple equally advantageous paths;
- they should be configurable because internal and external characteristics (e.g., the technology at hand, intellectual property, customer base, relationships with suppliers) may constrain a maturity model's applicability in its standardised version;
- they should not focus on a sequence of levels toward a predefined 'end state', but on factors driving evolution and change;
- we are witnessing a multitude of almost identical maturity models.

Below are the maturity models of Industry 4.0 and supply integration.

4.1 Industry 4.0 Maturity Model

Companies need to be motivated to participate in the digital transformation and in particular be supported in its implementation. Companies (and supply chains) are not immediately able to create a new smart factory, but they need to make targeted investments to move closer to Industry 4.0, in a step-by-step manner. Taking into account their current situation, companies must demonstrate the potential that they can achieve with Industry 4.0 measures. At the same time, measures must be derived to realise these potentials. Several Industry 4.0 maturity models have been developed to help business entities on this path. Kese and Tergtegen (2017) classified the majority of them, Table 2.

The most comprehensive approach to assessing maturity is represented by three models (Table 2) thematically oriented to the entire value chain: the Industry 4.0 Maturity Index, the '4i'—maturity model, and the Quickcheck Industrie 4.0 Maturity Model. Among them, most of the information was available on the Industry 4.0 Maturity Index, as a result of the Acatech study. Consequently, we took it as a model for finding intersections with the supply chain integration level.

The approach offers manufacturing companies practical guidance for developing an individual Industry 4.0 implementation strategy that is aligned with their business strategy. The idea behind a model is that implementation of new technologies, mostly ICT, must be accompanied by changes in organisation and culture as key aspects of implementation. The acatech Industrie 4.0 Maturity Index comprises a six-stage maturity model in which the attainment of each development stage delivers additional benefits (Schuh et al. 2017). It focuses on four key areas; each of them has two fundamental principles attached to it. The transformation process is a continuous journey of many successive steps that are taken incrementally and may not be perfectly synchronised across businesses, plants, lines, and cells.

The Acatech Industrie 4.0 Maturity Index assesses companies from a technological, organisational and cultural perspective, focusing on the business processes of manufacturing companies. Digitalisation does not itself form part of Industry 4.0. Computerisation and connectivity, as initial stages of maturity model, are basic requirements for Industry 4.0 implementation. They are followed by six further stages in which the capabilities required for Industry 4.0 are developed. A short description of stages (summarised from Schuh et al. 2017):

- Stage One: Computerisation: Different information technologies are used in isolation from each other within a company; the basis for digitalisation is provided. Computerisation is primarily used to efficiently perform repetitive tasks. Examples include CNC machines, computer-aided design (CAD), other business application systems that are not connected to the company's ERP system;
- Stage Two: Connectivity: The isolated deployment of information technology is replaced by connected components. Widely used business applications are all interconnected and mirror the company's core business processes. Events and states can be recorded in real-time in individual areas, such as manufacturing cells. Parts of the operational technology (OT) systems provide connectivity and interoperability, but full integration of the IT and OT layers has not yet occurred. Connectivity means that once a design has been created in engineering; its data can be pushed to production so that production steps can be executed accordingly. Suppliers can perform remote tasks at customers' premises. Examples include

Table 2 Classification of maturity models

Themes	Maturity models/providers				Application
General I4.0 specific aspects	**Industry 4.0—Checklist** BMWi	**Digital maturity—analysis tool** HNU, innosphere	**Digitisation index** Deutsche Telekom		Online self-check
Technological aspects in the foreground	**Industry 4.0 readiness model** IMPULS—foundation of the VDMA	**Industry 4.0 maturity test** Connected production	**Guide to Industry 4.0** IHK München and Oberbayern		Online self-check
Technological aspects in the foreground	**Toolbox industry 4.0** VDMA	**Industry 4.0 readiness** H&D International Group	**Maturity Model Industry 4.0** OÖ Business Agency GmbH	**Digital acceleration index** Boston Consulting Group	Cooperative analysis of maturity level
Oriented on the entire value chain	**Industry 4.0 Maturity Index** Acatech	**‘4i’ - Maturity Model** WZL der RWTH Aachen	**Quick check Industrie 4.0 Maturity Model** Competence center for SMEs (NRW)		Cooperative analysis of maturity level

Source Kese and Terstegen (2017)

Internet Protocol (IP), CAD/CAM processes, Manufacturing Execution System (MES).

- Stage Three: Visibility: Sensors, microchips and network technology enable processes to be captured from beginning to end. Events and states can be recorded in real-time throughout the entire company and beyond. It is possible to keep an up-to-date digital model of factories at all times. This model is also called a company's 'digital shadow' that helps to show what is happening in the company at any given moment so that management decisions can be based on real data. Problems include the lack of single source or no centralised database, very little data about production, logistics and services, more extensive use of data is prohibited by system boundaries, captured data is mostly only visible to a limited number of people. It is possible to more rapidly determine the delivery rates variance caused by a particular problem by means of real-time KPIs and dashboards. Suppliers and customers can be kept informed. The combination of existing data sources with sensors on the shop floor can deliver significant benefits, for example, integrating PLM, ERP, and MES systems provide a comprehensive picture that creates visibility regarding the status quo, modular approaches and apps can help to build a single source of truth.
- Stage Four: Transparency: The company attempts to understand why something is happening and uses this understanding to produce knowledge by means of root cause analyses. In order to identify and interpret interactions in the digital shadow, the captured data must be analysed by applying engineering knowledge. The semantic linking and aggregation of data to create information and the corresponding contextualisation provide the process knowledge required to support complex and rapid decision-making. New technologies that support the analysis of large volumes of data; for example, big data applications are deployed in parallel to business application systems such as ERP or MES systems. The common platform can be used to carry out extensive stochastic data analyses in order to reveal interactions in the company's digital shadow. Transparency is a requirement for predictive maintenance.
- Stage Five: Predictive capacity: The company is able to simulate different future scenarios and identify the most likely ones. This involves projecting the digital shadow into the future in order to depict a variety of scenarios that can then be assessed in terms of how likely they are to occur. For example, it is possible to flag up carrier failure before it even occurs and prevent that by changing carriers. Forecasting and recommendations are enabled.
- Stage Six: Adaptability: The goal of adaptability has been achieved when a company is able to use the data from the digital shadow to make decisions that have the best possible results in the shortest possible time and to implement the corresponding measures automatically, without human assistance.

The primary challenge for companies wishing to implement Industry 4.0 is to put these principles into practice by developing various capabilities. The goal is to generate knowledge from data in order to transform the company into a learning, agile organisation and enable rapid decision-making and adaptation processes throughout every part of the business and across all business process areas. Currently, decision-making processes can take weeks and decisions are often based on an intuitive feeling rather than hard data. Furthermore, the customer's needs are not comprehensively understood; the potential of new knowledge is partly lost through limited modifications to the business processes; the right information is sought and expected.

4.2 *Supply Chain Integration Maturity Model*

Throughout the discussions on stage-based supply chain integration, many contributory factors have been proposed by researchers to provide a typology on the subject (Aryee et al. 2008). Contributory factors to supply chain integration are differentiated into the 'hard' issues, such as technology, and the 'soft' issues, such as relations and collaborative strategies. Proposing that a fit must exist between the hard and soft issues in order to predict performance, Shah et al. (2002) put forward a model on supply chain integration. A similar model is suggested by Anderson and Lee (2000) but with more emphasis on the use of the internet and e-business to link customers and suppliers. Lockmay and McCormack (2004) presented an SCM maturity model based on a business process orientation concept. This model conceptualises how process maturity relates to the SCOR framework. Different levels are associated with process maturity. The model is structured in five levels. 'Ad hoc level' means when processes are unstructured and ill-defined. In the second level, basic processes are defined and documented. 'Linked level' means the breakthrough level when managers employ process management with strategic intent. In the integrated level, vendors and suppliers take cooperation to the process level. The highest level is extended when competition is based upon multiform networks (Lockmay and McCormack 2004).

In 2007, Aryee et al. proposed a new maturity model based on four previous models (Shah et al. 2002; Stevens 1989; Anderson and Lee 2000; Frohlich and Westbrook 2001). The model takes as its starting point a three-element-based supply chain structure of (1) business processes, (2) network structure and (3) management components (Lambert et al. 1998). These are then mapped onto three levels of process integration, which are defined as (1) optimisation, (2) integration and (3) synchronisation of processes. This perspective is taken further into the supply chain to embody customers and suppliers. While Aaryee and Naim's integration maturity model is extended to include ideas from all previous models, it has not proved to be directly applicable in our case. Despite being presented as a maturity model, the paper does not specify the maturity level for reuse in detail. In a recent study, Childerhouse et al. (2011) examine the uptake of SCI principles internationally and the resultant integration maturity, using Stevens' (1989) integration maturity model. This has given us an additional incentive to use it despite its earlier date of publication for our reasoning. It was supplemented with recent discoveries by the author of this very often cited model (Stevens and Johnson 2016).

Ho et al. (2016) developed the model based on CMMI, which has two parts including five maturity levels. These levels are:

- Initial: collaborative supply chain process is ad hoc and chaotic;
- Managed: collaboration process is planned and managed;
- Defined: supply chain collaboration process is well characterised, understood and described;
- Quantitatively managed: collaboration process is quantitatively managed in accordance with agreed-upon metrics;
- Optimising: collaboration process performance is continually improved through innovative technological improvement.

Correia et al. (2017) reviewed maturity models in supply chain sustainability. They determined that most of the reviewed papers have as their primary objective the development of maturity models, but only a few papers focus solely on the environmental dimension. They (ibid.) also found that sustainability brings new challenges and more complexity for the companies and their supply chain.

5 Results

For further consideration, Industry 4.0 Maturity Index and Stevens and Johnson's Supply chain integration maturity model were used. On Fig. 4, relations between levels of both maturity models are presented, namely in terms of information flow characteristics and the IT tools used. The most meaningful combinations of maturity levels are described hereinafter.

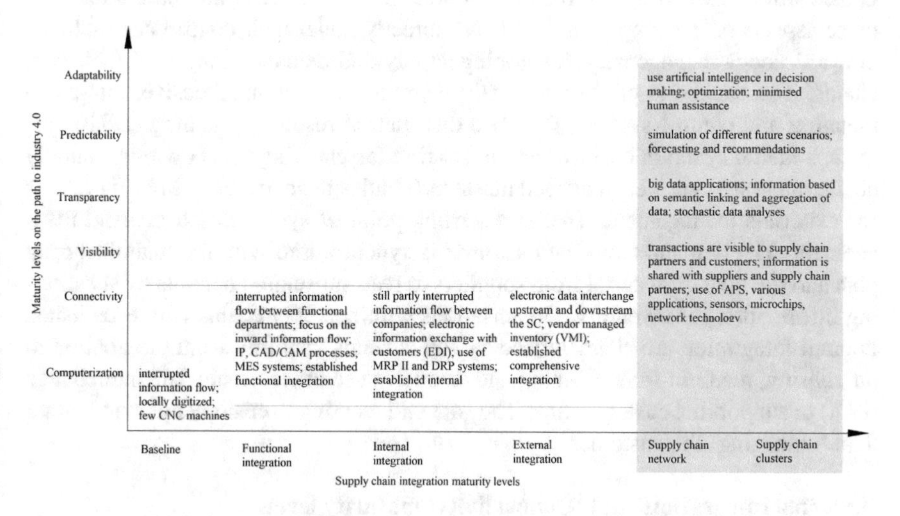

Fig. 4 Information flow characteristics and use of IT tools. *Source* Authors

'Baseline' and 'Computerisation' maturity levels

Several overlaps are observed between the descriptions of 'Baseline' and 'Computerisation' maturity levels. Organisationally viewed responsibility for different activities in the supply chain are vested in separate, almost independent, departments, which are equipped with different information technologies. Control systems and procedures covering sales, manufacturing, planning, material control, and purchasing are not integrated and unsynchronised. Manufacturing and production control would cover raw material through capacities and in-process inventories to finished goods. Further along the supply chain, sales and distribution divide the outward supply chain and inventories. Companies prepare mostly short-term plans if any. Operations are inefficient in individual companies as well as in the supply chain as a whole. Isolation of functions in a company can lead to inefficiency, and the consequences can spread even to the entire supply chain (Topolšek et al. 2010). The vulnerability of companies to the effects of changes in supply and demand patterns is significant.

'Functional integration' and 'Connectivity' maturity levels

Functional integration focuses principally on the inward flow of information and goods. Cost reduction is more important than performance improvement. Business functions are organised discretely. Customer service is reactive with poor visibility of real customer demand. Companies typically apply time-phased planning using MRP or MRPII techniques. The isolated deployment of information technology is replaced by connected components. The Internet Protocol is slowly becoming widely used. The process from product design to confirmation of completion of production can be fully IT supported (CAD/CAM, MES).

'Internal integration' and 'Connectivity' maturity levels

Organisations are mostly not focused on incoming flows. They are integrating only those aspects of the supply chain that are directly under their control and embrace outward goods management, integrating supply and demand along with their own chains. The backbone of their internal integration is a comprehensive, integrated planning and control system. They use distribution resource planning (DRP) systems, a method used in business administration for planning orders within a supply chain, integrated via well-managed master schedules to an MRPI or MRPII systems for materials management. That is a starting point of synchronised demand management. The demand from the customer is synchronised with the manufacturing plan and the flow of material from suppliers to reap substantial benefits by substituting information for inventory. Organisations within supply chains with established internal integration are characterised by full systems visibility from distribution to purchasing, medium-term planning and extensive use of electronic data interchange (EDI) to support the customer link. They are still reacting to customer demand rather than 'managing' the customer.

'External integration' and 'Connectivity' maturity levels

Full supply chain integration is achieved by extending the scope of integration outside the company to embrace suppliers and customers. In order to be attuned to the customer's needs and requirements, organisations are more or less customer-oriented. They are willing to understand the products, culture, market, and organisation. The manufacturer-suppliers relationship is progressing to mutual support and cooperation from the product development phase. Cooperation is based on long-term commitment.

'Supply chain network' and 'Visibility'
Supply chain networks are a result of the finding that supply chain is a network of relationships (Harland 1996) not a sequence (or chain) of transactions. They tend to increase the degree of maturity in terms of developing themselves in on-demand supply chains (IBM Institute for Business Value 2007). Customer order management is improved in a direction that order transactions and movements are visible to supply chain partners and customers, information is shared with suppliers and supply chain partners. This functionality is possible only with improved visibility based on the use of sensors, microchips, and network technology. Progression from 'connectivity' to 'visibility' is obligatory. Customer-driven manufacturing is enabled with the use of advanced planning and scheduling (APS) systems, automated product quality control, demand-pull manufacturing, real-time inventory control and seamless shift operations. Superior customer fulfilment is possible because of logistics excellence. Logistics establishes an integrated distribution network, enables visibility to the entire order-to-cash cycle; commitments are demand-driven with managed replenishment.

'Supply chain clusters' and 'Transparency'/'Predictability'/'Adaptability'
Stevens and Johnson (2016) consider that the centralised organisation structure and the underlying need for formality to support the central control of a Goal-Directed Networked Supply Chain gave rise to a rigid, inflexible structure unable to cope with the turbulent environment. Consequently, they are convinced that the future integrated supply chain model will be devoted, collaborative, supply chain clusters, based on a series of self-governing clusters comprising a network of suppliers and/or subcontractors associated by type, product structure, or flow. Planning in the described context takes place at a strategic level. Tactical and operational planning has already been established in previous levels of the integration level. Information, knowledge and insight are used better than in lower developed SC structures, mentioned before. Big data is a reality in the management of supply chain clusters. 'Visibility' itself is not enough for the efficacy of the network, and it must be stepwise upgraded to 'Transparency', 'Predictability' and 'Adaptability'. All these levels of Industry 4.0 maturity have been previously described.

6 Discussion and Conclusions

Industry 4.0 capabilities help manufacturing companies to dramatically reduce the time between an event occurring and the implementation of appropriate responses

(Schuh et al. 2017). In practice, this means that changes in customer requirements based on field data can be incorporated even during a product's manufacturing process because the company possesses the agility to adapt to the new situation. As a result, the customers can be supplied with a product tailored to their exact requirements in a significantly shorter period and in higher quality. The idea gained momentum with digitisation and digitalisation. Industry 4.0 is currently most frequently connected with a particular type of company: a production company. However, the core idea is quickly spreading to other sectors. It is now quite clear that in the near future it will be present throughout the supply chain. Additionally, Industry 4.0 mature companies are more green and sustainable. Several authors have already proved this link.

Supply chain integration has a much longer history than Industry 4.0 does. The real impetus to its development is also based on digitisation and digitalisation. The idea of its mature state is not clear yet. Perhaps this will be supply chain clusters. Maturation in the direction of Industry 4.0 is a trigger that will spark a leap to a higher level of supply chain maturity, whatever that will be. A maturing in the direction of Industry 4.0 and a maturing in the direction of supply chain integration are mutually creating the conditions for advancing on both maturity scales, but this progress is not linear.

A mature Industry 4.0 supply chain is also a mature integrated supply chain. Visibility, as the first level on the Industry 4.0 maturity scale, is required, when the supply chain identifies itself as a supply chain network and also acts one. In this way, an overview of transactions is in place, followed by confrontations with big data. The shareholders are sooner or later aware of the potential of the collected data of which they have an overview, the informative value of which is rather low. Application of analytical tools as a logical next step that will result in transparency and revealed interactions in the supply chain network. In this phase, the supply chain is mature enough to consider whether it is necessary to organisationally transform itself to supply chain clusters or maintain the existing organisation. Independent of the result of this, the supply chain can progress on the Industry 4.0 maturity scale to the levels of 'Predictability' and 'Adaptability' with the introduction of forecasting, decision algorithms, and a minimised need for human assistance in guiding the supply chain.

We can conclude that Industry 4.0 is an enabler of more mature supply chain integration and sustainable supply chains. Industry 4.0 accelerates the transition from supply chain networks to more advanced supply chain organisational structures or perhaps paves the way to the collaborative and sustainable supply chain clusters as a future state of supply chain integration.

Exploring the interactions between Industry 4.0, the green concept, supply chain integration and information flow has certainly become an interesting research area that cannot be avoided. More applications in practice will appear, and it will be easier to validate the links and effects between constituents. Until then, there is an opportunity for research based on simulations of the behaviour of complex environments of future supply chains.

References

Acatech-Plattform Industrie 4.0 (2014) Industrie 4.0—Whitepaper FuE Themen. http://www.acatech.de/fileadmin/user_upload/Baumstruktur_nach_Website/Acatech/root/de/Aktuelles___Presse/Presseinfos___News/ab_2014/Whitepaper_Industrie_4.0.pdf. Accessed 30 Oct 2018

Ackoff R (1989) From data to wisdom. J Appl Syst Anal 16:3–9

Al-Hakim L (2008) Modeling information flow for surgery management process. Int J Inf Qual 2(1):60–74

Anderson D, Lee H (2000) The internet-enabled supply chain: from first click to the last mile. Asset, 2

Aryee G, Naim MM, Lalwani C (2008) Supply chain integration using a maturity scale. J Manufact Technol Manage 19(5):559–575

Atani M, Kabore MP (2007) African index medicus: improving access to African health information. South Afr Family Pract 49(2):4–7

Barrat M (2004) Understanding the meaning of collaboration in the supply chain. Supply Chain Manage Int J 9(1):30–42

Becker J, Knackstedt R, Pöppelbuß J (2009) Developing maturity models for IT management—a procedure model and its application. Bus Inf Syst Eng (BISE) 1(3):213–222

Beynon-Davies P (2009) Business information systems. Palgrave MacMillan, Basingstoke

Bowersox DJ, Closs DC (1996) Logistical management: the integrated supply chain process, McGraw-Hill Series in marketing. The McGraw-Hill Companies, New York

Brettel M, Friederichsen N (2014) How virtualization, decentralization and network building change the manufacturing landscape: an Industry 4.0 perspective. Int J Mech Aerosp Ind Mechatron Manufact Eng 8(1), 37–44

Broy M (2010) Cyber-physical systems: innovation durch software intensive eingebettete Systeme. Deutsche Akademie der Technikwissenschaften

Bundesverband Informationswirtschaft, Telekommunikation und neue Medien e.V (2015) Industrie 4.0—Volkswirtschaftliches Potencial für Deutschland. Fraunhofer

Burstein MH, Diller DE (2004) A framework for dynamic information flow in mixed-initiative human/agent organizations. Appl Intell 20(3):283–298

Cachon G, Fisher M (2000) Supply chain inventory management and the value of shared information. Manage Sci 46(8):1032–1048

Capaldo A, Petruzzelli AM (2014) Partner geographic and organizational proximity and the innovative performance of knowledge-creating alliances. Eur Manage Rev 11(1):63–84

Carvalho JC, Rocha A, Abreu A (2016) Maturity models of healthcare information systems and technologies: a literature review. J Med Syst 40:131

Checkland P (1988) Information systems and system thinking: time to the unit? Int J Inf Manage 8(4):239–248

Chen H, Daugherty PJ, Landry TD (2009) Supply chain process integration: a theoretical framework. J Bus Logistics 30(2):27–46

Childerhouse P, Deakins E, Boehme T, Towill DR, Disney SM, Banomyong R (2011) Supply chain integration: an international comparison of maturity. Asia Pac J Mark Logistics 23(4):531–552

Chin TA, Tat HH, Sulaiman Z (2015) Green supply chain management, environmental collaboration and sustainability performance. Procedia 26:695–699

Chopra S, Meindl P (2001) Supply chain management: strategy planning and operation. Prentice Hall, New York

Columbos L (2015) Roundup of cloud computing forecasts and market estimates. https://www.forbes.com/sites/louiscolumbus/2015/01/24/roundup-of-cloud-computing-forecasts-and-market-estimates-2015/#3c34d3b7db7a. Accessed 15 Oct 2018

Cooper MC, Lambert DM, Pagh JD (1997) Supply chain management: more than a new name for logistics. Int J Logistics Manage 8(1):1–14

Correia E, Carvalho H, Azevedo S, Govindan K (2017) Maturity models in supply chain sustainability: a systematic literature review. Sustainability 9(64)

Dallasega P, Rauch E, Linder C (2018) Industry 4.0 as an enabler of proximity for construction supply chains: a systematic literature review. Comput Ind 99:202–225
De Bruin T, Rosemann M, Freeze R, Kulkarni U (2005a) Understanding the main phases of developing a maturity assessment model. Proceedings of the Australasian conference on information systems (ACIS). http://aisel.aisnet.org/cgi/viewcontent.cgi?article=1220&context=acis2005
De Bruin T, Rosemann M, Freeze R, Kulkarni U (2005b) Towards a business process management maturity model. In Proceedings of the thirteenth European conference on information systems, pp 26–28
De Olivera MP, Ladeira MB, McCormack KP (2011) The supply chain process. Management maturity model—SCPM3. IntechOpen, London
Denyer D, Tranfield D (2009) Producing a systematic review. In: The sage handbook of organizational research methods. London, pp 671–689
Dickmann P (ed) (2007) Schlanker Material fluss: mit Lean Production, Kanban und Innovationen. Springer, Berlin, Heidelberg
Dragan D, Keshavarzsaleh A, Jereb B, Topolšek D (2018) Integration with transport suppliers and the efficiency of travel agencies. Int J Value Chain Manage 9(2):122–148
Durugbo C, Tiwari A, Alcock JR (2013) Modeling information flow for organizations: a review of approaches and future challenges. Int J Inf Manage 33:597–610
Ellram LM, Cooper CM (1990) Supply chain management, partnerships, and shipper-third-party relationship. Int J Logistics Manage 1(2):1–10
Erlach K (2013) Value stream design: the way towards a lean factory. Springer, New York
Erol S (2016) Where is the green in industry 4.0? Or how information systems can play a role in creating intelligent and sustainable production systems of the future. Conference paper. Vienna
European Commission (2016) The fourth industrial revolution. https://ec.europa.eu/digital-single-market/en/fourth-industrial-revolution. Accessed 14 Oct 2018
Flynn B, Huo B, Zhao X (2009) The impact of supply chain integration on performance: a contingency and configuration approach. J Oper Manage 28:58–71
Frohlich MT, Westbrook R (2001) Arcs of integration: an international study of supply chain strategies. J Operat Manage 19:185–200
Gimenez C, van der Vaart T, Pieter van Donk D (2012) Supply chain integration and performance: the moderating effect of supply complexity. Int J Oper Prod Manage 32(5):583–610
Gimenez C, Ventura E (2005) Logistics-production, logistics-marketing and external integration: their impact on performance. Int J oper Prod Manage 25(1):20–38
Global Logistics Research Team at Michigan State University (1995) World class logistics: the challenge of managing change. Council of Logistics Management, Oak Brook
Hanfield R, Nichols EL (1999) Introduction to supply chain management. Pearson
Harland CM (1996) Supply chain management: relationships, chains and networks. Br J Manag 7(S1):S63–S80
Hearnshaw EJ, Wilson M (2013) A complex network approach to supply chain network theory. Int J Oper Prod Manage 33(4):442–469
Henczel S (2001) The information audit: A practical guide. In: Munich KG (ed) Saur: Information services management series
Herlyn WJ (2011) PPS in der Automobilindustrie: Produktionsprogrammplanung von Fahrzeugen und Aggregaten. Carl Hanser, München
Hinton CM (2002) Towards a pattern language for an information-centered business change. Int J Inf Manage 22(5):325–341
Ho D, Kumar A, Shiwakoti N (2016) Maturity model for supply chain collaboration: CMMI Approach. In: IEEE international conference of industrial engineering and engineering management
Houlihan JB (1988) International supply chains: a new approach. Manag Decis 26(3):13–19
IBM Institute for Business Value (2007) Follow the leaders—scoring high on the supply chain maturity model—Mainland China perspectives on supply chain fulfillment. IBM Global Services, Somers

Iversen J, Nielsen PA, Norbjerg J (1999) Situated assessment of problems in software development. Database Adv Inf Syst 30(2):66–81

Jones K (2016) Five ways the construction industry will benefit from augmented reality. https://www.linkedin.com/pulse/20140805163603-41493855-five-ways-the-construction-industry-will-benefit-from-augmented-reality/. Accessed 10 Oct 2018

Kembro J, Dag Näslund D, Olhager J (2017) Information sharing across multiple supply chain tiers: A Delphi study on antecedents. Int J Prod Econ 193:77–86. https://doi.org/10.1016/j.ijpe.2017.06.032

Kese D, Terstegen S Industrie 4.0—Reifegradmodelle (2017) Institut für angewandte Arbeitswissenschaft https://www.arbeitswissenschaft.net/uploads/tx_news/Tool_I40_Reifegradmodelle.pdf. Accessed 15 Sept 2018

Kiel D, Müller J, Arnold C, Voight K (2017) Sustainable industrial value creation: benefits and challenges of industry 4.0. Int J Innov Manage 21

Kim SW (2009) An investigation of the direct and indirect effect of supply chain integration on firm performance. Int J Prod Econ 119:328–346

Klitou D, Conrads J, Rasmussen M (2017) Germany: Industrie 4.0. Digital transformation monitor. http://ec.europa.eu/growth/tools-databases/dem/monitor/sites/default/files/DTM_Industrie%204.0.pdf. Accessed 7 Sept 2018

Kolberg D, Zühlke D (2015) Lean automation enabled by industry 4.0 technologies. IFAC-Papers OnLine 48(3):1870–1875. https://doi.org/10.1016/j.ifacol.2015.06.359

Kumar R, Chandrakar R (2012) Overview of green supply chain management: operation and environmental impact at different stages of the supply chain. Int J Eng Adv Technol 1(3)

Lambert DM, Cooper MC, Pagh JD (1998) Supply chain management: implementation issues and research opportunities. Int J Logistics Manage 9(2):1–20

Lasi H, Fettke P, Kemper HG, Feld T, Hoffmann M (2014) Industry 4.0. Bus Inf Eng 6(4):239–242. https://doi.org/10.1007/s12599-014-0334-4

Lee E (2008) Cyber-physical systems: design challenges. In: International symposium on object/component/service-oriented real-time distributed computing

Lee HL, Kut C, Tang C (2000) The value of information sharing in a two-level supply chain. Manage Sci 46(5):626–643

Lee KH, Kim JW (2011) Integrating suppliers into green product innovation development: an empirical case study in the semiconductor industry. Bus Strategy Environ 20(8):527–538

Lockmay A, McCormack K (2004) The development of a supply chain management process maturity model using the concept of business orientation. Supply Chain Manage Int J 9(4):272–278

Luthra S, Mangla SK (2018) Evaluating challenges to Industry 4.0 initiatives for supply chain sustainability in emerging economies. Process Saf Environ Prot 117:168–179. https://doi.org/10.1016/j.psep.2018.04.018

Madenas N, Tiwari A, Turener CJ, Woodward J (2014) Information flow in supply chain management: a review across the product lifecycle. J Manufact Sci Technol 7:335–346

Maier AM, Moultrie J, Clarkson PJ (2009) Developing maturity grids for assessing organisational capabilities: practitioner guidance. In: Proceedings of the 4th international conference on management consulting, Vienna, Austria

Mason-Jones R, Towill DR (1997) Information enrichment: designing the supply chain for competitive advantage. Supply Chain Manage 2(4):137–148

Mentzas G, Halaris C, Kavadias S (2001) Modeling business process with workflow systems: an evolution of alternative approaches. Int J Inf Manage 21(2):123–135

Mentzer JT, DeWitt W, Keebler JS, Min S, Nix NW, Smith CD, Zacharia ZG (2001) Defining supply chain management. J Bus Logistics 22(2):1–25

Ōno T (1988) Toyota production system: beyond large scale production. Productivity Press, Cambridge

Paulk M, Curtis B, Chrissis M, Weber C (1993) Capability maturity model for software, Version 1.1. IEEE Softw 10(4):18–27. https://pdfs.semanticscholar.org/85f1/ef3df4f12b8d50663f96f2f41cfde99423a2.pdf

Poirier C, Quinn F (2003) A survey of supply chain progress. Supply Chain Manage Rev 7:40–74
Pöppelbuß J, Röglinger M (2011) What makes a useful maturity model? A framework of general design principles for maturity models and its demonstration in business process management. ECIS Proceedings. http://aisel.aisnet.org/ecis2011/28
Prajogo D, Olhager J (2012) Supply chain integration and performance: the effects of long-term relationships, information technology and sharing, and logistics integration. Int J Prod Econ 135(1):514–522
Premkumar G, William R (1994) Organizational characteristics and information systems planning an empirical study. Inf Syst Res 5(2):75–119
Sahin F, Robinson EP (2002) Flow coordination and information sharing supply chains: review, implications and directions for future research. Decis Sci 33(4):505–536
Schuh G, Anderl R, Gausemeier J, ten Hompel M, Wahlster W (2017) Industrie 4.0 maturity index. Managing the digital transformation of companies (acatech STUDY). Herbert Utz Verlag, Munich
SCOR—Supply Chain Operations Reference Model. Revision 11.0 (2012) Supply Chain Council
Shah R, Goldstein SN, Ward PT (2002) Aligning supply chain management characteristics and inter-organizational information system types: an exploratory study. IEEE Trans Eng Manage 49(3):282–292
Spacey J (2017) 8 types of information flow. Simplicable. https://simplicable.com/new/information-flow. Accessed 10 Oct 2018
Sridhar K, Jones G (2013) The three fundamental criticism of the triple bottom line approach. An empirical study to link sustainability reports in companies based in the Asia-Pacific region and TBL shortcomings. Asian J Bus Ethics 2:91–111
Stevens GC (1989) Integrating the Supply Chains. Int J Seven Phys Distrib Mater Manage 8(8):3–8
Stevens GC, Johnson M (2016) Integrating the supply chain …25 years on. Int J Phys Logistics Manage 46(1):19–42
Szalavetz A (2017) The environmental impact of advanced manufacturing technologies: examples from Hungary. Central Eur Bus Rev 6(2):18–29
Topolšek D, Čižman A, Lipičnik M (2010) Collaborative behaviour as a facilitator of integration of logistic and marketing functions—the case of Slovene retailers. Promet Traffic Transp 22(5):353–362
Topolšek D (2011) The impact of collaboration or collaborative behavior on the level of internal integration: case study of Slovenian retailers and motor vehicle repair companies. Afr J Bus Manage 5(26):10345–10354
Trkman P (2000) Business success and informatization. Faculty of Economics, Ljubljana
Trkman P, Indihar Štemberger M, Jaklič J, Groznik A (2007) Process approach to supply chain integration. Supply Chain Manage Int J 12(2):116–128
Tyndall G, Gopal C, Partsch W, Kamauff J (1998) Supercharging supply chains: new ways to increase value through global operational excellence. Wiley, New York
Vickery SK, Jayaram J, Droge C, Calantone R (2002) The effect of an integrative supply chain strategy on customer service and financial performance: an analysis of direct versus indirect relationships. J Oper Manage 21:523–539
Wagner BA, Fillis I, Johansson U (2003) E-business and e-supply strategy in small and medium-sized businesses (SMEs). Supply Chain Manage Int J 8(4):343–354
Westrum R (2004) A typology of organisational cultures. Qual Saf Health Care 13(2):22–27
Wintraecken JJVR (1990) The NIAM information analysis method: theory and practice. Springer, Netherlands
Womack JP, Jones DT, Roos D (2007) The machine that changed the world: the story of lean production—Toyota's secret weapon in the global car wars that is revolutionizing world industry. Free Press, New York
Wu GC (2013) The influence of green supply chain integration and environmental uncertainty on green innovation in Taiwan's IT industry. Supply Chain Manage 18(5):539–552
Yazici HJ (2002) The role of communication in organizational change: an empirical investigation. Inf Manage 39(7):539–552

Yousefi N, Alibabaei A (2015) Information flow in the pharmaceutical supply chain. Iran J Pharm Res 14(4):1299–1303
Zhou H, Benton W (2007) Supply chain practice and information sharing. J Oper Manage 25(6):1348–1365

Digital Transport Management in Manufacturing Companies Based on Logistics 4.0 Concept

Piotr Cyplik and Marcin Hajdul

Abstract Manufacturing companies operating in turbulent environments require constant adaptation to changing conditions. Current expectations of customers are not only to buy product of acceptable quality, but also comprehensive logistics service. European Union research indicate that more than 40% of the total logistics costs are the costs of transport. Therefore, transport is susceptible to innovations and improvements. Authors proposed implementation of selected elements of logistics 4.0 concept for transport management as an integrator of supply chain. The article used proprietary methodologies for assessing the transport system, through which diagnosed root causes of problems of the traditional approach to transport management. Later in the research authors analysed the concept of logistics 4.0 in the context of the possibility of offset diagnosed root causes. As a result reference model for continuous improvement of the transport management system based on the concept of logistics 4.0 in a production company were created.

Keywords Logistics 4.0 · Internet of things · Transport management systems · Manufacturing companies · Supply chain integration

1 Introduction

1.1 Situation of the Manufacturing Companies in the European Union

Manufacturing companies are operating in large national and international networks in order to enlarge market reach. The challenge lies in the smooth management of data, proper execution of trade and transport with its complexities of distance and

P. Cyplik (✉) · M. Hajdul
Poznan School of Logistics, Estkowskiego 6, 61-755 Poznan, Poland
e-mail: piotr.cyplik@wsl.com.pl

M. Hajdul
e-mail: marcin.hajdul@wsl.com.pl

A. Kolinski et al. (eds.), *Integration of Information Flow for Greening Supply Chain Management*, Environmental Issues in Logistics and Manufacturing, https://doi.org/10.1007/978-3-030-24355-5_17

time, language and cultural barriers, and its dependency of national and international rules and regulations. This turbulent environment requires constant adaptation to changing network conditions. Moreover, current expectations of customers are not only to buy product of acceptable quality, but also comprehensive logistics service (Benoit et al. 2012; Moneimne et al. 2016).

For manufacturing companies reduction of logistics costs and in particular transport ones is the most important. EU research conducted by the European Logistics Association Alliance for Logistics Innovation through Collaboration in Europe (ALICE) in 2015–2016 indicates that more than 40% of the total logistics costs are the costs of transport. Therefore, transport is susceptible to innovations and improvements (Zijm and Klumpp 2016a, b).

According to ALICE, one of the major challenges for the companies is the lack of fast and safe real-time access to structured information and low utilization of solutions allowing communication between a forwarder, any subcontracts, customers, and physical resources. It stems from the fact that both the applicant and their partners (clients and subcontractors) operate in scattered systems (http://www.etp-logistics.eu/) (Mes and Iacob 2016).

Most of the Enterprise Resource Planning (ERP) and Transport Management (TMS) systems focus on a single transport modality (e.g. solutions like Transporeon or Transwide which are operating as a road transport marketplace). As a result, all actors from supply chain are not connected, and their transport processes are not synchronized. As a result, current efficiency and integration is minimal. This can be confirmed by statistics. Average load factor for road transport is 33%, empty runs for road transport (GUS 2016).

This leads to reducing efficiency, effectiveness, and safety of services rendered to clients by the applicant. Profitability of services offered by the applicant is also reduced, which results in lower competitiveness in both the national and international market (Redmer 2016).

1.2 Major Transport Problems of Manufacturing Companies

Authors made a panel study where major barriers in transport and logistics sector were defined. In total 45 senior logistics staff took part in the panel. Based on the panel, a list of barriers limiting the company development in terms of transport and logistics was developed. The major problem in organising transport processes indicated by the manufacturing companies participated in the survey is the lack of fast and safe real-time access to structured information and low utilization of solutions allowing communication between a forwarder, subcontracts, customers, and physical resources. It stems from the fact that both service providers and customers operate in scattered systems. The effect is that the same data must be entered manually in several or even a dozen or so systems. Additionally, the logistics resource and staff management itself is ineffective, as the respondent companies are using several non-integrated IT systems.

Organising transport, the staff of the companies are forced to use traditional e-mail and phone communication with customers, marine container terminals, agents, railroad carriers, road carriers, and administration representatives, which is not help in integration of supply chain (Montreuil et al. 2013). This leads to reducing efficiency, effectiveness, and safety of services rendered to customers by the applicant. Profitability of services offered by the companies is also reduced, which results in lower competitiveness in international markets (Nettsträter et al. 2015; Zarli et al. 2016).

Therefore, during the panel, the following barriers for the manufacturing companies in terms of transport and forwarding processes organisation were identified:

- lack of real-time access to information on the status of services rendered, which reduces the profitability of the service;
- work in a scattered IT environment resulting in the lack of access to analytical data regarding the use of the fleet of subcontractors (carriers acquired from the freight exchange);
- low efficiency and effectiveness of processing a customer order caused by the necessary manual acquisition of data on the current status of an order;
- the lack of electronic and structured communication with a customer and automatic notification on the status of an order to the customer;
- communication with a driver to provide them with information on current tasks—only via telephone, non-structured, and unarchived;
- the lack of automatic information sent by a vehicle regarding the location, the status of an order, the distance to a destination, estimated time of delivery—the concept of the Internet of Things (IoT) not used in practice;
- no system allowing the improvement of the safety of services rendered through, for example, identification whether a vehicle with a load is performing an order in accordance with the requirements or whether it was stolen or it changed its route—the concept of the IoT not used in practice,
- the lack of a solution allowing precise estimation of the time of goods delivery to a customer, considering the current road traffic—the concept of the IoT not used in practice.

2 Transport Problem Identification Methodology

2.1 *Classification of the Problem Identification Methods/Tools and the Analysis of the Causes of Their Occurrence*

In literature, one can find many different methods and tools for identifying problems and analysing the causes of them. The most commonly used are: Ishikawa diagram, Current Reality Tree, Conflict diagram, Why-Why diagram, Case Study, 8D's problem solving, ABCD (Suzuki method), Six Thinking Hats, Scenario analysis, Brainstorming.

Each of methods/tools assumes a course of action methods, some common features may be noticed. For the purpose of the classification of the identified methods/tools, seven analysis criteria have been distinguished, which include:

(1) The need for establishing a team of experts;
(2) The need for appointing a facilitator;
(3) Type of inferring;
(4) Subject range of the method;
(5) The method's formalization degree;
(6) The scalability of the problem's determining factors;
(7) Visual presentation of the results.

Within the first criterion, we can distinguish methods which, by definition, require establishing a team of experts. They include: the 8D's problem solving method, the ABCD method (the Suzuki method), the Six Thinking Hats method, and brainstorming. As for the Ishikawa diagram, the Current Reality Tree analysis, the Conflict diagram, the Why-Why diagram, and Scenario Planning establishing a team of experts is recommended but not necessary from the viewpoint of the characteristic of these methods. A case study does not require establishing a team of experts.

It is compulsory to appoint a facilitator (the second criterion) with the following methods: the 8D's problem solving method, the Six Thinking Hats method, and brainstorming. As for the ABCD method, the presence of a facilitator is recommended only, and the other methods do not require one at all.

The analysed methods differ from one another in different ways of inferring. Deduction is used in the case of the Ishikawa diagram, the Conflict diagram and the Why-Why diagram. Induction is used in the Current Reality Tree method, case studies, the ABCD method, the Six Thinking Hats method and the Scenario Planning. Brainstorming and the 8D's method both use induction and deduction.

The identified methods also differ in their scope of application (subject range). The Ishikawa diagram, the Current Reality Tree method, the Why-Why diagram and the ABCD diagram (the Suzuki method) identify primal and secondary problems and the cause of those problems. The other methods additionally have structured mechanisms of working out solutions to the identified problems and their root causes.

From the viewpoint of the methods' formalization degree (criterion 5), we can distinguish those which have:

- strictly formalized methodology of conduct,
- formalized framework conduct (only conduct stages are determined),
- non-formalized character.

The first group includes: the Ishikawa diagram, the Current Reality Tree method, the 8D's problem solving method, the ABCD method, and the Six Thinking Hats method. The methods, which are characterised by formalised framework conduct include: the Why-Why diagram (a variation of the Why-Why diagram, the 5 Why's method belongs to the group of strictly formalised methods), the Scenario Planning and brainstorming (the 635 Brainwriting technique and Phillips 66 Buzz Session are parts of the strictly formalised methods). The case study method is non-formalised. It leaves plenty of room or freedom to managers.

An important criterion in the classification of problems identification methods and the analysis of the causes of their occurrence is the scalability of the identified problems. In problem solving methods—the 8D's and ABCD, problems need to be classified in accordance with the adopted problem validity scale. With the Ishikawa diagram, it is recommended to scale problems. In the other methods, this criterion is not marked as compulsory (Cyplik and Hadas 2011).

The final distinguished criterion is the visualisation of the received results. Visualisation in the form of a diagram or a chart is obligatory with the Ishikawa diagram, the Current Reality Tree method, the Conflict diagram, and the Why-Why diagram. When summarising the 8D's problem solving method, it is possible to show the results in a visual form (depending on the selection of the problem solving tools), the ABCD (the Suzuki method) visualises results in the form of a table. The other methods, in their authors' opinion, do not require a visual presentation of the results. A summary of the above consideration is comprised in Table 1.

2.2 *The Transport System Virus Analysis*

The transport system virus analysis is a method to identify and analyse the reasons for various problems. The final result of this method is to elaborate a transport system problem virus. To illustrate the problems, one applied the medicine-based logic. The virus, which affects healthy tissues (system elements), makes them die or be transformed into hybrids. In turn, the hybrids do not achieve the system aims. "Sick" tissues cause disturbances in the system functioning. The key to eliminate the system virus (including the reasons for source problems) is its unambiguous identification. The elimination process is performed by carrying out dedicated improvement actions or, at least, by limiting the virus action range. This results in an improvement of the system functioning method. The methodology of the transport system virus analysis provides that the following stages and 10 consecutive measures will be executed (Cyplik and Hadas 2015).

STAGE I. Diagnosis of the transport system problems

1. Determining the system objective.
2. Designating an expert group.
3. Problem identification.
4. Statistical analysis of the identified problem.
5. Present state analysis (AS-IS).
6. Elaborating the transport system virus.

Table 1 Classification of the identified qualitative methods and problem identification tools and the analysis of their causes

Methodology/classification criteria	Requires establishing a team of experts	Requires appointing a facilitator	Type of inferring (induction vs. deduction)	Span (subject range) of a method	Method's formalisation degree	Scalability of problem determining factors	Visual presentation of results
Ishikawa diagram	+	–	Deduction	A, B	X	+	++ (fishbone diagram)
Current Reality Tree	+	–	Induction	A, B	X	–	++ (diagram)
Conflict diagram	+	–	Deduction	A, B, C	Y	–	++ (diagram)
Why-Why diagram	+	–	Deduction	A, B	Y (X)	–	++ (diagram)
Case Study	–	–	Induction	A, B, C	Z	–	–
8D's problem solving	++	++	Induction/deduction	A, B, C	X	++	(+) (selected tools)
ABCD (Suzuki method)	++	+	Induction	A, B	X	++	+ (table)
Six Thinking Hats	++	++	Induction	A, B, C	X	–	–
Scenario analysis	+	–	Induction	A, B, C	Y	–	–
Brainstorming	++	++	Induction/deduction	A, B, C	Y (X)	–	–

where
++ conduct required by methodology
+ possible/recommended, but not compulsory
– not required
A the method identifies primary and secondary problems
B the method identifies problem causes
C the method generates ideas and solutions
X strictly formalised methodology
Y framework formalisation—method's stages determined only
Z non-formalised

STAGE II. Implementing corrective actions

7. Corrective actions (solutions to the problems) selection and classification.
8. Elaborating a matrix to fix the identified problems with the suggested corrective actions and recommending final corrective actions.
9. Implementing the suggested corrective actions.
10. Implementation result assessment.

The transport system virus analysis is one of the methods which require appointment of an expert team and a moderator. This is motivated by the criteria of classifying the problem identification method and the analysis of their causes. The transport system virus analysis is characterised by both inductive (stage I) and deductive (stage II) ways of reasoning. This method identifies and graduates initial and secondary problems, their source causes, and generates new ideas and solutions. This method has a very precisely defined proceeding methodology and accurately visualised form of the transport system assessment (see Table 2).

2.3 Transport Virus of Manufacturing Company Operating in Turbulent Environment

The transport system virus analysis was used for identifying and analysing problems in manufacturing company from FMCG (Fast Moving Consumer Goods) sector during ECR Forum 2017 in Warsaw. Manufacturing company which was involved in the identification process is in many aspects a typical representative of companies operating on the market. Booth own means of transport as well as external fleet are in used. The main objective of the transport department within this company is to organize effectively and efficiently deliveries. As a result key transport problems of the company was identified. Elimination of identified problems was carried out in the next step. Such a measure was intended to improve the financial condition of the company by better utilization of available assets (human and physical). Figure 1 presents results of the diagnosis of the transport system in evaluated company.

3 Logistics Paradigms and Its Implication for Transport Management in Supply Chain Integration

3.1 Logistics 4.0 Concept

Rapid development of new technologies has escorted industry development from the early adoption of mechanical systems, to support production processes, to today's Industry 4.0 concept where machines are connected in a collaborative community

Table 2 The criteria fulfilment grade of the identified qualitative methods and problem identification tools and the analysis of their causes by the transport system analysis

Classification method/criteria	Requires an expert team	Requires a moderator	Reasoning type (induction/deduction)	Span (method material scope)	Method formalisation grade	Gradation of the factors which determine the problem	Result visual presentation
The transport system virus analysis	++	++	induction/deduction	A, B, C	X	++	++

where
++ required by the proceeding methodology
+ possible/recommended but not compulsory
– not required
A the method identifies both initial and secondary problems
B the method identifies the reasons for problems
C the method generates ideas and solutions
X precisely formalised methodology
Y frame formalisation—only particular stages determined
Z non-formalised

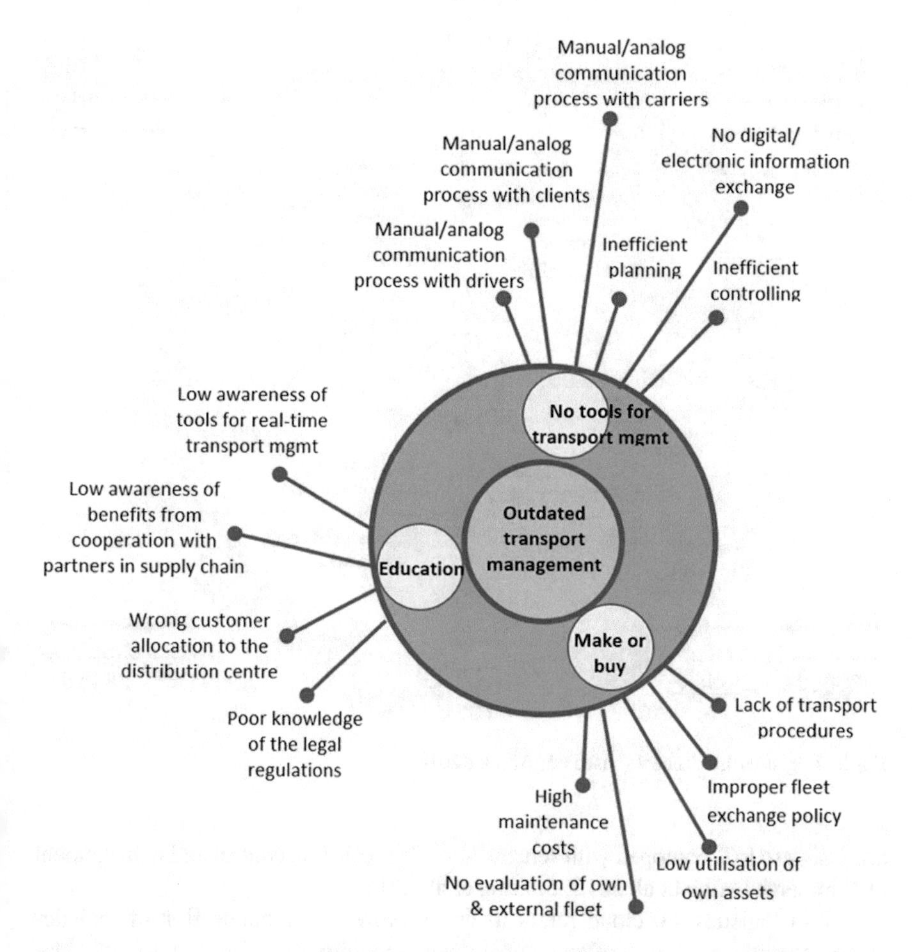

Fig. 1 The transport system virus of analysed manufacturing company

(Lee et al. 2014). Based on that researchers from Fraunhofer Institute defined assumptions of Logistics 4.0 concept. One of the pillars of Logistics 4.0 is the IoT, that allows connection of all logistics assets into one collaborative ecosystem (Fig. 2).

3.2 *Vision of Logistics 4.0 Cloud*

Based on the Logistics 4.0 concept authors create assumption of Logistics 4.0 reference model for manufacturing companies. In that model all partners in the supply chain are connected to Logistics 4.0 cloud where they are able to communicate in standardized digital way. Moreover, all assets used for transport execution are,

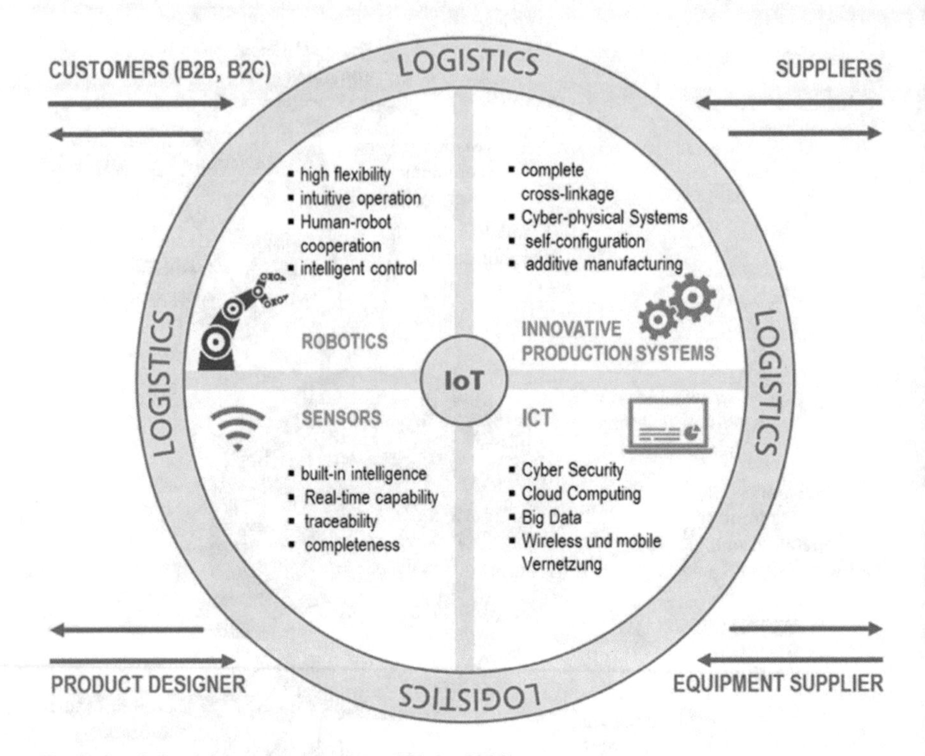

Fig. 2 Logistics 4.0 concept (based on: Akinlar 2014)

according to IoT, equipped with sensors allowing real-time control and management of shipments (Ming et al. 2012; Sharma et al. 2015).

IoT in logistics 4.0 cloud refers to the networked interconnection of logistics assets (objects), which are often equipped with ubiquitous intelligence. IoT increase the ubiquity of the Internet by integrating every logistics asset for interaction via embedded systems, which leads to a highly distributed network of devices communicating with human beings as well as other devices (Gubbi et al. 2013; Xia et al. 2012).

Finally, truck drivers are using mobile version of the Logistics 4.0 cloud which are available on smartphones and can be download from GooglePlay shop or AppStore. Figure 3 presents vision of Logistics 4.0 cloud.

As a whole Logistics 4.0 cloud is a comprehensive service aimed at integrating and improving data exchange and communication between partners in a supply chain and means of transport, which is made possible with mobile technologies based on the IoT concept. The service is to automate the process of cooperation with transport services providers, communication with drivers, as well as downloading information on the temperature or other cargo parameters in the loading space of a vehicle (with any frequency).

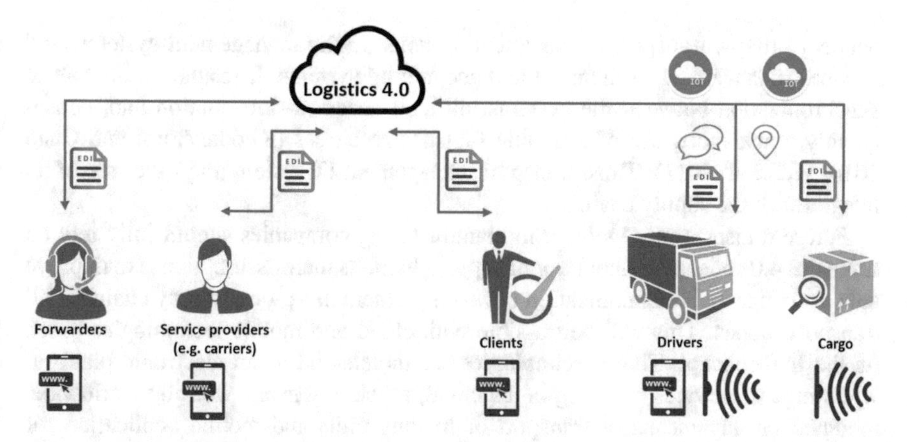

Fig. 3 Reference model of Logistics 4.0 cloud for manufacturing companies

As a result of implementing Logistics 4.0 concept manufacturing companies will be able to offer their customers and subcontractors a new quality of cooperation based mainly on electronic data exchange and management of all the processes and physical resources in real time, 24 h/day. The customers of companies will be able to track the status of their order from its placement, through physical transport, to delivery. The manufacturing company can also increase its competitiveness by offering to its national and international partners access to a cloud system. Subcontractors (e.g. carriers from the freight exchange) will have 24/7 access to the orders from manufacturing company. Customers will also be able to place orders 24/7. A system working in the cloud can also considerably facilitate the cooperation and exchange of data with contractors. Thus, the applicant will gain advantage over the competition which cooperates with customers and subcontractors in a traditional way (e.g. telephone, e-mail, no system implemented).

A substantial innovation is the automatization of the exchange of information between the manufacturing company and the drivers performing transport orders. The communication platform and mobile application available for each driver allows fully electronic communication with a driver, automatic sending of the list of tasks to be completed by a driver, electronic monitoring of the progress of deliveries, managing the statuses of deliveries, sending electronic copies of a customer's goods receipt acknowledgment.

4 Conclusions and Future Research

Current approach to transport management is not efficient anymore. This confirms transport virus of manufacturing companies identified by the authors. Moreover, current transport management IT systems are being developed under three independent groups: transport management systems, traffic management systems, and services management + financial and accounting systems. It results in the lack of synchronization between the systems, difficult access to information and, consequently, unbalanced use of available logistics resources (Khodakarami and Chan 2014; Wu et al. 2011). Thus, transport management IT system plays the role of an integrator of the supply chain.

Future transport management in manufacturing companies should fully rely on Logistics 4.0 concept as integrator of supply chain. Its main assumption is to improve data exchange and communication between partners in a global supply chain for all transport modes. This will be possible with cloud and mobile technologies based on the IoT concept. Those technologies are understood as an electronic platform with range of services, working in the cloud, wireless sensors, working worldwide, mounted on any means of transport or loading units and mobile application for drivers executing first/last mile delivery. The main innovation underlying the solution involves providing standardised electronic and automatic communication between a manufacturing companies, its customers, and any of its subcontractors (e.g. carriers acquired from the freight exchange) and objects (means of transport, loading units), which leads to improved safety and quality of deliveries and sustainable use of the available transport recourses, as well as improved effectiveness of transport and logistics services rendered by the applicant both in the country and abroad.

Cloud platform for transport management in manufacturing companies should meet the following requirements:

- information systems to enable/enhance global and multimodal freight transport and logistics planning and management integration with traffic management systems;
- multimodal journey planners for freight;
- tracking and tracing tools;
- intelligent cargo applications;
- tools for the implementation of single transport documents;
- IT infrastructures supporting information exchange and user authorisation/authentication: e-Freight;
- connectivity infrastructure;
- support for interoperability between different standards.

Authors conclude that it is necessary to develop a detailed reference model of cloud solution allowing integration of the IT systems and their further synchronized development so that, during planning, the user has real-time access to the data on road traffic, various services available, and transport resources.

Such a solution should provide a range of benefits for the users, such as:

- higher availability of goods thanks to more effective organization of transports;
- transport cost reduction;
- higher customer satisfaction;
- better use of loading space of means of transport;
- more intermodal transports;
- reduction of empty runs;

- reduced transport time;
- reduced time spent at loading/container terminals;
- lower CO_2 emission;
- supply chain integration.

Acknowledgements This paper has been the result of the study conducted within the grant by the Ministry of Science and Higher Education entitled “Development of production and logistics systems” (project No. KSL 2/17) pursued at the Poznan School of Logistics.

References

Benoit M, Rougès JF, Cimon Y, Poulin D, (2012 June) The physical internet and business model innovation. Technol Innov Manag Rev, pp 32–37

Cyplik P, Hadas L (2011) Production system virus analysis tool (PSVA)—problems identification and analysis framework—case study. LogForum 7(1):1–14

Cyplik P, Hadas L (2015) Transformation of a production-logistics system in the enterprises of broad assortment offer and a varied customer service strategy. In: Premises, methodology, evaluation. Polish Scientific Publishers, Warsaw

Gubbi J, Buyya R, Marusic S, Palaniswami M (2013) Internet of Things (IoT): a vision, architectural elements, and future directions. Future Gener Comput Syst 29(7):1645–1660. https://doi.org/10.1016/j.future.2013.01.010

GUS, Transport—wyniki działalności w 2016 r. http://stat.gov.pl/obszary-tematyczne/transport-i-lacznosc/transport/transport-wyniki-dzialalnosci-w-2016-r-,9,16.html (access: 08.09.2018)

Khodakarami F, Chan YE (2014) Exploring the role of customer relationship management (CRM) systems in customer knowledge creation. Inf Manag 51(1):27–42

Lee J, Kao HA, Yang S (2014) Service innovation and smart analytics for industry 4.0 and big data environment. Procedia CIRP 16:3–8. https://doi.org/10.1016/j.procir.2014.02.001

Mes MRK, Iacob ME (2016) Synchromodal transport planning at a logistics service provider. In: Zijm H, Klumpp M, Clausen U, Ten Hompel M (eds) Logistics and supply chain innovation: bridging the gap between theory and practice. Lecture notes in logistics. Springer International Publishing, pp 23–36. https://doi.org/10.1007/978-3-319-22288-2

Ming F, Zhu X, Torres M, Anaya L, Patanapongpibul L (2012) GSM/GPRS bearers efficiency analysis for machine type communications. In: IEEE 75th vehicular technology conference (VTC Spring), pp 1–5. https://doi.org/10.1109/vetecs.2012.6240122

Moneimne W, Hajdul M, Mikołajczak S (2016) Seamless communication in supply chains based on M2M technology. Logforum 12(4):3. https://doi.org/10.17270/J.LOG.2016.4.3

Montreuil B, Meller RD, Ballot E (2013) Physical internet foundations. In: Service orientation in holonic and multi agent manufacturing and robotics. Volume 472 of the series Studies in computational intelligence, pp 151–166, https://doi.org/10.1007/978-3-642-35852-4_10

Nettsträter A, Geißen T, Witthaut MED, Schoneboom J (2015) Logistics software systems and functions: an overview of ERP, WMS, TMS and SCM systems. In: Ten Hompel M, Rehof J, Wolf O (eds) Cloud computing for logistics. Lecture notes in logistics. Springer International Publishing, pp 1–11. https://doi.org/10.1007/978-3-319-13404-8_1

Redmer A (2016) Strategic vehicle fleet management—the replacement problem. Logforum 12(1):2. https://doi.org/10.17270/J.LOG.2016.1.2

Sharma A, Kansal V, Tomar RPS (2015) Location based services in M-commerce: customer trust and transaction security issues. Int J Comput Sci Secur (IJCSS) 9(2):11–21

Wu G, Talwar S, Johnsson K, Himayat N, Johnson KD (2011) M2M: from mobile to embedded internet. IEEE Commun Mag 49(4):36–43. https://doi.org/10.1109/MCOM.2011.5741144

Xia F, Yang LT, Wang L, Vinel A (2012) 2012, Internet of things. Int J Commun Syst 25:1101–1102. https://doi.org/10.1002/dac.2417
Zarli A, Bourdeau L, Segarra M (2016) REFINET: REthinking Future Infrastructure NETworks. Transp Res Procedia 14:448–456. https://doi.org/10.1016/j.trpro.2016.05.097
Zijm H, Klumpp M (2016a) Future logistics: what to expect, how to adapt. In: Freitag M, Kotzab H, Pannek J (eds) Dynamics in logistics: proceedings of the 5th international conference LDIC, 2016 Bremen, Germany. Lecture notes in logistics, 2017. Springer, Bremen, pp 365–379. https://doi.org/10.1007/978-3-319-45117-6_32
Zijm H, Klumpp M (2016b) Logistics and supply chain management: developments and trends. In: Zijm H, Klumpp M, Clausen U, ten Hompel M (eds) Logistics and supply chain innovation: bridging the gap between theory and practice. Lecture notes in logistics. Springer International Publishing, pp 1–20. https://doi.org/10.1007/978-3-319-22288-2

Online References

Akinlar S (2014) Logistics 4.0 and challenges for the supply chain planning and it. Präsentation Fraunhofer IML, September 24th. https://www.iis.fraunhofer.de/content/dam/iis/tr/Session%203_5_Logistics_Fraunhofer%20IML_Akinlar.pdf [access: 08.05.2017]
http://www.etp-logistics.eu/ [access: 08.05.2017]

The Model for Risk Management and Mastering Them in Supply Chain

Borut Jereb

Abstract Every organization aims for successful and continued operations, therefore efficient supply chain management is essential as risks in supply chain represent one of the major issues in today's businesses. In this chapter, on the one hand, a new risk management model is proposed in such a way that it takes into account different segments of the public and time-varying internal and external parameters from the observed system environment. On the other hand, a freely accessible catalog of risks in supply chains is described, which is based on the described new model. This model complies with the general risk management (ISO 31000) and supply chain security (ISO 28000) standards, while it also covers and includes recent findings from the risk management sphere of influence. The central concept of the model is an individual public exposed to individual risks. In the model, we determine what kind of a relationship a specific public has to some specific risk. Different publics have a different attitude to a specific risk—different publics have different exposures and different subjective uncertainties about a certain risk, even though objective uncertainty is the same for all public. In our model, we are assuming that risk is ultimately a characteristic of human beings and not of things or concepts as is the case with most of the existing risk management models. Human is not an object, but a self-aware subject with their own will, a power to do things and with his own exposure and risk acceptance. In addition, the model introduces functions which calculate new values of parameters and output on the basis of the given input and the accumulated history of the past processes' life cycles. Later on, the model resolves if calculated tolerance levels for risks, impacts and process parameters are adequate for every determined segment of the public. Calculated results are based on the provided tolerance levels, while the parameters, functions and levels are assumed to be non-deterministic (i.e. parameters, functions and levels may change in time). Contemporarily to the model, a freely accessible experimental risk catalog was formed and published online based on practical research in authentic organizations. Here, risks that have been identified thus far are presented and described, while the catalog also enables joining and cooperation of the experts from the field into a community. Hence, the catalog is

B. Jereb (✉)
Faculty of Logistics, University of Maribor, Mariborska cesta 7, 3000 Celje, Slovenia
e-mail: borut.jereb@um.si

A. Kolinski et al. (eds.), *Integration of Information Flow for Greening Supply Chain Management*, Environmental Issues in Logistics and Manufacturing, https://doi.org/10.1007/978-3-030-24355-5_18

considered as an ever growing list of hypothetical risks that enables an insight into the model, its value and practice while it can also be used as a listing and/or starting point for establishing supply chain risk management in organizations.

Keywords Supply chain · Risk management · Risk catalog · Requisite holism

1 Introduction

Risks are an integral part of our lives and it appears that people have never devoted as much attention to the challenges of risks as we do today. Risks are addressed by numerous articles, comments, and conversations. Perhaps expectedly, there are virtually countless conceptions and definitions of the term "risk". Even if a particular community agrees upon a single definition of risk, it is still anything but certain that such a community will reach uniform opinions or answers to questions such as (Bluhm et al. 2002; Hallikas et al. 2004; Benedetti et al. 2008; Gordy 2003; Scott 2005): How to perceive risks? How to measure them? Which risks are we most exposed to in a given moment? What are the consequences of exposure to risks—what is the impact of risks? Which risks are acceptable and to which magnitude or extent? Who are the risks acceptable to and who are they not acceptable to? How do risks change in time? What is their impact when observed individually and when taken together? What is their mutual effect and what are the consequences of these interactions? How should risks be managed? How to assess the amount of assets required to reduce, or eliminate the risks? The myriad of questions that have remained unanswered to this day points to the complexity of the problem imposed when one contemplates on a quest to address and manage the risks in a comprehensive manner.

From the point of view of internationalization and globalization, risk management becomes even more important. Globalization certainly brings increasing external threats and based on this, together with today's irregular market situation, we can conclude that certain measures need to be taken in order to guarantee stability of operations in an organization and its supply chain. Swoboda et al. (2010) have characterized this topic in more detail by precisely outlining the benefits of cooperation and alliance establishment between participants of supply chain.

ISO 31000:2009 establishes a number of principles that need to be implemented to make the risk management efficient (ISO 2018a). A risk assessment as the key activity of the risk management is the overall process of the risk identification, the risk analysis and the risk evaluation (ISO 2009) (see Fig. 1). It is the topic of this research and it requires a multidisciplinary approach since risk may cover a wide range of causes and consequences. It is important to understand that risk is a complex phenomenon, therefore making decisions in complex systems is a challenging task to accomplish (Colombo 2019).

Segments of the public are groups of people that have been identified by their current interest in, attitude to, or current behavior around, a particular issue, rep-

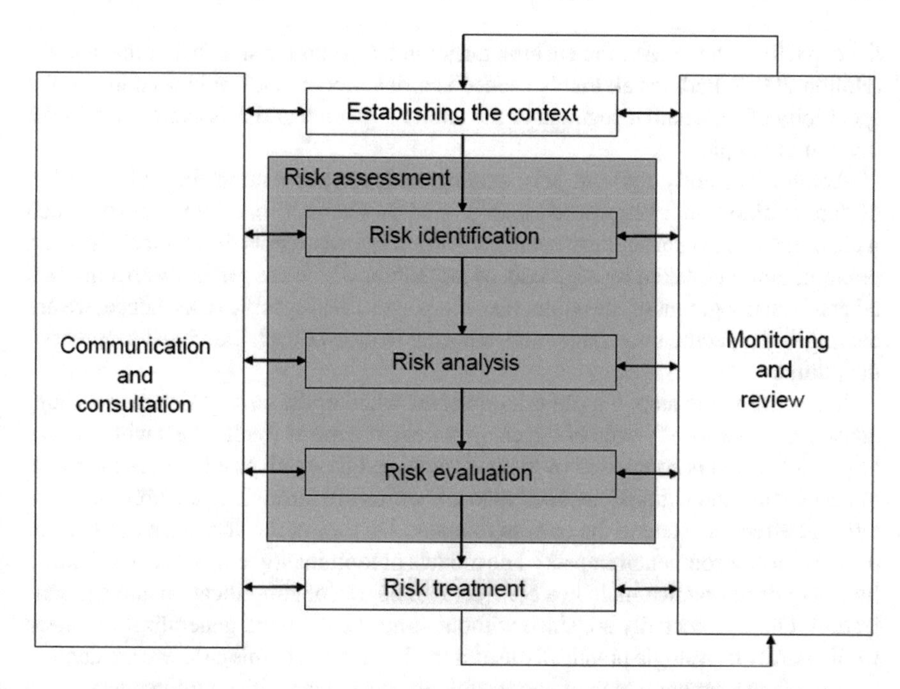

Fig. 1 Contribution of a risk assessment to the risk management process, ISO 31010

resenting the most important part of the environment which is considered in risk management. Such an approach in which segments of the public play the central role in risk management is new in scientific technically oriented literature.

As every human being is unique, different from all others, our relations to a certain risk encountered with regards to a particular situation can also differ greatly. Hence, people have a different view on and a relation to the same risk, which may be a result of different exposure as well as of different levels of uncertainty. Chopra and Sodhi (2004) already accentuated the difficulty of managing risks or uncertainties as actions have a great impact on individual risks—some actions may mitigate some risks while others might worsen them and vice versa. The problem is most commonly addressed not in relation to individuals, but in relation to groups of people, i.e. segments of the public that share a common stance with regard to a particular risk.

In scientific literature, as well as in practice, it is quite common to address risks as something intrinsic to any object, even inanimate, although only humans have the capacity of self-awareness. In his article, Glyn A. Holton (2004) addresses the question of the level at which risk is actually taken: can an organization actually be at risk, or is it in fact the individuals, i.e. the employees, who are the risk takers. In this context, they can either be regarded as individuals or as a specific segment of the public, within the organization. It should be widely accepted as a fact that in case of an undesirable event, an incident, a crisis, or a disaster, every community (segment of the public) generally bears its own level of risk. If we concede that only humans have the capacity to be at risk, the ensuing question is: "Whose risk is being managed?" (Holton 2004). Perhaps all that is needed is a risk model that would account for the specificity of a particular segment of the public—given that risk is exclusively in the domain of people.

Another currently relevant area dealing with the accounting for and inclusion of "uncertainty" and "exposure" in risk models seems to open up. Namely, such inclusion becomes particularly complex as soon as one accepts the fact that risks can predominantly be taken by segments of the public which are generally specific risk takers—each segment of the public (or each person) is at specific risks; hence, we are dealing with specific uncertainty and exposure in case of each individual segment of the public.

In the following step, we can ask ourselves whether the current risk models adequately account for the state of the environment in general (including a wide variety of public) which is comprised by such models, and in which past facts (as the result of past events and actions) are accumulated, which are intrinsic to the observed system and affect the state in the current moment. Do they at the very least account for the current environmental impact? The models predominantly employed in scientific literature or in practice include a considerable degree of simplification and generalization. Quite expectedly so, since without simplification and generalization, there would hardly be a single practically useful model created. In this case, we are dealing with the development which, if successful, always begins with simplification.

An approach considering the state of environment requires more complex modeling and risk management, thereby earning more trust of individuals involved in risk assessment activities than of top-ranking officials. Thus, governance is significantly improved—the model is attributed its importance, which has been confirmed by the results of surveys conducted during the implementation of risk assessment activities in organizations.

On the other hand, risks should be understood in order to be identified or perceived. We should be able to assess and measure their impacts, to monitor them, and ultimately, to manage them. In recent decades, the latter activity—i.e. risk management—has increasingly employed simulations (Houston et al. 2001), the reason being that in practice risks include the use of highly complex models (Pritsker 2006), in which particular risks, in addition to their mutual interdependence, also depend on the environmental parameters of system processes.

This research describes an integrated risk model that takes into account the aspects of risk at which segments of the public are considered and consequently. This research describes an integrated risk model that takes into account the aspects of risk at which segments of the public are considered and consequently, areas of risk management, formerly not being deliberated, are pondered. According to Silva et al. (2019) traditional risk management has been evolving towards a holistic perspective which can be helpful in increasing firm value from a strategic perspective. Many authors strive for a holistic approach to risk management because of its high relevance, as Bohnert et al. (2019) stated in their research. The aim is not quantification, but to break ground for future quantitative models. Our study is based on the risk management models described in the most important risk ISO standards (ISO 2009, 2018a, c). This model was partly validated in real life; the pilot testing was done on an actual logistics company that focuses mainly on warehousing. The output we got from this preliminary test and subsequent testing is a catalog of identified risks, where each risk is also defined or categorized according to different parameters or dimensions that will be explained later in the paper. As this test was well accepted by the test company we have reason to believe that we are on the right path to achieving our goal, which is to develop a widely usable model for supply chain risk assessment. Moreover, our goal is to manage a web-based catalog of supply chain risks, which is published under the Creative Common License, allowing everyone to use the catalog as a reference and to propose changes and additions to it. And while there might be particular models that took one or another dimension into consideration or maybe even a few that account for all dimensions mentioned in this article, our review of the scientific literature did not confirm any examples of such a model in a comprehensible way.

By testing the risk catalog, we did not make a complete validation of the proposed risk management model in full—we have shown the importance of segmenting the public and consequently taking into account the specificity and the subjective risk management approach dictated by the individual public. The use of other properties of the model can be used primarily in the study of risks through the simulation of stochastic event system undertakes several dimensions of the simulated risks.

However, in addition to setting up an accessible catalog of risk, the contribution of newly developed model can be seen also in improved manner of characterizing, assessing and managing risks.

The rest of the article is organized, as follows: The first part in continuation of the article will introduce theoretical background of setting organizations to manage external and internal risks, continuing with the risk management modeling in theory. Later on, the new model for risk management will be described and presented in detail. The prolongation of the article will be focused on the partial application of the developed model in practice by the description of the development and usage of risk catalog.

2 Problem Background

The term risk is used in many spheres of our lives. We all believe to understand its meaning, yet there are numerous different interpretations. Some are listed in the following Internet references (Kaplan and Garrick 1981; Houston et al. 2001; Business Dictionary 2019; InvestorWords 2019). Each field tends to interpret it in its own way: even within a single field, opinions often clash on various interpretations and even in an individual case, views on the risk arise that are often different and even opposing. The only conclusion that appears beyond debate is that risk, probability or possibility are inseparable terms. When addressing the former, the latter is always included in the discussion; the reverse, needless to say, is not true.

In principle at least, there also appears to be some level of consent as to what one can do about risks. They can be avoided, reduced, accepted as they are, or even transferred to others (e.g. to an insurance company). Each field forms its own definition of risks, or assumes the existing one. These definitions are not perfect, since one deals with a complex term—and the very number of different definitions is evidence underpinning this assertion. The use of respective definitions that tend to reduce the complexity of risks is inevitable for exact scientific disciplines, as they are the only way to enable the use of the concept, i.e. to operationalize it. On the other hand, risks attract interest and the issues around them are currently very relevant. A number of people deal with risk management methods for analyzing and managing risks, such as VaR (Jorion 2007), SARA (Corporation (2010), which are increasingly more complex and which include increasingly more risk properties or parameters. Standards have been defined and risk management frameworks have been established. Some of them are: AS/NZ 4360 (Australia 2004), ISO/IEC (ISO 2018a), The Risk IT (ISACA 2009).

Recently, the quest for a definition of risk has lead to a point when experts at the international institution ISO could not reach an agreement on some of the key terms that define risk—and thus to define risk itself. Hence, the standard ISO/IEC 27005:2008 Information technology—Security techniques—Information security risk management (ISO 2018b) lacks the precise definition of terms such as threat, vulnerability, probability of an occurrence, and, last but not least, risk. Soon after the publication of this standard, Ross (2009) published an article paying attention to this problem.

According to Mun (2010), the terms uncertainty and risk are different, yet related. Risks are the results of uncertainty, something that is undertaken by (or intrinsic to) somebody (or something), as a result of uncertainty. The same author submits that at the beginning, there is always uncertainty and the risks related to it, and that through time in which certain actions and events take place, these risks turn to facts. Furthermore, this author asserts that one can also encounter uncertainty that does not involve risk at all. This is described in the case of an airplane coming down to a certain crash, with only two passengers and one old parachute the functioning of which is doubtful at best. Both passengers are faced with the same uncertainty as to whether the parachute will open or not. If the object of uncertainty is the old parachute and the two passengers agree upon who will use it, then the person to use the parachute will assume the entire risk related to opening of the parachute from the moment he/she jumps out of the airplane until the very moment when he/she pulls the string and the parachute either opens or not. Meanwhile, the second person, having agreed not to use the parachute, does not assume any risk with regard to the functioning of the parachute; at the same time, the second person is also quite certain to die.

According to Holton (2004), risk includes only two essential components:

- uncertainty,
- exposure.

Uncertainty and exposure are, however, the most difficult concepts to define and account for. Our research shall henceforth employ Holton's definition. As such, this definition may fundamentally challenge the currently prevailing view on risks and the way they are addressed. The following example may clarify the issue. A bridge as a building does not undertake any risk, regardless of how poorly it may be built. Risks are only taken by stakeholders (people) related to this bridge in one way or the other. This is important, particularly because the bridge itself does not include a dimension of exposure as it will be defined shortly. Furthermore, the interpretation of uncertainty is assumed by (intrinsic to) the bridge.

To simplify, "a person" can be understood in particular cases either as a natural person (an individual) or a juristic person (a legal entity), although the latter is readily translatable into the specific community, or a group, of natural persons. Furthermore, such simplification soon leads to a dead end. There are very few examples in which only stakeholders in companies and organizations are the exclusive risk takers; rather, the risk also includes employees, stockholders, investors, the local community, etc. Any one of these stakeholders (or any group of them) indulges in their own uncertainty and exposure (Holton 2004).

Since probability is a constituent part of risk, I shall now briefly address this issue, as well.

Knight (1921) distinguished between two types of risk:

- "real or objective risk" which includes logic, probability and statistical methods,
- "uncertainty or subjective risk" where the idea of quantifying probability is hardly helpful—when probabilities are defined by individuals based on their beliefs, or when the system of values is established, based on opinions in order to describe their uncertainty.

Hence, it can be said with regard to risk that probability may be used as risk metrics; however, its use may be bounded and deficient. What is missing is the measure of "uncertainty", at least (Knight 1921).

Uncertainty
Uncertainty is a condition when one does not know whether a proposition or an assertion is true or false. Probability is the metrics that is most commonly used to express uncertainty; however, its applicability is bounded. At best, it can assess the uncertainty we are able to perceive. In addition, Dunovic et al. (2016) found out in their research that a vast majority of methods and techniques in different research spheres do not support uncertainty management concept.

However, what Knight designates as risk (objective risk and subjective risk) will in this research be referred to as uncertainty. In the following text the following two terms shall be used:

- objective uncertainty,
- subjective uncertainty.

Exposure
As Holton said, the litmus test for exposure is "Would we care?". People are thus exposed when we care about whether a certain proposition is true or false (Holton

2004). In general, a person is exposed when an event has some material or non-material consequences for that person, and caring is not a synonym for exposure. However, in our model, we will take into account different public in relation to a certain risk, so that each public will in principle have its own exposure. It is a subjective assessment of each individual public regarding its exposure to a certain risk. This assessment is based on Holton's view of exposure.

For example: we can be exposed to risk and be fully aware of it (balancing on the fence of a high bridge) or not be aware of it at all (balancing on the same fence while sleepwalking). Risk can be taken very seriously (speed limits in a village where a police patrol is always on duty), or act quite indifferently to it (speeding through the village in the middle of the night, knowing that the police patrol is not there and assuming that everyone is asleep). Thus, exposure introduces additional indistinctness, or undefinability, which depends primarily on the individual or a certain segment of the public and its perception of exposure and, consequently, of risk. Hence, we are not only dealing with the problem of metrics of uncertainty (see: Hubbard 2009), but rather with a problem of the metrics of exposure.

Risk

Risk can be described as exposure to uncertainty, therefore it follows from the above definition of exposure that risk depends on the attitude of persons (segment of the public). Since both uncertainty and exposure are difficult to define, risk is not easily definable, either. As a result, risk is difficult to model. It then follows that it is impossible to operationally define risk in a way that would allow its effective management (Holton 2004). At best, individuals and/or communities (the segments of the public) can define their perception of risk, which is mostly highly simplified. For example, a well known simplified approach is multiplying probability by potential loss. Problems arising from using such a simplified approach are described in Taleb's Black Swan (Taleb 2007) or in Hubbard's The Failure of Risk Management (Hubbard 2009).

The rest of this research is based on the assumption that risk is composed of:

- uncertainty, which should be divided into: Objective uncertainty and Subjective uncertainty,
- exposure for every defined public (at least one public should be defined) as a necessarily defined parameter of the observed risk.

3 A Model Considering the Segments of the Public

The described model pursues the ambition to be sufficiently general in order to be able to use in various situations and in various fields where risk is encountered—perhaps as suggested by Holton (2004), who, in his examples, refers to trading natural gas, launching a new business, military adventures, as well as romance. Although the model described in this article can be used in a wide array of fields, the example of a business process model is provided in the following subsection.

Depending on the particular field at hand that we wish to model, the importance of a particular part of the model (various public, internal vs. external, dynamic behaviour in time, etc.) may differ (Holton 2004); however, it can seldom happen that an individual part of the model is completely negligible in a particularly used case.

The terms and definitions described in the model in this research are based on the ISO 31000 and ISO 31010 (ISO 2009), denoting a high-level model of risk management. In this research the model is expanded in a way that nothing from the ISO standard is changed. In the mentioned standard there are many hints indicating the awareness of different public with their own properties, but the idea of the segments of the public (according to their interests) is not used.

The following text in this chapter is focused on those ideas of ours which are different com- pared to the ISO 31000. More consistent relations between risk, the consequence, the process state, risk, the consequence and the process state criteria (borders of accepting) are described in a mathematical form in the appendix. This theoretical consideration is underpinned by a simple example of a business process using documents. The described model is easy to use in simulations which are mostly employed to optimize processes when a mathematical model is not available (when a mathematical formula is too complex) and we ask ourselves "What if …?"

Process, Its State and Time Dimension

Business processes are represented by process graphs, i.e. mathematical structures in which the nodes represent a particular process and the link between two nodes represents their relation.

Example 1 (ISO 2018a) Clerk A regularly receives documents of two types: document X and document Y. Upon receipt, clerk A performing the business process A establishes whether documents are adequate for further processing. If any document is not, Clerk A rejects it, producing the explanation Z, including a request for the amendment of the document. If the document is adequate for further processing, it is recorded in Incoming Mail and forwarded to other clerks: type X documents are forwarded to the clerk B, performing the business process B; and type Y documents are forwarded to the clerk C, performing the business process C. Figure 2 illustrates this simplified example of business processes.

The state of each process is described by parameters—the process state depends on its specific properties which are represented by its parameters. Some examples of such parameters are: process time parameters, the maturity level, sensibility to some types of risks, the period of the year in which its importance may be low or high, the risk acceptance, the impact acceptance, etc. The model does not define what each parameter actually represents, nor does it define the number of parameters.

Example 2 In our business processes example, the process A parameter could be the number of delays involved in forwarding or rejecting any document by the clerk A

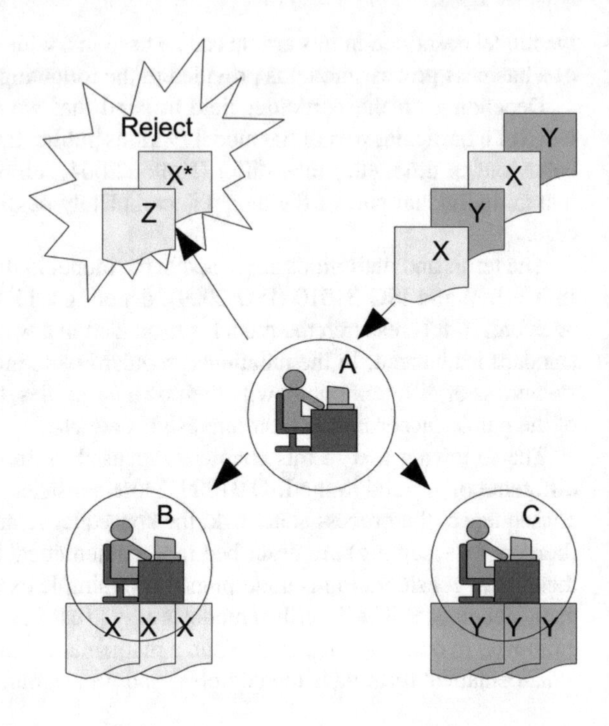

Fig. 2 A simplified business process in which the clerk A reviews and sorts/classifies the received documents and forwards them to the business processes B (clerk B) and C (clerk C), own study

(the clerk acts later than required by the respective regulations). If the clerk A never makes a mistake, the type X documents are sent to the clerk B. However, the clerk could make a mistake and send a wrong document to the clerk B. A document may also be ambiguous and it may only later become evident that it is of a different type than initially believed by the clerk A. In the first or second case, the document sent to the clerk B is of the wrong type. Within the process B, the number of wrong type documents received can be measured and recorded in a particular parameter of the process B.

The most important aspect of process parameters is that they allow for the past life cycle of each business process to be "accumulated" within them; this accumulated information is then used to accumulate the impacts and new business process parameter values. In this way, modelling also comprises the "history" of the modelled system. These parameters include the accumulated history of past moments and accordingly, the past combinations of risks and other impacts relevant to the business process.

Example 3 In the above example, each individual delay could be insignificant while a number of delays could have adverse consequences. It is, therefore, not only necessary to record individual delays, but also the total sum of all delays. This is an example of an additional process parameter.

The model should include the dimension of time, which introduces non-determinism. In many real situations, some or all processes include the time dimension in their input, output, or in the manner in which the following state of a process is calculated (see the Appendix).

In this paper, business processes are represented by process graphs, i.e. mathematical structures in which the nodes represent a particular process and the link between two nodes represents their relation.

The Process graph PG is defined as a directed graph described by Eq. (1).

$$PG = \{P, E\} = \{P, (P_k, P_l), (P_m, P_n) \ldots (P_q, P_r)\};$$
$$k, l, m, n, q, r = \{1, 2, 3, \ldots, |PG|\} \quad (1)$$

where P represents a set of processes; E represents a set of edges representing the flow of information, in which particular processes from P are the sources and destinations, respectively, of such flows. E is a set of ordered pairs, in which the pair (P_x, P_y) is considered to be directed from the process P_x to the process P_y. It represents the output information flow for the process P_x and the input information flow for the process P_y. Each pair (P_x, P_y) represents the information on the mutual relationship between the process P_x and P_y. P_x is a direct predecessor of P_y and vice versa, P_y is a direct successor of P_x. In our model, both P and E are finite sets.

The behavior of the process P_k is influenced by its input denoted by *Input* (P_k). The output of the process P_k is denoted by *Output* (P_k) and it is generated according to the following items:

1. its current status (or state in which the process is),
2. its current input, and
3. the rules for generating the output according to the status and input.

Calculation of the process states described by parameters is further explained.

The definitions of the process P_k input and output are, as follows (see Eqs. 2 and 3):

$$Input(P_k) = \{(P_x, P_k)\} = \{Inp_{k,1}, Inp_{k,2}, \ldots, Inp_{k,n}\} \quad (2)$$

$$Output(P_k) = \{(P_k, P_y)\} = \{Out_{k,1}, Out_{k,2}, \ldots, Out_{k,n}\} \quad (3)$$

The state of the process P_k is described by Eq. (4) in which $Param_{k,x}(t)$ denotes the value of the parameter x of the process P_k in time t.

$$State(P_k, t) = \{Param_{k,1}(t), Param_{k,2}(t), \ldots, Param_{k,m}(t)\} \quad (4)$$

In addition, there is the function Φ_{SC} that calculates new values of the process parameters (i.e. the new state) in each discrete (temporal) moment, based on:

1. Business process input *Input*(P_k, t);
2. Current values of business process parameters *State*(P_k, t).

Equation (5) represents the state of the process P_k, which is changing through time.

$$\begin{aligned} State(P_k, t + \Delta) &= \{Param_{k,1}(t + \Delta), Param_{k,2}(t + \Delta), \\ &\quad \ldots, Param_{k,m}(t + \Delta)\} \\ &= \Phi_{SC}(Input(P_k, t), State(P_k, t)) \end{aligned} \tag{5}$$

The *State*(P_k, t) comprises all accumulated influences spread from P_k in the future. These influences are based on the past combinations of inputs and states of the P_k. In other words: it represents a kind of accumulated history of the P_k, that could be reflected in the future by generated impacts.

In the above explained equations we still do not consider the following described segmentations including the risks and segments of the public.

Internal and External Context

In general the external context represents the external environment while the internal context represents the internal environment, in which the organization strives to achieve its objective (ISO 2018a).

The "world" of system processes is represented by the combination of all known inputs and outputs. The "world" is the environment in which system processes "live". Processes in the "world" depend on the stream of inputs—information flowing—through them. In their life cycle, they change their own "world" by their output, which, at the same time, is a part of the information stream.

All inputs, outputs, risks, and consequences of a process, and consequently, the entire "known world", should be segmented into "internal" and into "external". The observed system, composed of processes with all their parameters and mutual "output-input" relationships between processes, defines the internal world, while the external world is defined by everything else. In the model, only risks as part of the external input, and consequences as part of the external output of the observed system processes, are of our interest (see Fig. 3).

Usually, we do not have the exact knowledge of processes of the external world with all their parameters; however, we do know the input (and risks) from the external world to the observed system, as well as the output (and consequences) sent to the external world from the observed system.

In a real situation, it is difficult or even impossible to have any influence on the external risks entering the observed system; on the other hand, we have the power to minimize or even to avoid the internal risks (which are consequences from yet another process of the system). Consequently, the ability to influence the internal risks (or consequences) is the reason why the internal world should be distinguished form from the external one.

Example 4 Let us illustrate this with our business processes example with clerks: the Company has the power to reduce the clerk A's risks related to incorrectly forwarded documents to the clerks B and C by adopting appropriate internal rules and

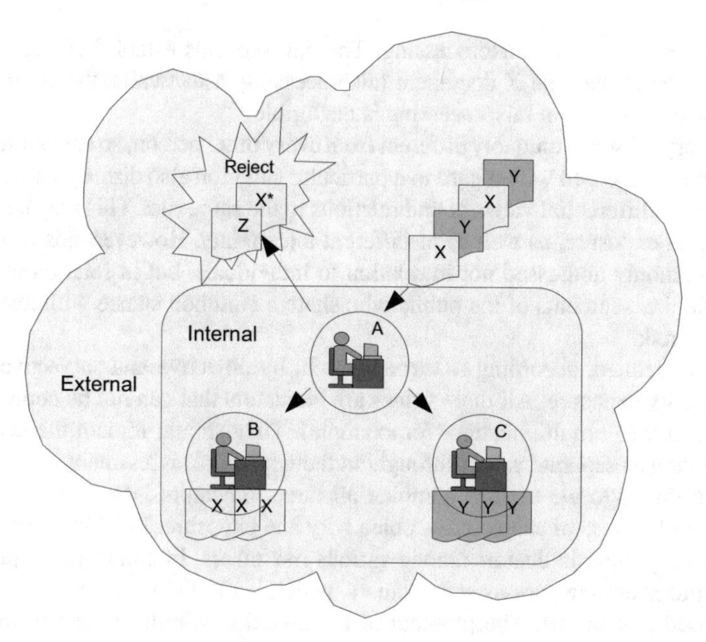

Fig. 3 Segmentation of inputs and outputs of a system of processes according to whether the origins and terms are internal or external, own study

procedures, thus affecting the internal output (including internal impacts) to some extent. However, the Company has no power over the print quality of the incoming documents; hence, the print quality is an external risk.

Risks inevitably cause consequences (output of the observed system or process); however, in addition to risks, consequences also depend on the process state and input in general.

Example 5 One example of two risks in the process described above is arrival of a poorly legible document—perhaps a poor photocopy of the original document. Poor legibility of a document can pose a threat to the correctness of its further processing. The clerk A may confirm such a document as being correct and forward it for further processing; however, it may turn out later in the process that an essential part of this document is illegible or not legible enough to allow for the certainty of its particular contents. This, in turn, can lead to even bigger material or non-material damage with legal consequences. Thus, processing a copy of a document (this can include bad print due to a worn out printer cartridge or toner) always includes an increased risk of damage incurred later in the process. Similarly, damage with legal consequences may result from an unjustified rejection of a document.

In addition, a detailed analysis of the example of risk may lead to a conclusion that misprocessing of the type X documents may cause considerable damage while the damage due to misprocessing of the type Y documents is quite negligible. Hence, in practice, the extent of potential damage would have been determined according to the share of the type X documents and the expected (given the known data from the past) occurrence of misprocessing. The damage thus established represents a consequence of the type X document misprocessing. Meanwhile, the consequence of the type Y document misprocessing is negligible.

As every person is uniquely different from every other person, so can our relations to a certain risk posed with regard to a particular situation also differ greatly. Hence, people have differential views of and relations to the same risk. This may be a result of different exposure, as well as of different uncertainty. However, this problem is most commonly addressed not in relation to individuals, but in relation to groups of people, i.e. segments of the public who share a common stance with regard to a particular risk.

Risk is defined, according to our approach, by objective and subjective uncertainty and by exposure. All three values are indicators that can not be comprised in one indicator (by multiplication, for example). They should remain the subject of investigation as separate values throughout the whole risk assessment.

Some time ago we tried to combine all three indicators into a single unit and thus created a mess of subjectivity, objectivity and exposure. In addition to this, we tried to unify these indicators among various risk takers. In doing so, we proposed compromises among various stakeholders, which led to the problem of confidence in the used risk model. The problem of the model credibility is one of the chief problems in risk management.

In practice, we have more confidence in calculations of objective than subjective uncertainty, irrespective of the fact that we are faced by a relatively small statistical samples. Such a point of view is explained in Hubbard's book (Hubbard 2009). Subjective evaluation poses the problem described by the Prospect Theory (Kahneman and Tversky 1979). In our practice we were mostly confronted with subjectivity in risk assessment (which consists of the phase of risk identification, risk analysis and evaluation).

Finally, we should compare our approach to the standard, in which it is written, as follows (ISO 2018a): "Organizations of all types and sizes face internal and external factors and influences that make it uncertain whether and when they will achieve their objectives. The effect this uncertainty has on organization's objectives is risk." This definition still does not provide for explicit differentiation between objective and subjective, and there is no word about exposure. But later in the text there are many hints about divergence among various people, between objectiveness and subjectiveness. Further, there are explicit hints (ISO 2009), for example, expressed in "perceptions and values of external stakeholders" or "the way in which probabilities are to be expressed", etc. Also (ISO 2009) explicit awareness of different objective ("reviews of historical data") and subjective methods ("identify risks by means of a structured set of prompts of questions") for risk identification is mentioned. The standard identifies qualitative, semi-quantitative or quantitative methods used in risk analyses, it refers to the estimation of risks and consequences—however, there is just a slight awareness of the objective and subjective nature of uncertainty and about exposure, but not an explicit recognition of the risk indicators structure. The standard assumes only likelihood as the one, all-inclusive indicator of uncertainty, but without exposure.

If risk is exposure to uncertainty and if uncertainty is divided into the objective and subjective uncertainty as described above, the set of risks of the process Pk in the proposed model is denoted by Risk (P_k, t) as Eq. (6) expressed.

$$\begin{aligned} Risk(P_k,t) &= \{R_{k,1}(t), R_{k,2}(t), \ldots, R_{k,m}(t)\} \\ &= \left\{ \begin{matrix} (Uncertainty_{k,1}(t), Exposure_{k,1}(t)), \\ (Uncertainty_{k,2}(t), Exposure_{k,2}(t)), \\ \ldots, \\ (Uncertainty_{k,m}(t), Exposure_{k,m}(t)) \end{matrix} \right\} \\ &= \left\{ \begin{matrix} (ObjUncertainty_{k,1}(t), SubUncertainty_{k,1}(t), Exposure_{k,1}(t)), \\ (ObjUncertainty_{k,2}(t), SubUncertainty_{k,1}(t), Exposure_{k,2}(t)), \\ \ldots, \\ (ObjUncertainty_{k,m}(t), SubUncertainty_{k,1}(t), Exposure_{k,m}(t)) \end{matrix} \right\} \end{aligned} \tag{6}$$

The function Φ_{RC} of risk calculating should be written as Eq. (7).

$$\begin{gathered} Risk(P_k,t) = \{R_{k,1}(t), R_{k,2}(t), \ldots, R_{k,m}(t)\} \\ \Phi_{RC}\begin{pmatrix} Uncertainty(P_k,t), \\ Exposure(P_k,t) \end{pmatrix} = \Phi_{RC}\begin{pmatrix} ObjUncertainty(P_k,t), \\ SubUncertainty(P_k,t), \\ Exposure(P_k,t) \end{pmatrix} \end{gathered} \tag{7}$$

whereby in Eqs. (6) and (7):

1. P_k is process k.
2. *Uncertainty* (P_k, t) is uncertainty in the process P_k at time t; in the second step, uncertainty is further divided into objective (*ObjUncertainty*) and subjective (*SubUncertainty*) uncertainty.
3. *Exposure* (P_k, t) is exposure in the process P_k at time t.
4. Particular risks for the process P_k are represented by a set of m risks $\{R_{k,1}(t), R_{k,2}(t), \ldots, R_{k,m}(t)\}$ at time t.
5. Function Φ_{RC} calculates risks.

In this subsection we still do not consider the segments of the public. They are topic of the next subsection, in which the most important equations are repeated.

According to the definition of risks, regardless of the segment of the public parameter it follows Eq. (8).

$$Risk(P_k,t) \subseteq Input(P_k,t) \tag{8}$$

General input in time t of the process P_k is denoted by *GeneralInput* (P_k, t) and is defined as Eq. (9).

$$GeneralInput(P_k,t) = Input(P_k,t) - Risk(P_k,t) \tag{9}$$

In our model, consequences are calculated with a function with properties similar to the function employed to calculate parameter values. Function parameters, too, are the same, but the calculation differs. Consequences are the result of combined effects of inputs and internal states of processes, while internal states of a process can also be changed. As we already mentioned above: when a consequence terminates in the system, such a consequence becomes a risk originating in the system in the next period of time and it is of special importance for risk management, because it is within the scope of our influence and target actions.

The function Φ_{CC} calculates the consequences (impacts) by applying the output generation rules, for a given combination of Input (P_k, t) and State (P_k, t). The set of consequences of the process P_k is denoted by Consequence (P_k, t) and it is calculated for every time slice Δ, as is expressed in Eq. (10).

$$\begin{aligned} Consequence(P_k, t+\Delta) &= \{C_{k,1}(t+\Delta), C_{k,2}(t+\Delta), \ldots, C_{k,m}(t+\Delta)\} \\ &= \Phi_{CC}\begin{pmatrix} Input(P_k, t), \\ State(P_k, t) \end{pmatrix} \\ &= \Phi_{CC}\begin{pmatrix} Risk(P_k, t), \\ GeneralInput(P_k, t), \\ State(P_k, t) \end{pmatrix} \end{aligned} \tag{10}$$

Considering Eqs. (5) and (10), consequences should also be calculated as Eq. (11) shows.

$$\begin{aligned} Consequence(P_k, t+\Delta) &= \left\{C_{k,1}(t+\Delta), C_{k,2}(t+\Delta), \ldots, C_{k,m}(t+\Delta)\right\} \\ &= \Phi_{CC}\begin{pmatrix} Risk(P_k, t), \\ GeneralInput(P_k, t), \\ \Phi_{SC}\begin{pmatrix} Input(P_k, t-\Delta), \\ State(P_k, t-\Delta) \end{pmatrix} \end{pmatrix} \end{aligned} \tag{11}$$

As presented in Eq. (11), the consequence following this moment (the time of observing the P tt), denoted by Consequence (P_k, t + Δ), depends on the following:

1. Input (P_k, t) which is composed of:
   (a) Risk (P_k, t) and
   (b) tteneralInput (P_k, t);
2. State (P_k, t) composed of:
   (a) Input (P_k, t − Δ) and
   (b) State (P_k, t − Δ)
3. Functions Φ_{CC} and Φ_{SC}.

According to the definition of impacts, regardless of the segment of the public parameter it follows Eq. (12).

$$Consequence(P_k, t) \subseteq Output(P_k, t) \tag{12}$$

The general output of the process Pk at time t is denoted by GeneralOutput(Pk,t) and is defined as Eq. (13).

$$GeneralOutput(P_k, t) = Output(P_k, t) - Consequence(P_k, t) \tag{13}$$

Public Segmentation

Risks should be conducted for each segment of the public separately. The given view adopted throughout this article, however, justifies the calculation of risk, impact and process states for each particular segment of the public.

In Eq. (9) GeneralInput of the process Pk does not depend on a segment of the public. General input does not include exposure or uncertainty—it is just data. However, for the sake of generality, it should depend on a segment of the public regardless that it is the same for all segments of the public. According to this approach Eq. (9) should be expressed as Eq. 14.

$$GeneralInput(P_k, Public_l, t) = Input(P_k, Public_l, t) - Risk(P_k, Public_l, t) \tag{14}$$

Equation (7) for calculating risks conducting the segment of the public is expressed as Eq. (15).

$$\begin{aligned} Risk(P_k, Public_l, t) &= \left\{R_{k,l,1}(t), R_{k,l,2}(t), \ldots, R_{k,l,m}(t)\right\} \\ &= \Phi_{RC}\begin{pmatrix} Uncertainty(P_k, Public_l, t), \\ Exposure(P_k, Public_l, t) \end{pmatrix} \\ &= \Phi_{RC}\begin{pmatrix} ObjUncertainty(P_k, Public_l, t), \\ SubUncertainty(P_k, Public_l, t), \\ Exposure(P_k, Public_l, t) \end{pmatrix} \end{aligned} \tag{15}$$

Equation (5) for calculating processes considering 14 the state conducting the segment of the public and segmenting input to risks, uncertainty and exposure is expressed by Eq. 16 (see also Eq. 15):

$$\begin{aligned} &State(P_k, Public_l, t + \Delta) \\ &= \left\{Param_{k,l,1}(t + \Delta), Param_{k,l,2}(t + \Delta), \ldots, Param_{k,l,m}(t + \Delta)\right\} \\ &= \Phi_{SC}\begin{pmatrix} Input(P_k, Public_l, t), \\ State(P_k, Public_l, t) \end{pmatrix} \\ &= \Phi_{SC}\begin{pmatrix} Risk(P_k, Public_l, t), \\ GeneralInput(P_k, t), \\ State(P_k, Public_l, t) \end{pmatrix} \end{aligned}$$

$$= \Phi_{SC}\left(\begin{array}{c}\Phi_{RC}\left(\begin{array}{c}ObjUncertainty(P_k, Public_l, t),\\ SubUncertainty(P_k, Public_l, t),\\ Exposure(P_k, Public_l, t)\end{array}\right),\\ GeneralInput(P_k, t),\\ State(P_k, Public_l, t)\end{array}\right) \tag{16}$$

Equation (10) for calculating consequences, considering Eqs. (15, 16) and conducting segments of the public, is expressed by Eq. (17).

$$\begin{aligned}
&Consequence(P_k, Public_l, t+\Delta)\\
&\quad = \left\{I_{k,l,1}(t+\Delta), I_{k,l,2}(t+\Delta), \ldots, I_{k,l,m}(t+\Delta)\right\}\\
&\quad = \Phi_{IC}\left(\begin{array}{c}Input(P_k, Public_l, t),\\ State(P_k, Public_l, t)\end{array}\right)\\
&\quad = \Phi_{CC}\left(\begin{array}{c}Risk(P_k, Public_l, t),\\ GeneralInput(P_k, t),\\ State(P_k, Public_l, t)\end{array}\right)\\
&\quad = \Phi_{CC}\left(\begin{array}{c}\Phi_{RC}\left(\begin{array}{c}ObjUncertainty(P_k, Public_l, t),\\ SubUncertainty(P_k, Public_l, t),\\ Exposure(P_k, Public_l, t)\end{array}\right),\\ GeneralInput(P_k, t),\\ State(P_k, Public_l, t)\end{array}\right)
\end{aligned} \tag{17}$$

Considering Eqs. (11, 14–16) and conducting segments of the public risks should be expressed by Eq. (18):

$$\begin{aligned}
&Consequence(P_k, Public_l, t+\Delta)\\
&\quad = \left\{C_{k,l,1}(t+\Delta), C_{k,l,2}(t+\Delta), \ldots, C_{k,l,m}(t+\Delta)\right\}\\
&\quad = \Phi_{CC}\left(\begin{array}{c}\Phi_{RC}\left(\begin{array}{c}ObjUncertainty(P_k, Public_l, t),\\ SubUncertainty(P_k, Public_l, t),\\ Exposure(P_k, Public_l, t)\end{array}\right),\\ GeneralInput(P_k, t),\\ State(P_k, Public_l, t)\end{array}\right)
\end{aligned}$$

$$\Phi_{CC}\left(\begin{array}{c}\Phi_{RC}\left(\begin{array}{c}ObjUncertainty(P_k, Public_l, t),\\ SubUncertainty(P_k, Public_l, t),\\ Exposure(P_k, Public_l, t)\end{array}\right),\\ GeneralInput(P_k, t),\\ \Phi_{SC}\left(\begin{array}{c}\Phi_{RC}\left(\begin{array}{c}ObjUncertainty(P_k, Public_l, t-\Delta),\\ SubUncertainty(P_k, Public_l, t-\Delta),\\ Exposure(P_k, Public_l, t-\Delta)\end{array}\right),\\ GeneralInput(P_k, t-\Delta),\\ State(P_k, Public_l, t-\Delta)\end{array}\right)\end{array}\right) \tag{18}$$

Equation (15) shows how to calculate risk, being the input to a business process based on objective and subjective uncertainty and exposure at a point in time. Equation (16) shows how to calculate process states based on known risks, general input and process states recorded for a prior time slice at a certain point in time. Equation (17) explains the calculation of the impact based on the same inputs as for internal process states. Equation (18) gives the calculation of impacts using a transitive relation for the calculation of internal process states in a prior time slice by taking into consideration risks, general input and internal process states in the time slice prior to the last time slice. All equations include business processes and segments of the public.

These equations constitute the foundation of the algorithm for the calculation of the impacts in a model. The impact calculation is central to risk management modeling, and it is illustrated by Fig. 4.

Risk Criteria

Furthermore, risk criteria representing acceptance borders should also be defined for risks, im- pacts, and process states.

Each time the consequence and parameter values are calculated, the calculated values must be compared against the tolerance values for the following:

1. risks that present specific inputs in a particular process,
2. calculated values of particular business process parameters, and
3. values of calculated effects for each segment of the public, respectively.

If any of the tolerance values is exceeded, the analysis of the causes leading to such a condition should be commenced.

Example 6 For the business process A (see Fig. 2) and for all segments of the public it is true that risks and acceptance borders do not change over time. The risks that accompany business processes should be:

1. R1—poorly legible received document.
2. R2—delays resulting from untimely forwarding or rejection of a document by the clerk A.
3. R3—wrong type of the document sent from the clerk A to the clerk B. The following individual segments of the public have been observed:

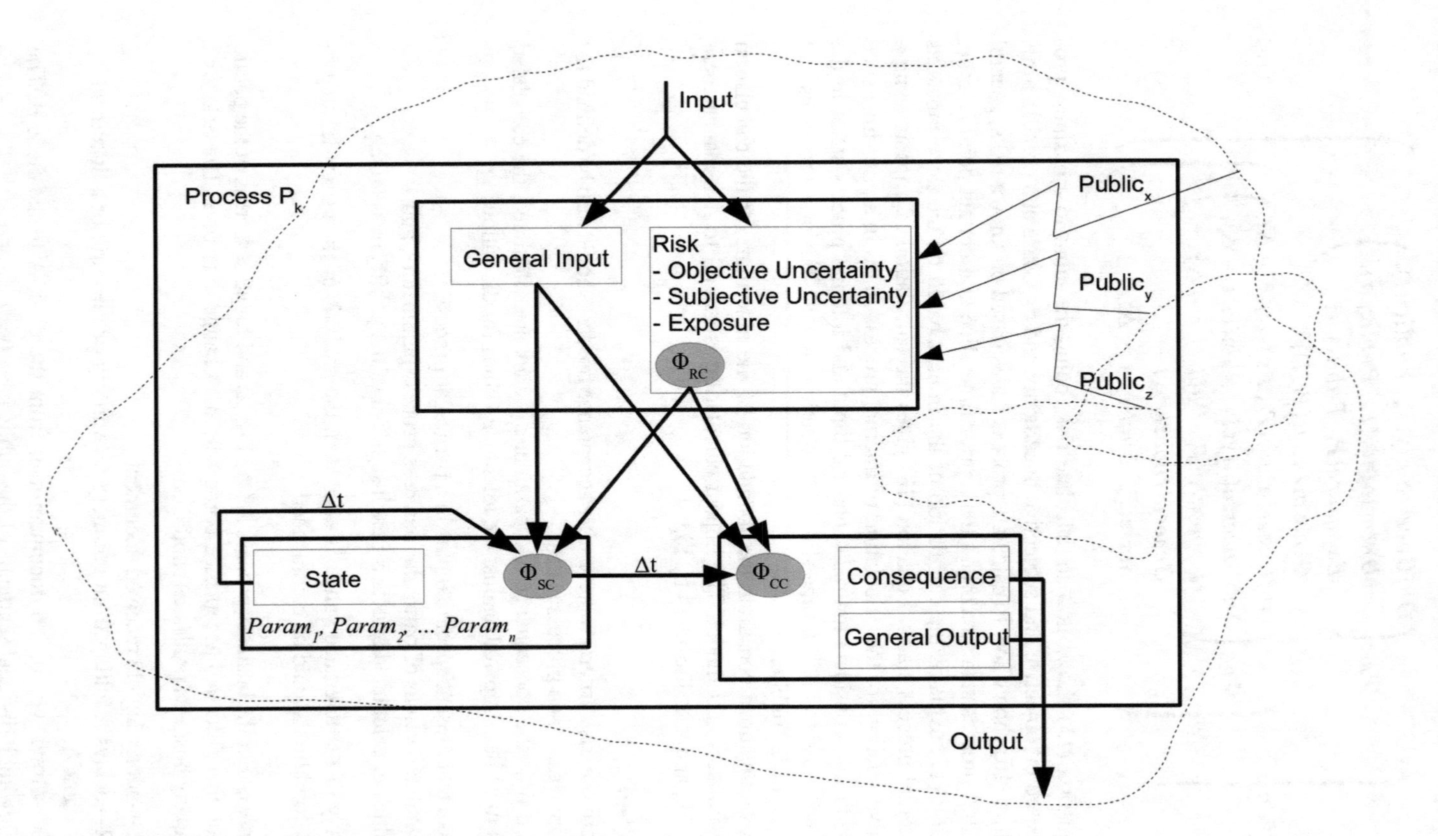

Fig. 4 The main elements of the risk management model, own study

Table 1 Objective uncertainty as to the individual risk and segment of the public

	SJ_1	SJ_2	SJ_3
R_1	M	M	Ø
R_2	S	S	Ø
R_3	M	M	Ø

Source Own study

1. SJ1—employees who carry out the business process A.
2. SJ2—owners of the business process A.
3. SJ3—users of the business process A.

Objective and subjective uncertainty, exposure and risks have the following set of four values: {Ø—zero value, S—relatively small values, M—middle values, H—relatively high values}. Although we use the same designation of values, they have different implications for uncertainty, exposure and risks. Tables 1,2, and 3 show values that change in simulations.

Table 4 shows the calculated risks by using a function (see Eq. 15 in Appendix for example). In this case the function is simplified in order to calculate risk as the worst option in the Cartesian product between objective and subjective uncertainty, and the exposure.

Table 2 Subjective uncertainty as to the individual risk and segment of the public

	SJ_1	SJ_2	SJ_3
R_1	Ø	M	M
R_2	Ø	V	V
R_3	Ø	S	V

Source Own study

Table 3 Exposure to the individual risk and segment of the public

	SJ_1	SJ_2	SJ_3
R_1	M	S	M
R_2	M	S	V
R_3	S	V	V

Source Own study

Table 4 Calculated risks for an individual segment of the public

	SJ_1	SJ_2	SJ_3
R_1	M	S	M
R_2	S	V	V
R_3	S	V	V

Source Own study

Table 5 Accepted risks for an individual segment of the public

	SJ_1	SJ_2	SJ_3
R_1	M,S	M,S	M,S
R_2	M,S	M,S	M,S,V
R_3	M	M	M,S

Source Own study

If the acceptance borders were such that acceptable risks are as described in Table 5, the risk R_3 would be unacceptable to all segments of the public and the risk R_2 would be unacceptable to SJ_2, while the remaining risks are acceptable.

In practice, we need to decide what to do with these risks. If we want to reduce them, it is necessary to take steps towards reducing uncertainty and/or exposure. In a similar way we should calculate and assess the business processes states and the corresponding impacts.

For risks the acceptance border is calculated in Eq. (19), using the function Φ_{RAB}; the acceptance border for the impacts is defined with Eq. (20) by the function Φ_{IAB}; and the acceptance border for the process states is defined with Eq. (21) by the function Φ_{SAB}.

$$\begin{aligned} & RiskAcceptanceBorder(P_k, Public_l, t) \\ & \quad = \Phi_{RAB}(Risk(P_k, Public_l, t)) \\ & \quad = \{RAB_{k,l,1}(t), RAB_{k,l,2}(t), \ldots, RAB_{k,l,m}(t)\} \end{aligned} \tag{19}$$

$$\begin{aligned} & ImpactAcceptanceBorder(P_k, Public_l, t) \\ & \quad = \Phi_{IAB}(Impact(P_k, Public_l, t)) \\ & \quad = \{IAB_{k,l,1}(t), IAB_{k,l,2}(t), \ldots, IAB_{k,l,m}(t)\} \end{aligned} \tag{20}$$

$$\begin{aligned} & StateAcceptanceBorder(P_k, Public_l, t) \\ & \quad = \Phi_{SAB}(State(P_k, Public_l, t)) \\ & \quad = \{SAB_{k,l,1}(t), SAB_{k,l,2}(t), \ldots, SAB_{k,l,m}(t)\} \end{aligned} \tag{21}$$

In Eqs. (22–24), tolerable, or acceptable values for risk, impacts, and values of the process states are defined according to the given acceptance borders.

$$\begin{aligned} & AcceptedRisks(P_k, Public_l, t) \\ & \quad = \{R_{k,l,x}(t); x = 1, 2, \ldots, m \wedge R_{k,l,x}(t) < RAB_{k,l,x}(t)\} \end{aligned} \tag{22}$$

$$\begin{aligned} & AcceptedImpacts(P_k, Public_l, t) \\ & \quad = \{I_{k,l,x}(t); x = 1, 2, \ldots, m \wedge I_{k,l,x}(t) < IAB_{k,l,x}(t)\} \end{aligned} \tag{23}$$

$$AcceptedStates(P_k, Public_l, t) = \{Param_{k,l,x}(t); x = 1, 2, \ldots, m \wedge Param_{k,l,x}(t) < SAB_{k,l,x}(t)\} \quad (24)$$

Equations (25–27) define the unacceptable (intolerable) values which represent a set of values that is equal to the set of all possible values minus the set of acceptable values.

$$NotAcceptedRisks(P_k, Public_l, t) = Risk(P_k, Public_l, t) - AcceptedRisk(P_k, Public_l, t) \quad (25)$$

$$NotAcceptedImpacts(P_k, Public_l, t) = Impact(P_k, Public_l, t) - AcceptedImpact(P_k, Public_l, t) \quad (26)$$

$$NotAcceptedStates(P_k, Public_l, t) = State(P_k, Public_l, t) - AcceptedState(P_k, Public_l, t) \quad (27)$$

The other collected data are important, but they are not the topic of this research. For the sake of completeness they shall be listed, as follows:

- Sources need to operate the business process. If this process is the IT process, then we choose among information, application, IT infrastructure and the people responsible to execute the IT process (they are the IT sources); in the case of a logistics process we choose among the flow of goods or services, information, logistics infrastructure and suprastructure and people (they are logistic sources), etc. Any risk, occurring in an observed process, can have an effect on one or more of these sources.
- The nature or type of goods or materials (flammable or frozen materials, for example) needed in the life cycle of a business process in order to produce the process output, which is a product or service.
- Segments of the public. These public are risk takers (for example: owners, employees, management, union representatives, residents in the 20 km-circle around the nuclear plant, etc.).
- The level at which risk arises. Usually we distinguish between business and technological levels, but sometimes we introduce intermediate levels. These levels are usually in connection with one or more specific segments of the public. There are four levels proposed as example (Kahneman and Tversky 1979).

4 Risk Catalog

No company today can operate in a completely secure environment without risk, deriving from supply chains, particularly considering trends of globalization and global sourcing. Supply chain risks have become a main concern in today's logistic and other business processes in any company. Therefore we can say that the process of risk management is crucial for uninterrupted operations of companies in all fields of business.

Risks need to be understood in order to begin their efficient management. Perhaps they can be most easily grasped through the example of investments. Investments are the foundation of any business activity—investments enable maintenance, increase of the scope of business operations, or changing the business activity (ISACA 2012)—and involve risks and their management as a vital part of operating activities; there are virtually no investments without risks.

This chapter proposes a general principle for risk model based on ISO 31000 and on the proposition of segmenting the risks into any given number of dimensions.

When considering risk management in organizations and in the supply chains they form, following certain guidelines is advised to ensure the process is thorough and efficient. We use of ISO 31000 family of international standards, which provides a framework for risk management in all types of organizations as is described in upper section. It takes into account different aspects of an organization and its risk management, including internal and external context, structures, processes, functions etc. The basic risk management process, as is defined in ISO 31010:2009, can be seen on Fig. 1 (ISO 2009).

Risk Assessment

The processes included in risk assessment, especially risk identification and analysis, are the most crucial in the whole risk management process. We have to be aware that risks that are not identified and defined in the first stages of risk assessment are not later treated and therefore go unseen and unmanaged. Because of that, a model for efficient supply chain risk assessment in organizations was developed.

The first step in risk assessment is always risk identification. This process should be carefully approached and as extensive as possible in order to identify as much potential risks as possible to avoid overlooking crucial risks.

ISO 31010 (ISO 2018a) proposes numerous techniques and methods for risk assessment. This standard "deal with risk assessment techniques and attempt to catalog a set of general techniques and methods useful in the assessment and analysis of risks (Luko 2016)". Out of those, we selected three—free interviews, structured interviews and brainstorming, which we used in the phase of risk identification in the first steps of assessing risks in our pilot testing. During sessions between trained external personnel and organization's employees risks are identified and then later put into the description model. It has to be noted though that the use of our model and the catalog that is derived from it is in no way connected to the use of these three methods. Every organization should approach risk identification using methods they find most suitable in their context.

Every identified risk has its specific attributes, which we strive to describe with the use of our model. Since we believe that risk identification and analysis are the key activities in managing risks, several dimensions by which each identified risk in a company or supply chain should be described are included in the model and consequently in the risk catalog which serves as a base for risk analysis. These attributes of a certain risk can be general, where we can be quite certain that the same attributes are true in every organization, or they can be organization specific, where some attributes of a risk have to be defined in a specific organization that is undertaking risk assessment.

Each of the above mentioned attributes that can be generalized are infiltrated in our model in the form of dimensions, where each risk is described by being placed in a certain group within a dimension. With this we also provide risk segmentation and consequently some additional ease of manipulation with lists of identified risks. At the moment, our model proposes five dimensions of risk definition that are not dependent on a certain organization and can therefore be generalized:

- type of risk, which is in accordance with risk groups as defined in ISO 28000,
- logistics resources, on the use of which a certain risk can have an influence,
- publics that are highly exposed to a certain risk,
- risk origin according to the organization and its supply chain,
- domain of risk management in regard of business or technological area.

The possibility of introducing ISO 26000 (ISO 2010) to supply chain risk management is where all the originality of suggested model and risk catalog lies. The standard enabled a development of the next dimension model which includes elements of social responsibility from the mentioned standard as it suggests the organizations to "take into consideration societal, environmental, legal, cultural, political and organizational diversity, as well as difference in economic conditions, while being consistent with international norms of behaviour" (ISO 2010). These were supposed to be the main factors when it comes to detecting and understanding different risks in supply chains because "the expectations of society regarding the performance or organizations continue to grow" (ISO 2010).

As stated earlier, some dimensions of risk definition have to be additionally implemented to achieve a thorough understanding of risks, such as influences between risks, its consequences etc., but these risk attributes are usually mainly dependent on the organization's environment and therefore have to be defines specifically.

Dimensions that are included in our model are described in this article, and short descriptions of organization specific dimensions that need to be implemented are given.

Risk Segmentation According to ISO 28000:2007

This model and the catalog that derives from it are structured so that they complement an international standard on security in supply chains, ISO 28000. In this standard, several fields from where risks or security threats to a company or a supply chain can originate are defined. Because the standard defines these groups broadly enough and yet in a manner that includes all relevant aspects of potential risks, we use this groping as the base for our risk assessment process. In the first step each identified risk is placed in these groups (ISO 2007):

- physical failure threats and risks, such as functional failure, incidental damage, malicious damage or terrorist or criminal action,
- operational threats and risks, including the control of the security, human factors and other activities which affect the organizations performance, condition or safety,
- natural environmental events (storm, floods, etc.), which may render security measures and equipment ineffective,
- factors outside of the organization's control, such as failures in externally supplied equipment and services,
- stakeholder threats and risks such as failure to meet regulatory requirements or damage to reputation or brand,
- design and installation of security equipment including replacement, maintenance, etc.; 8.information and data management and communications,
- a threat to continuity of operation.

The description of a risk based on the group from ISO 28000 is also the first dimension of risk definition in the risk catalog. Since some risks are more complex than others, some cannot be defined simply by one group; therefore some risks also have a secondary group placements.

Risk Segmentation According to the Affected Logistics Resources
As we analyze risks we need to be aware that there are different resources of logistics operations in supply chains. These resources represent fundamental resources which are used in logistic processes and consequently in supply chain management processes. Supply chain risks can have a significant effect on the use of this resources and therefore this interaction has to be recorded, which we achieve by defining, on which logistics resource or its use a certain risk can have an effect. The idea behind resources definition and their use in risk management comes from the field of IT, where risk management is based on interactions between resources and IT risks, as are defined in COBIT 4.1 (ISACA 2012). Based on our research of different definitions of logistics and also consultations with logistics expert, we defined four primary logistics resources, without which logistics processes cannot take place. We believe that the implementation of logistics is based on the following logistics resources:

- Flow of goods and services should be managed from the point of origin to the point of use in order to meet the requirements of customers.
- Information, which cause a change in the state of a dynamic system, if the system was able to decode data and to attribute them with a relevant meaning, and also deliver a change of knowledge in accordance with certain rules where the system has access to them.
- Logistics infrastructure and suprastructure as basic physical and organizational structures needed for the operation of logistics.
- People are the personnel required to plan, organize, acquire, implement, deliver, support, monitor and evaluate the logistics systems and services. They may be internal, outsourced or contracted as required.

Any consequence of risk, occurring in a supply chain, can influence one or more of these resources. If we wish to effectively manage risks, we need to be aware of logistics resources that a specific risk and its consequences possibly affect. That is why the second dimension of defining risk in our model is to ascertain which resources of logistics can be affected by an identified risk. Again, as with ISO 28000 grouping, some risks are complex and have wider influences; therefore they have to be defined as influential on more than one resource of logistics.

Risk Segmentation According to Risk Takers—Public

Segments of the public are groups of people that have been identified by their current interest in, attitude to, or current behavior around, a particular issue, representing the most important part of the environment which is considered in risk management. Such an approach in which segments of the public play the central role in risk management is new in scientific technically oriented literature.

As every human being is unique, different from all others, our relations to a certain risk encountered with regards to a particular situation can also differ greatly. Hence, people have a different view on and a relation to the same risk, which may be a result of different exposure as well as of different levels of uncertainty. The problem is most commonly addressed not in relation to individuals, but in relation to groups of people, i.e. segments of the public that share a common stance with regard to a particular risk.

When defining risks and their influences, we can take a different approach as that of most today's literature on the subject. If we assume that only people can perceive themselves and inanimate things cannot, we can also assert that finally, a certain risk can only influence people, who are susceptible to perceptions. According to this theory we segment all people, involved in a supply chain and its surroundings, to different publics, that is different groups of people with same interests or functions according to the individual risk. When defining risks in our model, we say that this dimension of risk identification is exactly that—defining, which publics are affected by a certain risk. This is also in accordance with ISO 31000, where one of the main principles for effective risk management is that "risk management takes human and cultural factors into account. It recognizes the capabilities, perceptions and intentions of external and internal people that can facilitate or hinder achievement of the organization's objectives" (ISO 2018a).

Also, the standard defines the importance of communication and consultation with stake- holders, which our model achieves by segmenting them into publics. ISO 31000 describes this importance: "Communication and consultation with stakeholders is important as they make judgments about risks based on their perceptions of risk. These perceptions can vary due to differences in values, needs, assumptions, concepts and concerns of stakeholders. As their views can have a significant impact on the decisions made, the stakeholders' perception should be identified, recorded, and taken into account in the decision making process" (ISO 2018a). Specific shareholders, as the standard names them, are publics as are defined in our model. We chose to use the term publics based on the knowledge from public relations, which is a field that uses segmenting of the public with best results and where this segmentation is most widely used in practice.

Risk Segmentation According to the Origin from the View of the Supply Chain
A supply chain is a complex system of several organizations that work together in a specific environment, where they "face internal and external factors and influences that make it uncertain whether and when they will achieve their objectives" (ISO 2018a). Based on the extent of risk consequences regarding the supply chain, we can define risks according to another dimension in our model. A risk can come from three different origins:

- from a company that is included in the supply chain,
- from the whole supply chain (but not from the observed company),
- from outside of the supply chain, in its environment.

Every company has dependencies on multiple third parties. As a part of a supply chain, a company is usually tightly connected with parties in the supply chain, more than with other companies from "outside". Therefore any company should suppose that companies, involved in a specific supply chain, have some kind of influence between themselves. However, Andrew Steward wrote that dependencies are risks, because, by definition, if you depend on someone then they could act in a way that negatively impacts you (Stewart 2004). Steward also recognized that dependency is a crucial dimension of risk that is often not considered as part of risk assessment or is ignored for political reasons; these risks tend to be more subtle and only emerge when analyzing business processes and not the technology components or infrastructure.

Risk Segmentation According to Business or Technological Significance
All organizations' activities can be characterized as technological or commercial. In accordance we can also define risks as mainly technological, commercial or universal. This is another dimension of our risk definition model.

Together, a list of identified risks, their definitions by dimensions and additional descriptions where needed form a base for the risk catalog, published on the Internet.

Further Definitions During Risk Assessment
As stated earlier, in the process of risk identification, analysis and evaluation in a specific organization, we have to implement additional dimensions of risk definitions in order to completely understand risks, their connections and impact.

As we know, supply chains are as diverse as today's consumer markets. Based on the type of a supply chain or goods that are supplied in a specific chain, we can define risks according to another dimension in our model. Some risks can occur in all types of supply chains, but some are specific to a certain type of a chain, for example cold chains, production of flammable materials etc.

For evaluating risks we also have to define their impact (or influence) to a specific public during the assessment process. We have to be aware that every specific public is influenced by a certain risk in its own way and responds to risks differently. By analyzing the impact with aspect to publics, we can gain a better insight into the consequences of a risk. This is not the same as only defining which public is affected, it is an expansion of that previous dimension; here possible effects of the risk are analyzed in more detail.

In many real situations, some or all risks and impacts depend on time. It is the reason why the model should include the dimension of time, which introduces non-determinism. In some time frames a single risk can be minor and in some a major influence on the organization. These time frames, if present, have to be defined in the process of risk assessment to gain a perspective over changes with time.

For every risk an acceptability level has to be defined. We also have to consider the time component of the risk when applicable in order to fully acknowledge all levels of potential impact and to correctly define the acceptability level. With this, a frame is set where we can assess to which extent and if even a risk needs to be managed.

We have to acknowledge that no process in a company can exist without links to other processes. The same goes for any risk—not a single risk can be isolated, not having any effect on other processes and also risks in a company or in the supply chain as a whole. Because of that, we need to define connections between all identified risks, and that is the next dimension in our model.

A general idea of risk management is that every risk should have a person or group, designated for its management, usually named risk owner. ISO 31000 defines a risk owner as a "person or entity with the accountability and authority to manage a risk", and that "the organization should ensure that there is accountability, authority and appropriate competence for managing risk, including implementing and maintaining the risk management process and ensuring the adequacy, effectiveness and efficiency of any controls (ISO 2018a)." By defining a specific person for every risk we achieve a higher level of awareness with those who need to partake in risk management.

The final product of the conventional risk identification and risk analysis is a risk catalog, which contains all the identified and defined risks in a single organization. We have collected these results in a risk catalog, extending it with the whole range of the supply chain risks and making it publicly available as a valuable resource in this field. Since the process of risk assessment is slow and can be insufficiently accurate, our idea of a publicly available catalog provides organizations with the option to use previously gained knowledge of the risk management process. This risk catalog contains supply chain risks as defined in different companies from different branches of operations, which can therefore be an excellent resource for any manager dealing with risks to use as a guideline and a check list. The use of a check list as a tool for risk identification is also strongly recommended by ISO 31010, which defines it as a "list of hazards, risks or control failures that have been developed usually from experience, either as a result of a previous risk assessment or as a result of the past failures" (ISO 2018a). Based on that we believe our risk catalog is in accordance with the ISO risk management set of standards, which also takes the frameworks proposed in the standards to a higher level with the inclusion of more supply chain risk management experts and through disseminating the knowledge throughout the community.

The need for a risk catalog can be recognized from many perspectives. Even ISO 31000 (ISO 2018a) defines the output of risk identification as "a comprehensive list of risks based on those events that might create, enhance, prevent, degrade, accelerate or delay the achievement of objectives". An organization can undertake the process of risk management by itself, but because of the daunting scope of this project many decide not to manage their risks all together. By using the catalog as a resource and a check list, the major step of risk management has already been undertaken, allowing the organization to approach risk management more prepared and with fewer complications. We can see that a risk catalog of this scope, which to this day has not yet existed as a publicly accessible source of information, is much needed in today's business environment. Even if the catalog is used only as a check list of possible risks in supply chain operations, it represents a crucial next step in the evolution of supply chain risk management worldwide.

However, since our philosophy is that the catalog is an ever growing publication, we believe that all users should be able to contribute, comment or add to the catalog. This is to be done by submissions of ideas to the editorial board, which shall assess the contributions and include them in the catalog when appropriate. With this we hope to achieve a widespread interest in the use of the catalog among professionals from the supply chain field and to additionally increase its scope and quality. As supply chain risk managers we have to be aware of the importance of cooperation between companies. One single company or its employees can never identify as many risks as a group of companies can. Our aim is to connect experts across supply chains all over the world and establish a community with a common goal—to provide an insight into risk assessment and the risk catalog.

The catalog is available online at http://labinf.fl.uni-mb.si/risk-catalog/. An extensive list of supply chain risks is provided, and the risks are described by the categories listed above. Additionally, and explanation of the dimensions is provided, accompanied with the list of catalog codes. For every dimension code, a list of risks under that code is also given.

On the first page of the catalog website there is a short description of the model and the catalog, followed by the most important dimension of the risk definition, grouped by ISO 28000 categories of risks. Additionally, all dimensions of the risk definition are listed. At the bottom of this page you can also find a downloadable version of the catalog. Below a part of a page of the risk catalog is shown in Fig. 5.

When you wish to find out more about the catalog itself and also about the risk assessment process, we recommend that you should visit the sub-page named "Risk assessment". There you can find a short description of the risk assessment process and our proposals for it. Most importantly, here you can find links to the descriptions of different dimensions by which risks are defined in the risk catalog. Below a part of this page is shown in Fig. 6.

A certain dimension of definitions, for example "List of affected logistics resources", can be accessed easily by clicking on the title, then a sub-page would open with a short description of the dimension and with all the category codes and categories by which a risk can be described in this dimension.

Welcome | Data | Model | About | Application

Below, you can find all data currently a part of the risk catalog. The included risks were identified and analysed in accordance with the model, which you can find here.

The table below is sortable by different columns, you can sort a column alphabetically by clicking on the column header.

Risk	Group according to ISO 28000	Secondary group according to ISO 28000	Primary logistics resource	Secondary logistics resource	Primary public	Secondary public	Origin of risk	Level of logistics planning	Source of risk	Area of impact
Limited or no access to the key locker	a.PHY		ISL		OPE		COM	OPL		
Fall of wall/ceiling	a.PHY		ISL		IMP	OPE	OSC	TPL		
Collapse of tent	a.PHY		ISL		IMP	OPE	OSC	TPL		
Planted bomb or explosive	a.PHY		ALS		ALL		OSC	OPL		
Damage to the forklift ramp	a.PHY		ISL	FLW	OPE		COM	OPL		
Damage of cranes, lifts	a.PHY		ISL	FLW	MNG	OPE	COM	OPL		
Collapse of the roof (snow)	a.PHY		ISL	FLW	IMP	OPE	OSC	TPL		
Destruction or reduction of value of goods	a.PHY		ISL		MNG	CCU	COM	TPL		

Fig. 5 First page of the online risk catalog, own study

Since risk assessment according to ISO 31000 is comprised out of three different processes, we have made use of the same principle in our risk catalog and divided our processes into these three categories. Risk identification is the first process of the risk assessment. The risk catalog is a very useful tool for identifying risks, but in every specific organization, additional parameters of risk have to be defined in order to complete the risk identification phase according to ISO 31000—sources of risk, areas of impact, risk causes and their potential consequences. As these cannot be generalized, they are out of the current scope of this catalog. In most cases though, many organizations share similar sources of risk, risk consequences and impact. The list is currently under development. We hope that with more contributions by supply chain risk experts, this list will also be more complete.

The next stage is the risk analysis, which provides an input to the risk evaluation and to decisions on whether risks need to be treated, and on the most appropriate risk treatment strategies and methods.

Some risk descriptions are general, and some are organization specific. Since this risk catalog aims to be a resource for all organizations of all types and sizes, only general definition dimensions are included. Additional dimensions by which we recommend an organization to define and analyse a certain risk are proposed in this article in the chapter "Further definitions during risk assessment". In the "Risk

Dimensions of risk definition

List of groups by ISO 28000

This model is structured so that it complements an international standard on security in supply chains, ISO 28000. In this standard, several fields from where risks to a company or a supply chain can originate are defined. Each identified risk is placed in one of these groups.

Code	Description
PHY	Physical failure threats and risks, such as functional failure, incidental damage, malicious damage or terrorist or criminal action.
OPT	Operational threats and risks, including the control of the security, human factors and other activities which affect the organizations performance, condition or safety.
NAT	Natural environmental events (storm, floods, etc.), which may render security measures and equipment ineffective.
OUT	Factors outside of the organization's control, such as failures in externally supplied equipment and services.
STK	Stakeholder threats and risks such as failure to meet regulatory requirements or damage to reputation or brand.
SEC	Design and installation of security equipment including replacement, maintenance, etc..
IDC	Information and data management and communications.
CON	A threat to continuity of operations.

List of affected publics

When defining risks and their influences, we can take a different approach as that of most today's literature on the subject. If we assume that only people can perceive themselves and inanimate things cannot, we can also assert that finally, a certain risk can only influence people, who are susceptible to perceptions. According to this theory we segment all people, involved in a supply chain and its surroundings, to different publics, that is different groups of people with same interests or functions. When defining risks in our model, we say that one dimension of risk identification is exactly that – defining, which publics are affected by a certain risk. The publics, defined in our model so far, are shown below.

Code	Description
IMP	Infrastructure maintenance personnel
EMP	Equipment maintenance personnel
DRV	Drivers
FIS	Financial sector
PLN	Planning sector
ITP	IT personnel
MNG	Management
INP	Internal public

Fig. 6 Risk analysis page, own study

analysis" sub-page, a list of all risks is provided, and those risks are defined by different dimensions. Every categorization is performed with a code of a relevant category of a dimension, which is also a hyper-link, leading to a sub-page with the description of the category and a list of all risks that fall into that category of a certain dimension.

When you wish to know more about a certain category or you wish to see all risks that fall into the category, click on the code in the first column and a sub-page will open with its description and a list of relevant risks.

Risk evaluation as the final step of the risk assessment, as defined in ISO 31000, is the process of deciding about the risks in need of a treatment and the priorities for the treatment implementation. This step cannot be generalized and is therefore not included in this risk catalog, but is entirely dependent on specific organizations.

5 Conclusions and Further Research

Risk management is a process intended for improving, developing and intensifying security levels in organizations because it gives extensive overview and insight into possible risks which could disturb or in any way affect organization's performance and productivity. This enables organizations to make suitable decisions when managing risks. Therefore risk management should be a fundamental concern of every organization, included in every aspect of it. With this, efficiency and thoroughness would be guaranteed.

During our supply chain risk research we identified some key issues in the field, the major issue being the lack of standardization and models which can make risk management in an organization easier and more efficient. Consequently, we developed a model which captures and identifies risks in an organization and its supply chain based on the process approach.

In order to handle risks, organizations need suitable tools that should be convenient and cost effective. Most common, easy and relatively inexpensive approach used by organizations are simulations of models of their business. This approach requires before-established framework for making continuous assessments and simulations. Therefore, a proper model is a requirement for fruitful risk management that follows successful simulations.

Despite our model being complex, it is still constructed in a way that permits exclusion of particular dimension defined by the segmentation. It can also be simplified to the degree of frequently used models. De facto, building a model of complex tasks with all relations is for the most part too complicated to create a final model in a few or even in just one single step as the model needs to be adapted many times in order to make it more precise and more knowledge about the inputs and rules defined by the functions, calculating impact, process states needs and public segmentation to be added to make the model more realistic and useful.

If each process represents an identity, then several of these identities should be connected with their inputs and outputs to simulate the (whole or observed) stochastic system. Although this research was focused on the processes level of the model, it is important to keep in mind that the model should be resized from the level of processes to at least the level of activities if not further to the level of particular tasks. Our risk assessment model enables managers to approach risk management by detailing recommended steps and simultaneously providing them with a tool for risk assessment. We construct such a model of stochastic processes for testing purposes. However, the entire real system of processes has not been fully tested in practice yet.

On the other hand, we have prepared a catalog of supply chain risks based on our model that is published and accessible freely. In it we summarized many of the ideas from our model—it is primarily to take into account the various publics who have their own, subjective, attitude and their own exposure to an individual risk. It allows managers to approach risk management in a simplified manner, detailing recommended steps, and at the same time providing them with a tool for risk assessment. It is in accordance with the general risk management standard—ISO 31000, and also incorporates some relevant recent findings from general and supply chain risk management, especially from the viewpoint of public segmentation.

In addition, an online supply chain risk catalog presents a simple check list of risks as identified by experts from practice. Moreover, some general descriptions according to different risk parameters or dimensions are provided in this catalog. Organization specific aspects of risks should be added during the risk assessment process to ensure a thorough understanding of an organization's risks and to provide an extensive input into the process of risk treatment. For example, including social responsibility factors into the next dimension of the risk assessment model is therefore essential as it leads to holism, it assists the progress of systemic thinking and makes it easier to use it in practice, while still considering the environmental, societal, legal, political, legal and organizational diversity.

One of the main reasons our catalog of risks is free and easily accessible is our belief that an individual cannot provide the sufficient knowledge to compose a list of risks and make it as extensive as possible or as it should be to encompass every detected risk. For this our aim was to reach as many experts as possible, considering the fact that perfecting the catalog would be easier with creative inter-expert cooperation. We believe that this catalog, especially with its focus on people and the public, presents an excellent risk management resource in practice, especially in supply chains. With that in mind we encourage managers and experts from the field of risk management to use the developed model and provide us with new ideas for its improvement or extension of the catalog.

References

Benedetti L, Bixio D, Claeys F, Vanrolleghem PA (2008) Tools to support a model-based methodology for emission/immission and benefit/cost/risk analysis of wastewater systems that considers uncertainty. Environ Model Softw 23(8):1082–1091

Bluhm C, Overbeck L, Wagner C (2002) An introduction to credit risk modeling. CRC Press Company, Chapman Hall

Bohnert A, Gatzert N, Hoyt RE, Lechner P (2019) The drivers and value of enterprise risk management: evidence from ERM ratings. Eur J Financ 25(3):234–255

Burcar Dunovic I, Radujkovic M, Vukomanovic M (2016) Internal and external risk based assessment and evaluation for the large infrastructure projects. J Civ Eng Manag 22(5):673–682

Business Dictionary (2019) What is risk? definition and meaning—BusinessDictionary.com. URL http://www.businessdictionary.com/definition/risk.html

Chopra S, Sodhi MS (2004) Managing risk to avoid supply-chain breakdown. MIT Sloan Manag Rev

Colombo S (2019) The holistic risk analysis and modelling (HoRAM) method. Saf Sci 112:18–37

Corporation TAR (2010) The security auditor's research assistant (sara). URL http://www-arc.com/sara/

Gordy MB (2003) A risk-factor model foundation for ratings-based bank capital rules. J Financ Intermed 12(3):199–232

Hallikas J, Karvonen I, Pulkkinen U, Virolainen VM, Tuominen M (2004) Risk management processes in supplier networks. Int J Prod Econ 90(1):47–58

Holton GA (2004) Defining Risk. Financ Anal J 60(6):19–25

Houston DX, Mackulak GT, Collofello JS (2001) Stochastic simulation of risk factor potential effects for software development risk management. J Syst Softw 59(3):247–257
Hubbard DW (2009) The failure of risk management: why it's broken and how to fix it. Wiley
InvestorWords (2019) What is risk? Definition and meaning. URL http://www.investorwords.com/4292/risk.html
ISACA (2009) The risk of IT framework. ISACA, Rolling Meadows
ISACA (2012) COBIT 5. ISACA, Rolling Meadows
ISO (2007) ISO 28000 Specification for security management for the supply chain
ISO (2009) IEC 31010:2009 Risk management—Risk assessment techniques
ISO (2010) ISO 26000:2010 Guidance on social responsibility. ISO
ISO (2018a) ISO 31000: Risk management— Guidelines. ISO
ISO (2018b) ISO27005:2018 Information technology—security techniques—information security risk management. ISO
ISO (2018c) ISO/IEC 27000:2016(E) Information technology—security techniques—information security management systems. ISO
Jorion P (2007) Value at risk: the new benchmark for managing financial risk, 3rd edn. McGraw-Hill
Kahneman D, Tversky A (1979) Prospect theory: an analysis of decision under risk. Econometrica 47(2):263–291
Kaplan S, Garrick BJ (1981) On the quantitative definition of risk. Risk Anal 1(1):11–27
Knight FH (1921) Risk, uncertainty, and profit. Hart Schafner, and Marx, New York
Luko SN (2016) Risk assessment techniques. Qual Eng 26(4):379–382
Mun J (2010) Modeling risk: applying risk simulation, strategic real options, stochastic forecasting, business analytics, and portfolio optimization. Gunn, Wiley-Finance
Pritsker M (2006) The hidden dangers of historical simulation. J Bank Financ 30(2):561–582
Ross SJ (2009) IS security matters: four little words. Inf Syst Control J 1:9–12
Schlarman S (2009) IT Risk Exploration: the IT Risk Management Taxonomy and Evolution. Inf Syst Control J. 3:27–30
Scott HS (2005) Capital adequacy beyond basel, banking, securities and insurance. Oxford University Press Inc, Oxford
Silva JR, Silva AFD, Chan BL (2019) Enterprise risk management and firm value: evidence from Brazil. Emerg Mark Financ Trade 55(3):687–703
Standards Australia (2004) AS/NSZ 4360:2004—Risk management. Standards Australia
Stewart A (2004) On risk: Perception and direction. Comput Secur 23(5):362–370
Swoboda B, Pop NA, Dabija DC (2010) Vertical alliances between retail and manufacturer. URL http://papers.ssrn.com/sol3/papers.cfm? abstract{_}id = 2305261
Taleb N (2007) The black swan: the impact of the highly improbable. Random House USA Inc

Modern Marketing and Logistics Approaches in the Implementation of E-Commerce

Nataliia Mashchak and Oksana Dovhun

Abstract E-commerce is a very broad topic that has interesting and contemporary components that require thorough study. It is indisputable that internet trade promotes the development of enterprises. Not to the same extent for everyone, but also, depending on the desire to respond quickly to market demands and develop approaches to managing sales through the Internet, marketing and logistics. Taking into account this, in the work the researches and publications of scientists on the preconditions, trends and perspectives of the use of effective marketing tools, logistic management mechanisms that can promote e-commerce development are researched. The authors systematized the main market changes that are projected for their comprehensive understanding by enterprises. Market players should monitor trends, take into account barriers and threats to satisfy all consumers. Today, the speed and quality of logistics, and in particular, the logistics of e-commerce, are becoming an important competitive advantage. The integrated logistic management approaches that promote optimization and efficiency of decisions of production, distribution and sales of enterprise goods, and ensuring the positive characteristics of supply chains are relevant. The study uses a systematic approach to study the basic laws and mechanisms of systems, the variety of internal and external relations. The research is carried out at the empirical level with the selection of facts that reflect the main trends of e-commerce development.

Keywords Marketing · Logistics · E-commerce market · Internet trading · Fulfillment

N. Mashchak (✉) · O. Dovhun
National University "Lviv Polytechnic", Lviv, Ukraine
e-mail: nata.pomirko@gmail.com

O. Dovhun
e-mail: dovhun.oksana@ukr.net

A. Kolinski et al. (eds.), *Integration of Information Flow for Greening Supply Chain Management*, Environmental Issues in Logistics and Manufacturing, https://doi.org/10.1007/978-3-030-24355-5_19

1 Introduction

In the world, online sales are growing rapidly (82% of businesses confirmed sales growth in 2017), while 88% of retailers expect a further increase in the amount of orders in 2018. According to GFK research, the key factors for buying in an online store for users is the cost and speed of delivery of the order, as well as the ability to monitor the cargo and to open it before payment. And if the latter factors are easy to improve, the first factors relate to logistics purely, which accounts for a lion's share of the company's costs and influences the time of delivery of the goods to the client. According to the annual report "Ecommerce Report 2018", only 53% of respondents-participants in the e-commerce market are satisfied with their warehouse logistics and full financing processes, while 24% consider their problem as purchasing and forecasting. 63% admit that it is not always possible to deliver the necessary goods to the client in time, and the most often the reason for this is an absence of the necessary goods in the warehouse. 80% of respondents believe that they incur additional transportation, customer service and warehousing costs associated with delivery errors. 78% say about logistics problems in peak periods (pre-holiday period, seasonal sales), and 52% of enterprises are forced to attract additional employees to logistics services (from warehouse staff to couriers). 62% of companies use a specialized warehouse management system, and 38% plan to implement such a system in case of an increase in the number of errors (E-Commerce Fulfilment Report 2017).

Consumers are already so accustomed to delivery at the same or the next day, their requests are constantly growing, they are not ready to endure the mistakes of online stores that can not deliver on time. But so far, enterprises have the opportunity to solve this fundamental problem, which in the near future can threaten loyalty of customers, increase the number of returns, reduce profitability and lead to an increase in overhead costs.

2 Main Trends of E-Commerce

F. Piller explores processes in the consumer e-commerce market, N. Dholakya examines the issues of global e-commerce and online marketing collaboration, R. M. Lavrenyuk, Paleha Yu. I., Gorban Yu. I., Kuklinova T. V. are focusing on the creation, promotion and development of e-commerce systems (Savitska 2018).

It is advisable to distinguish between scientists, who are dealing with individual issues in the field of Internet marketing. S. Godin, author of studies and several books on changing concepts in business and consumer behavior, F. Kotler, H. Cartadgaya, A. Setiavan—proposed new marketing concepts 3.0 and 4.0 (Kotler et al. 2016), B. Loteborne—complex "4C", Ukrainian scientists who develop the theory of Internet marketing are Krykavskyy Y., Illyashenko S., Oklander T., Savytska N., Lylyk I.

(Krykavskyy and Yakymyshyn 2018; Illyashenko and Derikolenko 2016; Oklander 2017) and many other scholars who are studying other issues of Internet marketing.

Scientists are presenting such competitive advantages of Internet commerce (Savitska 2018; Krykavskyy and Yakymyshyn 2018; Oklander 2017):

- possibility of reducing operating costs, while increasing customer support through technology;
- expansion and globalization of markets;
- improving customer satisfaction with personalized offers and increasing efficiency of managing customer base;
- reduction of the number of personnel, intermediaries, outsourcing possibilities of individual services;
- creation of new products and services, for example, electronic delivery and support services, providing reference services, establishing contact services between customers and suppliers, etc.
- increasing trust in companies, since information about the introduction, company services and reviews is more open and transparent;
- creation of a reputation not only for clients, but also among potential clients, formation of interested communities;
- control, analysis of activities and the possibility of adjusting actions, requests in accordance with internal and external environmental changes, etc. (Kuklinova 2017).

According to data (Ukrainian Retail Association 2018) by 2019, b2b companies of the world will have spent more on developing e-commerce than online retailers. In 2017, 56% of b2b companies' customers bought half of their products and services online, in comparison to 30% in 2014. An average age of a b2b client is less than 35 years old. 89% of buyers are looking for b2b sector products and services online, 74% do it for more than 50% of purchases. 71% of online users are looking at an average of 12 contractors before choosing one company. In the US, b2b-e-commerce market is very well developed—by 2020 it can reach $1.1 trillion and $6 trillion all over the world. Amazon and Alibaba retailers are developing fast. Alibaba generates 80% of all sales in China. The B2b platform of the retailer brings together Western buyers and Chinese manufacturers. Amazon Business united more than 30,000 vendors and earned $1 billion in its first year of operation, increasing its revenue by 20% per month, representing 44% of total online sales and 4% of retailer sales in the United States. With the acquisition of Whole Foods Markets in Canada, Amazon became a serious market player with 400 stores and entered a grocery market with $770 billion (Ukrainian Retail Association 2018).

3 Major Tendencies of Logistics on the Ukrainian E-Commerce Market

Every year, Ukrainian e-commerce shows an increase of 35%, which makes related industries to offer new services for Internet business. According to a study that was conducted by Ukrainian E-commerce Expert, the share of e-commerce in the client's portfolio of postal services is 60–70%. The fact that the Ukrainian e-commerce market is the most dynamic is approved with the study of Europe B2C E-Commerce Report 2017 in Europe.

Internet purchases in the first half of 2018 exceeded last year's figures by one and a half times: the growth was 44%, and turnover—6 billion UAH. It is reported by the Ukrainian Interbank Association of EMA payment system members. According to EMA studies with Gemius Ukraine, 26% of Ukrainians regularly make online purchases. And the number of those who never buy online is steadily declining: from 24% in November 2017, to 15% in May 2018.

As research shows, that clothing, appliances and electronics were the "engine" of Internet commerce for several years, but in 2018 the spectrum of purchasing interests began to expand. Yes, traders mark an increase in demand for instruments, including those for professional activities, for example, for medical instruments and consumables. Also, a demand for highly specialized goods, for example, for fishing, extreme sports, painting, etc. has increased. Popularity of home delivery of food is growing. Cosmetics and medicines began to become more popular with Internet buyers.

The trend of need for handmade products is growing. It was defined several years ago as unique for the Ukrainian market and the most promising ones. "The online market of handmade is different, because even the smallest manufacturer finds its buyer on its own, through advertisement sites, social networks, etc."—experts note in the study (E-commerce: What and how Ukrainians buy in 2017, 2018).

However, the logistics of delivery of all the above-mentioned goods is a direction that many people leave behind till the end when creating an online store (Erceg and Damoska Sekuloska 2019). Teams focus primarily on the layout, photos and product descriptions, promotions, marketing, and even reporting on the data. Logistics is not considered, and this is a mistake. Ways of ordering, delivering and returning goods are critical, especially now, when multichannel is becoming more important. Buyers are impatient, that means that many of them are waiting for delivery even not the next day. If they do not get what they want in a particular online store, they will look for another one. The same can be said about the cost of delivery. During the consumer survey, the question "What obstacles do they come across when buying online?" There were two answers: high delivery costs and too long time of delivery (5 most important e-commerce trends in 2017, 2017.

At the annual eDelivery Day event (July 2018, Kyiv), logistics companies and players made the following conclusions regarding logistics in the Ukrainian e-commerce market (eDeliveryDay: news, trends and new logistic solutions for e-commerce 2018):

- logistics became more complicated, but the client does not see it and does not always feel it;
- the range of goods on the marketplaces is expanding, and the logistics also changes with it;
- marketplaces are focused on improving service quality, especially during peak periods;
- marketplaces are trying to optimize their costs, including logistics, due to the growth of competition;
- The market has accustomed the consumer to the fact that a client wants to get his goods tomorrow. In parallel, it is necessary to adhere strictly to the specified date of delivery;
- "day to day" delivery works perfectly. It reduces the percentage of refusals;
- the marketplaces gradually go to fulfilment service;
- an active development of address delivery is going on. In the future, we expect delivery with help of drones and post-office boxes at the entrances of residential buildings.

While providing e-commerce services, you have to understand the specifics of a segment, and not to "adapt" it to retail e-commerce or offline standards. E-commerce has a number of key features: a minimum size of a batch, being sent, extremely scanty information about the characteristics of goods, sometimes—the necessity to ban some goods from a supplier, the return of goods. Specific knowledge and fundamentally new approaches to the organization of warehouse processes are needed. Optimization of logistics is the basis of a successful development of the online store (E-commerce and logistics—how to overcome pass 2017).

4 Digital Marketing

Advertising campaigns and other communications of companies were formed on the basis of marketing strategies. What to advertise exactly, when and through which channels are important issues. Along with this, important concepts that provide a comprehensive opportunity to explore the market, to discover the potential and help to plan your actions in order to increase sales and development. Marketing is developing and its approaches, that aimed at the success of the company as a whole, are changing, developing, taking into account the innovation of technologies and digitalization.

One can discuss what part of the marketing complex is changing, taking into account the development of e-marketing. The authors (Swieczak 2017) believe that this is a price, as price comparison aggregators allow to compare prices for any goods, to analyze prices of different distributors, and also to get an ability to obtain information about prices abroad from different sites, to calculate the possible costs in different cases, to choose the most optimal. However, other factors should also be taken into account: the size of the goods, because it should be delivered, qualitative characteristics (if the production of goods is carried out in different countries), therefore, it is important to analyze various factors and changes in the components of the marketing complex as they may be larger or smaller, depending on the variety of goods. Perhaps this will be a place of sale or promotion, or may be the fifth component of the marketing complex, which is increasingly taken into account—people, influence on them is complicated by requirements, awareness, psychological and personal characteristics.

Currently, we are witnesses of the development of marketing 3.0, or the era of values. Now companies consider people not only as consumers, but as human beings, endowed with thoughts, emotions, and soul. For their part, consumers are increasingly looking for solutions that are designed to improve the environment, economy, life and the world as a whole. In the purchased goods and services people are looking for satisfaction not only their functional and emotional needs, but also needs of the human soul.

Like consumer-oriented marketing 2.0, marketing 3.0 is also designed to meet consumer demands. However, companies using this kind of marketing have a mission, a vision and values that are important for the world as a whole. Marketing 3.0 raises the concept of marketing to the level of human aspirations, values, emotions. There are modern marketing levels. Experienced marketers understand that not only the rules of the game are changing—customers are changing. According to information saturation, buyers are not able to devote enough time to a single brand—and competitors are trying to attract people's attention. Today people are getting used to that companies try to explain, to communicate with buyers through messengers, present personalized offers. Marketing 4.0 has in-depth and detailed tips on how to create messages that will be heard, how to gain customer's confidence and how to persuade them to spread positive propaganda about your brand (Swieczak 2017; Lukowski 2017; Marketing Journal 2017).

> Marketing 4.0—is an approach which more effectively takes into account the convergence of the offline and on line worlds of businesses and customers. The concept focuses on how, in the times of a digital economy boom, offline touch serves as a major differentiation in an increasingly online world. It also encompasses how style blends with substance, in that even as brands need to adopt flexible and adaptive styles in the view of fast-changing technological developments, the brand's core, authentic character is ever more important. (Marketing Journal 2017)

Marketing activity in e-commerce requires an integrated approach, it needs to be conducted both offline and online, and if you compare them, the company's strategy and sales department activities, then you can draw a map of the success of marketing in general.

Therefore, it's important that marketers with offline experience take into account the changes made by the social media to the online market every day. Many of the tools and experience that is used are also used offline.

In Królewski and Sala (2016) the secrets of online marketing for offline marketers are given:

- to replace or complement the usual advertising with SMM and SEO, that is, customers who are looking for goods will see the advertisements of those companies that use these tools;

- it is worth using e-mail marketing and mobile-marketing;
- the Internet page should be not only visual and aesthetically good, but also has to be functional for building trust;
- it is advisable to develop not only online presence but also online sales;
- it is expedient to add your clients to your community or blog;
- it is worth cooperating with companies that offer complementary goods;
- it is worth finding leaders of thoughts;
- it is worth to contribute to the formation of feedback, questions, and discussions from clients.

Secrets of offline marketing for online marketers (Królewski and Sala (2016):

- before starting to position the page, think about positioning of the brand and the corporate offer—it is necessary to understand and create a proposal, to engage in positioning, taking into account competitors, not only positioning in search engines;
- proven marketing strategies operate in different cases, regardless of whether the firm operates on the real market or virtual;
- not everyone is virtual—and this should be taken into account, especially if the company sells products that are bought by very different people of a social level and age;
- it should be remembered that there are issues that can not always be solved only with the help of online marketing tools;
- it is important to ensure that there is no significant difference in the messages that the company carries through online and offline communications, as customers may be disoriented;
- it must be kept in mind that unfortunately, in offline marketing, employees may not always find precise measurements and figures such as Google Analytics or other services;
- when working online, people get used that some things are free or not expensive: however, research, affiliate programs, transfers, and advertising are not cheap and often are more expensive.

Thus, by integrating online and offline marketing strategies, businesses can develop more effectively. For example, in addition to the usual outdoor advertising and print media advertising, companies distribute advertising through Internet channels and carry out: master classes, promotions, distribution of discount and savings cards, charity events, contests, flash mobs, ambient marketing, etc. Innovation helps to increase a product value for a customer. Customers are curious when they receive not only a standard set of services, but also modern, technical—additional options. For example, according to a research, clients of FedEx Express Polska (Skowron and Broda 2017), for 87% of respondents it is important to see contemporary decisions that are applied. The most valuable message is sms or e-mail about parcel (26.86%), real-time posting (22.05%) and receiving (11.49%). Customers expect good communication with postal companies—more than 77% believe that innovative solutions make for them valuable usage. 80% of clients would not use the services, if in addition to typical courier solutions, the company would not offer other additional solutions (on-line tracking, messages, etc.). Many customers are guided by the complexity of services.

ROPO Ratio (Research Online Purchase Offline) indicator shows the link between online advertising and offline shopping. 82% of smartphone users are looking for stores near the house and 18% of searches lead to a purchase overnight. Online advertising is one of the fastest growing Amazon projects, which will rank third in 2018 after Google and Facebook. Thus, online advertising in its various manifestations is a powerful impact for e-commerce.

In Bundeskartellamt (2018) "online advertising is defined as various forms of advertising which are delivered through the Internet, both desktop and mobile". There are few types of online advertising: display advertising (banner advertising), mobile advertising, video advertising (either placed before or embedded within a video), search engine advertising (SEA) and social media advertising (SMM).

The statistics of various indicators that characterize Internet activity in different channels confirms their positive dynamics. We will provide basic indicators that reflect the scope of Internet commerce and online advertising related to it. Projections of advertising costs, which according to the Global Internet advertising market TechNavio Analyst in 2017 made up—USD183.06 billion, in 2018—$204.89 billion. Forecasts in some other sources were higher, but along with this real data, these indicators are exceeding today. In fact already in 2017 the amount of expenditure that was projected for 2018 was bigger—amounted to 229.25 billion dollars, to the end of 2019 it is forecasted to be 304.34 billion dollars, and in 2020—$336 billion—this is more than twice as much as in 2015 (Smith 2018; Statista 2015).

It will be much more common for customers to buy b2b-companies services just like for b2c-sector customers—online. The share of Google's and Facebook's advertising costs in the United States is 38.6 and 19.9%, respectively.

Business receives on average $2 in revenue for every $1 they spend on Google Ads. An average click-through rate (CTR) in Google Ads across all industries is 3.17% for the search network and 0.46% on the display network. An average cost per click (CPC) in Google Ads across all industries is $2.69 on the search network and $0.63 on the display network. Within the Google Ads platform, the average CPC is over $2 higher for ads in the search results than for those run on the display network (Statista 2018).

By 2020 80% of the advertising process will have been automated. Mobile advertising costs will be getting higher significantly. Target ads are twice as effective as non-tagged. CTR video advertisements is 1.84%. This is the highest indicator among all digital advertising formats. The cost of native ads in the US should reach $21 billion in 2018 (Marketing Media Review 2018).

Multi-channel strategies for local businesses show about 80% higher traffic when people see ads that certain goods are nearby in the store. Social networks—the proportion of people who purchase exactly through social networks (get inspiration) is increasing. The Linkedin network is gaining popularity in the B2B market, this network provides information that the messages sent through InMail system have

Table 1 The volume of the e-commerce in Ukraine

Indicator	2013	2014	2015	2016	2017
Volume of e-commerce in Ukraine, billion UAH	7.0	12.3	25.5	38.4	50.0
The volume of e-commerce in Ukraine, billion euros	0.66	0.78	1.06	1.35	1.71
Annual index of growth of e-commerce in equivalents, euro, %	48.0	18.8	34.9	27.4	30.1
E-commerce in Ukraine, % (in the structure of all retail trade in Ukraine)	0.8	1.4	2.5	3.3	4.1

Source Prepared by authors based on Marketing Media Review (2018)

twice bigger opening rate than those delivered to users outside the network—via ordinary e-mail (StartUp 2018).

Videos on Facebook—64% of consumers say that viewing motivates them to make a purchase. American companies spent $13.23 billion on video advertising. E-mail distribution—brings about 21% of revenue from all income, despite the fact that advertising on the Internet channels is developing. 65% of marketers claim that dynamic content is effective. The level of opening letters is 14.31% higher in the segmented campaigns (Smith 2018).

According to data (Ukrainian Retail Association 2018) for 2017, Ukrainians spent $14.2 billion on the Prom.ua, Bigl.ua, Crafta.ua, Shafa.ua (EVO projects) marketplaces. It is 68% more than during the same period last year. "The number of orders increased by 61%. An average bill grew insignificantly—by 4% and makes up 962 UAH" says Denis Gorovyy, EVO Development Director. Often, in 2017, they bought clothes, shoes, accessories, appliances and electronics, household and garden goods, cosmetics and perfumes, gifts, books and goods for hobbies. The online technology and electronics segment is about 10%.

At the end of 2017, the company Rozetka, which successfully existed exclusively online, opened its first offline hypermarket. This is a reflection of the global trend. The reason for the trend's popularity is simple—brands want to get as many contacts as possible with potential buyers (Marketing Media Review 2018). The volume of the Internet trading in Ukraine is shown in Table 1.

What do consumers expect in Ukraine? 72% of consumers want to see discounts and promotions in social networks. Often advertising should not be intrusive, not directly for sale, and 58% of marketers say they "tell the story" instead of direct advertising, but in the meantime, only 37% of consumers want it. 84% of consumers want to receive promotional letters at least once a month. The level of use of the complemented reality is growing (Ukrainian Retail Association 2018).

Advertisers will buy ads that will be based on real data about users. According to Mind (2018), all sectors have increased their budgets at least by 10–20%. "But there are special clients who managed to build e-commerce, by achieving profitability and increasing of advertising costs by 25–30% per year." Last years, three "whales" were the drivers of growth of Internet advertising: video advertising, sales through RTB (Real-Time Bidding, that is, the purchase of real-time advertising on an auction basis) and advertising on mobile devices. According to the InAU, in 2017 the mobile segment took over 35% of the total turnover of the Internet advertising. In 2017, RTB-procurement through programmatic (a set of methods of buying advertising on the Internet using automated systems and algorithms for decision-making on a transaction without a human being) was one of the drivers of growth in Ukraine. “Although the world of RTB is not going through the best times” (Mind 2018; IAB report 2017.

5 Marketing Communications of STRYI Company (STRYI)

Let’s consider contemporary marketing communications on the example of the Ukrainian company of woodcarving—STRYI. First of all, it should be noted that inspiring information provided on the company’s website, that encourages clients is not only a fact. These details and vision of cooperation contribute to satisfying customers’ expectations that the company is interested not only in sales but also in the formation of long-term relationships.

This is how the story of STRYI company about the company begins on the site: “Each of us is in unbreathing astonishment admiring the beautiful woodcarving work—a painting, sculpture, back plate, candlestick, carved furniture or just patterns. When you look at such work—you see hundreds of hours of mistakes, new attempts and experiments, days, weeks, months of disappointment and search, and moments of sincere joy of what you’ve created. It is not just work—it is a part of heart, soul and moments of life of another person.”

Basic information, that describes the company’s activities: “STRYI is the first Ukrainian manufacturer of tools and accessories for woodcarving! More than 500 items are given to your attention—tools, limewood templates for creativity, sharpening and polishing accessories, woodcarving courses, round-the-clock advice, etc. Manufacturer STRYI successfully works already 10 years. To 55 countries this summer it shipped its products” (Fig. 1) (STRYI 2018).

The company also indicates benefits of a product itself: “Our main advantage is production of the made to-order instruments. All requests regarding the profile, size, form, handle, sharpening and other details are always taken into account. We don’t stick to standards which foreign producers follow. We listen to our customers first of all.”. Information about delivery: “The tools are carefully packed while delivering. Cutting edges are protected from the damages. Company guarantee the declared quality of the tools and proper condition while receiving”.

Even when a product is of high quality, it does not mean that customers will find the company themselves, and even will be more loyal to it. Today, it is worthwhile to expand the chances of cooperation, including through such tools that are successfully used by STRYI company. It is important that with help of communicating in the Internet, the company can find not only end users, but also companies or professionals

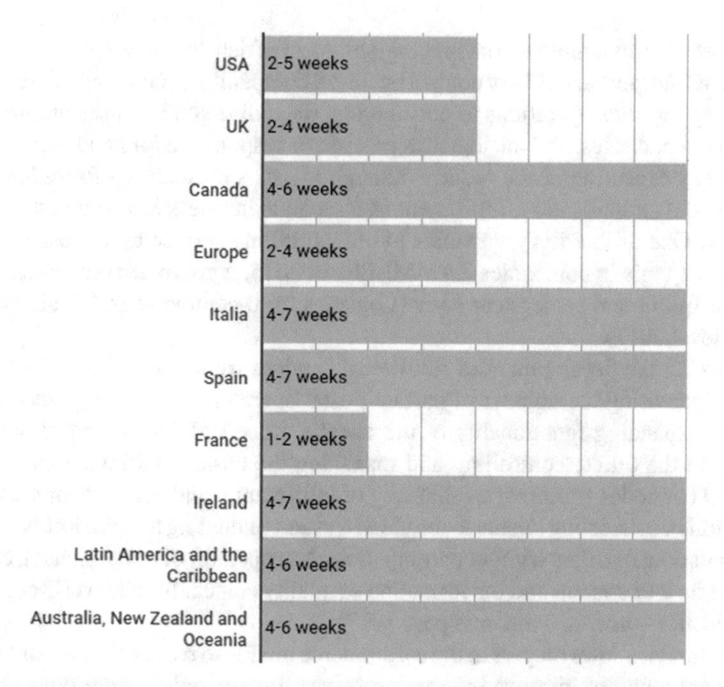

Fig. 1 Countries, in which a product is delivered and an average delivery time

who can become representatives of the company in different cities, countries. The company has a website (the main information is in Ukrainian and English), uses Facebook advertising, Google Ads contextual advertising, actively fills the Facebook page with information, has 9 representative offices in Ukraine and abroad, as well as distributors, places videos on youtube. Registration and short surveys provide the opportunity to conduct e-mail distribution taking into account the interests of a particular group of consumers, to implement loyalty programs.

6 Innovative Solutions for Logistics of E-Commerce Market

Taking into account global trends, every day new services appear in e-commerce logistics. In particular, a service with a name "fulfillment" is gaining popularity on the Ukrainian market. This service has long been in demand in the world. A year and a half ago, a Ukrainian market began to talk about a relatively new business service—fulfillment—which gained worldwide popularity thanks to revolutionary logistics solutions from Amazon. There is no single definition in the service. But market participants interpret fulfillment as a full complex of operational and warehouse processing of orders for online stores: reception of goods from suppliers, storage in the warehouse, assembly and packaging, shipment of delivery orders, delivery by own means of transport or delivery control by postal companies, processing of returns. So, by giving logistics operations to outsourcing, the online store focuses on promoting its product and sales, and the logistics provider is responsible for handling the goods and timely execution of the order. Although logistics outsourcing for online stores today is not an innovation, fulfillment as a separate business has started to be taken recently. One of the first companies to offer fulfilling service to its customers was a group of logistic companies ZAMMLER. In 2015, a group of companies opened the first fulfillment center near Kyiv (Logistics for e-commerce in Ukraine reaches a new level 2017).

Plans for the development of fulfillment services from ZAMMLER for the near future: increasing the number of partner postal operators, integrated with the CRM system, expanding functionality of the client's office and developing of analytical reports in the office, controlling and managing the customer's inventory, working with 2D bar codes (improving reliability of information and accuracy of warehouse operations), connecting clients to the CRM-system (including the possibility of direct import of bootable files without participation of an operator of a warehouse), controlling the flow of content and the formation of quality content for clients (E-commerce and logistics—how to overcome pass 2017).

Fulfillment allows players in the e-commerce market to reduce the cost of logistics significantly and to improve internal processes. Firstly, online store does not need to invest a lot of resources in a warehouse and staff training. Secondly, a logistics operator is able to process the full volume of orders quickly and efficiently, even in the high season of sales. Thirdly, logistics providers can make a return of unsold goods to the supplier much more effectively and more efficiently. The final consumer remains in the win, because if the logistics processes are built correctly, the time of delivery of the Internet order is considerably reduced (Logistics for e-commerce in Ukraine reaches a new level 2017).

Logistics company DHL introduced a new Parcel Metro delivery service, specially developed for online stores. For today, DHL Parcel Metro is being tested in Chicago, New York and Los Angeles and offers delivery on the same or the next day.

In fact, the new service represents itself as an aggregator: local systems of delivery, as well as freelance drivers are registered in the electronic system. They indicate the routes and prices, and also the conditions at which they are ready to deliver parcels. "Intelligent" system chooses couriers who offer optimal conditions of service itself.

After placing an order, a buyer selects destination of the delivery and the desired time of receipt: in two hours, the same day or the next day. A courier is appointed to him. Then in a special application, which can be embedded in the website of the online store, the buyer can track delivery in real time, instruct the courier and even change delivery time (Fig. 2).

According to studies conducted in the US, most consumers want their midday orders to be delivered to them on the same day. It is also important for shoppers to be able to track parcels in real time, to receive delivery notifications, to contact the courier.

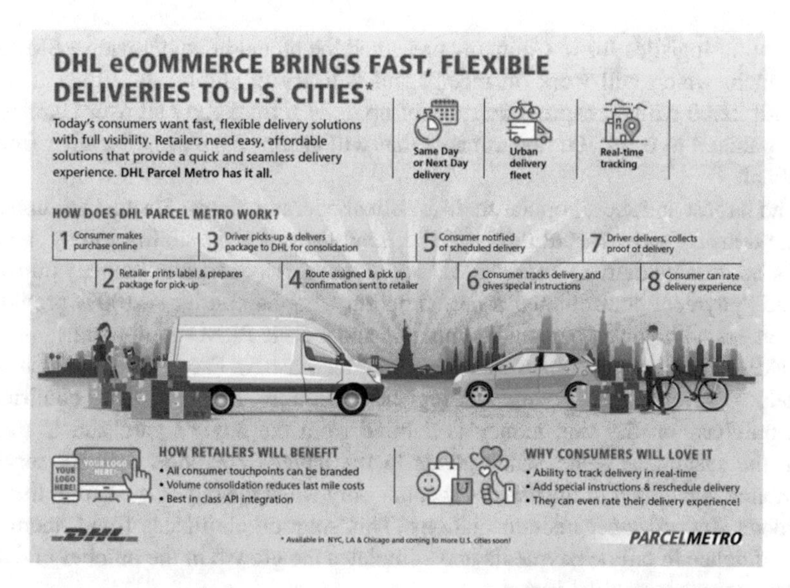

Fig. 2 Scheme of service delivery by DHL Parcel Metro. *Source* DHL launches craudsords platform for e-commerce service (2018)

DHL predicts that due to the new service, the number of "empty" courier departures will decrease significantly, service time will be reduced, and the need to deliver several times will disappear. In addition, in densely populated areas of metropolitan area, it will be possible to use hiking couriers or couriers on bicycles, which will provide fast, reliable, as well as economically and environmentally beneficial delivery at the last mile.

DHL is planning to start providing services at Parcel Metro in Dallas and Atlanta in the second quarter of 2018, and by the end of the year in San Francisco and Washington. 3.5 years ago a global leader of courier delivery Global Mail DHL conducted rebranding. For markets in North and South America, the company has become known as DHL eCommerce. The name was chosen on purpose, in DHL they claimed that a main direction of development during the coming years is the delivery of goods from online stores. The new name had to signal the business environment and the consumer about shifting of emphasis from the usual cargo transportation to expanding the list of courier services for the e-commerce market. Consequently, DHL is consistently executing its program (DHL launches craudsords platform for e-commerce service 2018).

At the above mentioned annual event, eDelivery Day, logistics companies and players of the Ukrainian e-commerce market presented their novelties and discussed the preparations for the new business season. Let's consider the main innovations for improving logistics in the Ukrainian e-commerce market (eDeliveryDay: news, trends and new logistic solutions for e-commerce 2018):

- Mailing-logistics Justin Company presented the model of an innovative SMART-branch, which will work on receipt and delivery of shipments 7 days a week until 22:00 with an express delivery of up to 24 h from Kyiv all over Ukraine. It is planned to open 700 departments that will ensure the closest location from a client.
- The largest and most popular mail-logistical operator "Nova Poshta" has updated the delivery processes at the last mile: now there is a possibility to pay in cash for addressing delivery services and additional services such as "money transfer" and "payment control" and a new function "mobile courier"—100% providing couriers with pocket personal computers and mobile PRRO equipment.
- UAPAY National Payment Service introduced a new Escrow service that will help to develop e-commerce and logistics operators' business. After confirming a purchase on the site, money is debited from the buyer's card and is frozen in the system. A seller sends goods to the buyer by express delivery service. Money is credited to the seller's account only when the buyer is ensured that the goods shipped meet his expectations. This solution eliminates fraud, increases confidence in online payments and stimulates the growth in the number of deals, and, consequently, deliveries.
- Delivery Group, which provides transport and logistics services for B2B on the territory of Ukraine and beyond its borders, has focused its efforts on the implementation of the electronic form of invoice. In their opinion, it will speed up the process of signing documents and making payments between a buyer and a seller, it will minimize an influence of a human factor on the processes, will increase the security of document circulation, strengthen control at all stages of transportation for both parties, guarantees timely receipt of information on the load on the road, moreover, it saves paper and energy resources, and, consequently, cost savings (a cost per paper document is 13.87 UAH, and an electronic one only 1 UAH).

The largest shoe retailer Intertop Ukraine introduced its omnichannel logistics, which boosted sales of the online store by 64%, and peak traffic in November 2017 reached 1,650,000 visits to the site. They use the following sales channels: Offline Network, Online Store, reserve, Pick-UP, Shop-to-shop, Store-to-Door, online store shelf. Their plans are to develop Intertop Ukraine as a marketplace, to cooperate with "Ukrposhta" and to optimize the process of returning goods.

According to experts, in the e-commerce market there is a serious technological gap. On the one side—needs and possibilities of Internet technologies, on the other—a courier in the old "Lanos". This is a technological gap between the "ideal world" and its physical reflection. But the above new technological solutions offered by logistics operators and market players should become a bridge that eliminates this abyss and provides any market participant (regardless of its magnitude and level of organization) with quality service and an ability to focus on its core competency (E-commerce and logistics—how to overcome pass, 2017).

7 Conclusions

As a conclusion, we have been systematized the main changes in marketing, which will affect the further development of e-commerce:

- a departure from a clear division of marketing and sales—it's one of the trends for the coming years. Through automation at all stages, marketing will be inextricably linked with sales as a single process;
- developing an internet marketing strategy, clearly defining a target audience and segmenting are important theoretical aspects of marketing, however, which should be also practical;
- Ukrainian companies have some problems in search and preparation of online marketers. The market will develop in this direction—to offer specialized courses, trainings, outsourcing services for different projects;
- business is required to create sites, and to develop sales—through sites or through social media pages (the value of medias as a platform of products increases);
- the role of offline stores in parallel with online product sales increases;
- continuous creativity. Billboards advertising may not change for several weeks, but advertising on the Internet requires detailed adjustment: the cost of money will not be redeemed if potential clients will receive the same advertising several times, therefore, advertising needs to be tested and modified in time, and depending on the segments;
- expansion of advertising possibilities. Sites that attract broad audience will continue to be popular and interesting. Technological capabilities will allow to improve the target and aggregate the audience in order to identify the end-user better and according to the target. Currently, large companies, as well as local small companies, can have benefit from it, if would respond quickly to the requests of new instruments;
- the role of video advertising and mobile advertising, respectively, the costs of these types of ads grow;
- influencer marketing will be integrated into all marketing activities within a few years, and marketers point out that it will become inter-functional discipline and go beyond marketing area;
- customers expect an integrated approach to sales and service.

Also we would like to note that when organizing any logistics system, including logistics in e-commerce, it is important to pay attention to all processes in the chain: storage of goods, payment system, timely delivery and attractive packaging of goods. The niche of e-commerce is now very promising, and with a proper organization of all business processes, even a start-up company can achieve high results. It is important to take into account some main threats, such as:

- poor inventory management, resulting in overselling and underselling;
- loss of loyal customer due to inaccurate delivery;
- the inability to identify/react quickly and profitably to new market opportunities/peak periods;

- increase of returns due to errors in ordering or delivery of goods;
- lost market opportunities through lean purchasing and decisions recasting
- low productivity due to ineffective picking routes/placing in storage;
- labor costs, storage and overhead costs.

Expanding the capabilities of e-commerce lies not only in the modern marketing and logistics tools, which can affect the wishes and encourage customers, but also in creativity and innovation, which needs further research.

References

5 most important e-commerce trends in 2017 (2017) https://logist.fm/publications/5-nayvazhlivishih-trendiv-e-commerce-v-2017-roci

7 fresh trends in 2018 (2018) http://mmr.ua/show/top-7_trendov_reklamy__kotorye_nelyzya_propustity_v_2018_godu#1893198022.1531813974

DHL launches craudsords platform for e-commerce service (2018) https://logist.fm/news/dhl-zapustila-kraudsorsingovuyu-platformu-dlya-obsluzhivaniya-e-commerce

Digital advertising spending worldwide from 2015 to 2020 (2018) https://www.statista.com/statistics/237974/online-advertising-spending-worldwide/

E-commerce and logistics—how to overcome pass (2017) https://logist.fm/news/e-commerce-i-logistika-kak-preodolet-propast

E-Commerce Fulfilment Report (2017) https://logist.fm/news/e-commerce-fulfilment-report-2017

eDelivery Day: news, trends and new logistic solutions for e-commerce (2018) https://interfax.com.ua/news/press-release/519428.html

Erceg A, Damoska Sekuloska J (2019) E-logistics and e-SCM: how to increase competitiveness. LogForum 15(1):155–169

How will the prices of internet advertising increase in 2018? (2018) https://mind.ua/publications/20181489-yak-zrostut-cini-na-internet-reklamu-u-2018-roci

IAB internet advertising revenue report (2017) https://www.iab.com/wp-content/uploads/2018/05/IAB-2017-Full-Year-Internet-Advertising-Revenue-Report.REV_.pdf

Illyashenko S, Derikolenko A (2016) Features of application of Internet marketing tools for promotion of products of Ukrainian industrial enterprises in the market. Marketing and logistics in the management system: theses of the reports of the XI international scientific and practical conference, Lviv Polytechnic Publishing House, Lviv, pp 85–87

Jak pracowac lepiej i madrzej niz konkurenci (2018) StartUp, No. 27, Grudzien-Luty 2018. Wydawca Colorful Media, Poznan, pp 21–26

Kotler P, Kartajaya H, Setiawan I (2016) Marketing 4.0. https://marketinginsidergroup.com/strategy/marketing-4-0-philip-kotler/

Królewski J, Sala P (2016) E-marketing. Współczesne trendy. Pakiet startowy. Wydawca: Wydawnictwo Naukowe PWN Warszawa, 1

Krykavskyy Y, Yakymyshyn L (2018) Complementarity of marketing and logistics strategies in the supply chain of daily demand. Mark Digit Technol 2(1) (2018). http://mdt-opu.com.ua/files/download/mdt2.1.2018.pdf

Kuklinova T (2017) Internet-commerce as a means of increasing the competitiveness of enterprises in a globalizing environment. http://uam.in.ua/upload/iblock/767/767f9eacb36cc09804037b17ebe5b502.pdf

Logistics for e-commerce in Ukraine reaches a new level (2017) https://www.epravda.com.ua/rus/publications/2017/05/19/624977/

Łukowski W (2017) Wpływ internetu rzechy (Internet of things) na wartość dodana Marketingu 4.0

Marketing 4.0. http://www.marketingjournal.org/marketing-4-0-when-online-meets-offline-style-meets-substance-and-machine-to-machine-meets-human-to-human-philip-kotler-hermawan-kartajaya-iwan-setiawan/

Oklander M (2017) Segmentation of online communities. http://oklander.info/wp-content/uploads/2017/11/202_statya.docx

Online advertising series of papers on "Competition and Consumer Protection in the Digital Economy" (2018). https://www.bundeskartellamt.de/SharedDocs/Publikation/EN/Schriftenreihe_Digitales_III.pdf?__blob=publicationFile&v=5

Savitska N (2018) Marketing in social networks: strategies and tools in B2C market. http://mdt-opu.com.ua/index.php/mdt/article/view/5

Site of the company "STRYI" https://stryi.ua/

Skowron S, Broda K (2017) Kreatywność i działalność innowacyjna jako czynniki tworzenia wartości dla klienta. Mark rynek 4(2017):333

Smith B (2018) 31 advertising statistics to know in 2018. https://www.wordstream.com/blog/ws/2018/07/19/advertising-statistics

Świeczak W (2017) Wpływ wpolczesnych technologii na zmianę działan marketingowych w organizacji. Marketing 4.0

Trends E-commerce (2018) https://rau.ua/ru/ecommerceuk/trends-e-commerce-2018-year/

Management Challenges of Smart Grids

Noémi Piricz

Abstract If we wish to integrate green energy into daily life we should sincerely discuss the management tasks of smart grids. Smart grids belong to hot topics in engineering journals but their management and other economic angles have received less attention. Dangers of cybercrime have increasing publicity meanwhile businesses consider the stable, reliable electric power supply to be self-evident. The modern electric power supply chains—more concretely the smart grids—show the near future. The technical and engineering parts of them seem to exist and there are more theories for their economic and financial operation, however, the integration of all these into one system may hide more unsolved problems. There is a lack of discussion of the consequences and management of the increased roles of end-users and the behaviour of them. Our critical review deals with the managing questions of smart grids based on the most cited relevant research papers. We also discuss the challenges of the management side of modern electric power supply chains, facing E-mobility and E-vehicles (G2V and V2G). In our view smart grids are not purely a technology, but a complex set of intertwined technologies, which requires drastic changes in both user behaviour and society. We see that papers dominantly use the scheme of entirely rational end-user behaviour which may cause unexpected difficulties in practice.

Keywords Electric power supply chains · Grid-to-vehicle · Vehicle-to-grid · Smart grids

1 The Changing Roles of Actors in Modern Electric Power Supply Chains

Amin and Giacomoni (2012) see smart grids (SG) as special networks which are self-healing, empower and incorporate the consumers, tolerant of attack, provide

N. Piricz (✉)
Keleti Faculty of Business and Management, Óbuda University, Tavaszmező str. 17, Budapest 1084, Hungary
e-mail: piricz.noemi@kgk.uni-obuda.hu

A. Kolinski et al. (eds.), *Integration of Information Flow for Greening Supply Chain Management*, Environmental Issues in Logistics and Manufacturing, https://doi.org/10.1007/978-3-030-24355-5_20

Table 1 General comparison between traditional electric power supply chains and smart grids (Own study based on Fang et al. 2012 and IEA (International Energy Agency), 2011)

	Traditional electric power supply chains	Smart grids
Electrical infrastructure and communication	One-way flow, one-way communication, these are separated	Two-way flow and two-way communication which are parallel and inter-connected
Source of energy	Traditional sources (e.g. from coal, oil, nuclear power)	Besides traditional sources alternative energy and energy reload by end-users
Roles of actors	Few customer choices	Many customer choices
Monitoring	Manual monitoring	Self-monitoring
In case of problem	Manual restoration	Self-healing

power quality needed by 21st-century users, accommodate a wide variety of supply and demand, and are fully enabled and supported by competitive markets. This broad explanation of smart grid may be applied to a range of commodity infrastructures, including water, gas, electricity or hydrogen. However, we focus here on electric supply chains and we use the following definition: "A smart grid is an electricity network combined with an ICT network, adapted to renewable energy sources. Its 'smartness' allows balancing the supply and demand of energy on the grid, thus making the electricity grid more sustainable, efficient and robust" (Planko et al. 2017:38).

In Table 1 we shortly list the main differences between traditional electric power supply chains and smart grids. Here we focus on those novelties which may relate to managing challenges.

Self-healing means that "For instance, once a medium voltage transformer failure event occurs in the distribution grid, the SG may automatically change the power flow and recover the power delivery service" (Fang et al. 2012).

Smart grid is the near future and worldwide there are more and more research projects, tests and directives at different levels to force its application in practice.

According to conventional logistics, the electricity supply is a typical pull system, without significant storage possibilities at the final product level. More concretely power plants and other generators produce the required electrical energy from moment to moment, otherwise the system loses its balance, and the service breaks off—generating high balancing costs and enormous costs of restarting (Bajor 2007).

So as we see in Table 1 modern electric grids involve more actors with developed interconnected connections. Additionally customers have more and very vital roles in smart grids: They are not just end users of electricity but will have also right to generate, store, and manage the use of energy (Fang et al. 2012). In these future electric networks those individual and business consumers who have electric vehicles (EV) may serve as active participants which is absolutely a new role comparing with traditional electric supply chains. The modern smart grid technology can enable EVs

Table 2 Operating features of traditional electric power supply chains and smart grids (Own study based on Fang et al. 2012 and http://eng.gruppohera.it/group/business_activities/business_energy/electricity/)

	Traditional electric power supply chains	Smart grids
Generation	– dominantly traditional power plants – centralized generation – free market activity	– traditional power plants (bulk generation) and renewable power plants – distributed generation (V2G, G2V) – free market activity
Whole-sale trade	– free market activity	– free market activity – increasing role of international trade among countries
Transmission and dispatching	– natural monopoly	– natural monopoly? – new role of V2G and G2 V
Distribution and metering	– natural monopoly	– natural monopoly? – new role of V2G and G2 V
Retail sales	– free market activity	– free market activity – new role of 'energy cooperatives'

to be used as distributed storage devices, feeding electricity stored in their batteries back into the system when needed (vehicle-to-grid supply or V2G) (Morgan 2012). V2G together with normal EV-charging (grid-to-vehicle or G2V) preferably in off-peak periods may balance the daily load.

In the last decades the energy supply showed a tendency of strengthening market features (Verbong et al. 2013). This trend probably continues in smart grids as well. In Table 2—which shows some operating features of traditional electric power supply chains and smart grids—we can see that the involved actors (e.g. producers and traders) operate both in the conditions of natural monopoly and free market. On the top of all that these actors represent individuals and organizations with public and private ownership. The question is how these really different members of smart grids will able to cooperate in high level and support e.g. resilience and self-healing.

In a modern smart grid, at the demand side we distinguish (a) large commercial and industrial, (b) small commercial and industrial, (c) residential buyers and (d) individuals who own electric vehicles, (e) fleet of plug-in electric vehicles (PEVs) (Albert et al. 2009). Group (a) own the modernist technology and act not just as end-users but wholesalers or retailer or they are able to offer any services (like IT or maintenance). Small commercial and industrial energy buyers (group b) have limited resources and capacities therefore their role is similar to residential, individual buyers (group c). So the members of group (c) may participate not just in consumption but retail (e.g. due to renewable energy sources), actually their financial possibilities and interest are restricted so they will not make essential investment in the system

(Mohsenian-Rad et al. 2010). The last two groups (individuals who own electric vehicles, fleet of PEVs) are absolutely new but they will have vital roles in demand-side management (e.g. load-shifting) (Mohsenian-Rad et al. 2010). The groups of c-d-e will be the main targeted end-users of demand-side management programs like dynamic pricing and other economic incentives additionally they will be able to participate in grid-to-vehicle (G2V) and vehicle-to-grid (V2G) processes.

There is already a clear tendency of growing international electricity trade. In Europe Germany is the most active and its energy export surplus increases year by year. Germany also receives electricity from France which is fully forwarded to other countries. In 2015 "Germany exported the most electricity to the Netherlands, who sent some of it on to Belgium and Great Britain. Second in line was Switzerland, who sent nearly all of the electricity from Germany directly on to Italy" (Burger 2016). They refuse to export at dumping price while both the import and the export price was the same.

Besides international energy commerce another trend ways up for smart grids, namely individual users establish 'energy cooperatives'. "A cooperative is an autonomous association of persons united to meet common economic, social, and cultural goals. They achieve their objectives through a jointly-owned and democratically-controlled enterprise" (European Commission). According to the Commission cooperatives can offer open, democratic and voluntary cooperation, fair distribution of results that's why they represent responsibility, equality, equity and solidarity. This form of organisation is traditionally popular in agriculture, forestry and banking but nowadays its number is dynamically increasing in the energy sector as well.

In Germany over thousand energy cooperatives (Balch 2015) and in the UK 2300 community energy organisations operate. This means that in the UK 500,000 people are involved in these communities which is roughly one tenth of all the households within the country (Department of Energy and Climate Change, UK 2014). "There are more than 500 energy collectives in the Netherlands, 220 of which have the legal capacity of cooperative" which means totally 35,000–40,000 involved people (The Dutch energy transition 2018).

In the USA energy cooperatives are popular as well. According to America's Electric Cooperatives more than 900 consumer-owned, not-for-profit electric cooperatives are active in 47 US states. This means that almost 11% of the total kilowatt-hours annually sold in the U.S. was produced by these cooperatives.

These activities can create value for the local community in several ways, such as creation of jobs in the region, financial benefits of local investors, ownership aspect, and development of environment consciousness in a direct way, practising possibilities for the youth.

2 Management Problems of Smart Grids

2.1 Management Problems of Grid-to-Vehicle (G2V) and Vehicle-to-Grid (V2G) Supply

In case of new electric power supply chains energy is volatile while it originates not only from stable traditional sources but alternative energy sources such as solar or wind energy and additionally from special storage facilities, like electric batteries of EVs.

In spite of that the electricity used in charging EV batteries will possibly not increase heavily but remain small relative to overall electricity demand for the foreseeable future, EVs could flatten peak load. Namely EVs can be used for storing electricity (from 20 up to 100 kWh) that's why grid-to-vehicle (G2V) does not mean just normal load from the electric power supply chain but also stored electricity in the battery of an EV (Fang et al. 2012). More scholars point out that thanks to the emergence of technologies such as plug-in hybrid (PHEV), fully electric (EV), vehicle-to-grid (V2G) and grid-to-vehicle (G2V) vehicles, the storage and management of electricity can be better assured (Hannan et al. 2017; Putrus et al. 2015; De Gennaro et al. 2014). If we consider daily use of cars—"in the U.S. the car is driven only one hour a day on average" (Fang et al. 2012:9 cite Kempton et al. 2009)—we see it as a possible solution.

Vehicle-to-grid (V2G) electric vehicles are able to reload electricity back to the power grid (when the EV is parking and connected to the electric power supply chain) and be scheduled to do so at any times as the grid needs it. However this supply requires payments to owners, and it reduces the life-cycle cost of owning an electric vehicle (Parsons et al. 2014; Hannan et al. 2017). This seems feasible but we should not neglect some technical and managing warnings. Several researchers call attention to the potential difficulties such as significant degradation of power system performance and efficiency, and even overloading or uncoordinated charging (Schneider et al. 2008; Hadley and Tsvetkova 2009; Roe et al. 2009; Sortomme et al. 2011). Fang et al. (2012) suggest the optimization of EVs charging profile which may lead to smaller and smaller peak power demand. Preparing for large smart grids and organizing numeral devices Fang et al. (2012) identify 2 options: top-down and bottom-up approaches.

However these ideas directly depend on the number of EVs. According to statistics EV sales achieved a new world record in 2016 (over 750,000). "In 2016, China was by far the largest electric car market, accounting for more than 40% of the electric cars sold in the world and more than double the amount sold in the United States. The year 2016 was also the first time year-on-year electric car sales growth had fallen below 50% since 2010" (Global EV Outlook 2017:5–6). So we can recognize dynamic growth together with dramatic increase in electric power which may outline the importance of V2G and G2V.

If we imagine that V2G and G2V processes operate in large scale we assume that consumption is heavily based on previous agreements between service providers, distributors and end-users so the energy use is scheduled as per daily, weekly and monthly bases. During V2G (vehicle-to-grid) the role of logistics companies using electric vehicles will increase because the possible higher number of owned EVs. Generally not just logistics firms but large corporations may receive an extra influencing power in smart grid. Till now we have not encountered a paper discussing or investigating this question.

2.2 Managing Challenges of Smart Grids

Demand-side management (DSM) involves initiatives and technologies that encourage consumers to optimise their energy use. This set of programs is not so fresh however its application for homes is a new trend (Verbong et al. 2013). Due to the increased roles of end-users the demand-response feature of smart grids seems to be a vital economic pillar. This ability of smart grids also contributes to efficient operation and resilience. That's why the high-level state support and legal background is understandable. In the USA the state-administration tries to fulfil this need when US President George W. Bush signed the Energy Policy Act in 2005 and large-scale studies about demand-response and advanced metering have been completed by the US Department of Energy and the US Federal Energy Regulatory Commission (both in 2006).

For a better understanding let us have a look at Fig. 1 which makes visible the challenges of the supply and demand of electricity in smart grids. It is very difficult to secure electricity at stable disposal if both supply and demand have elastic parts. On the supply side elasticity may come from renewable energy sources (Gelazanskas and Gamage 2014), G2V supply or certain strategic reasons of maintaining resilient system. The elastic demand originates from business and individual end-users' uncertain consumption. The level of elasticity and uncertainty depend on efficiency of economic incentives (including dynamic pricing), the behaviour of end-users, the level

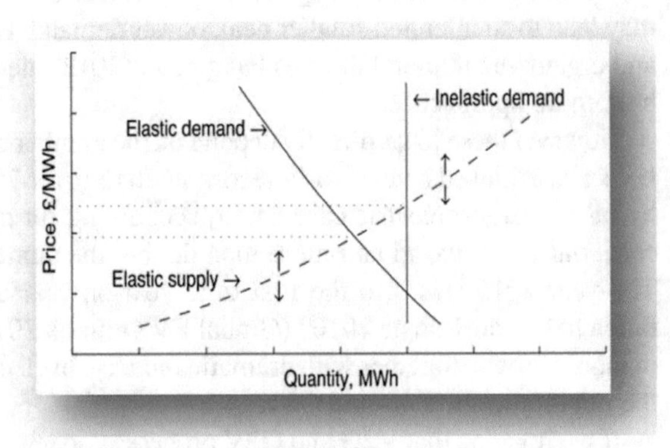

Fig. 1 Supply and demand of electricity (Gelazanskas and Gamage 2014:24)

of V2G and G2V, the general economic situation (volume of orders and production), the weather (heating in winter, cooling in summer) and several other conditions.

With regard of the demand side the level of uncertainty can be decreased by demand-side management. During DSM both the peak and low consumption should be balanced as it is illustrated in Fig. 2. We understand from this table that flattening of elastic consumption means a complicated task which may bring doubts if this system can operate by any pricing system. The most frequently suggested pricing methods of demand respond strategies are discussed in the next subchapter.

Palensky and Dietrich (2011) make the following groups of DSM programs: the market demand-side programs are e.g. real-time pricing, price signals and incentives which are under control of economists. On the other hand grid management and emergency signals belong to the so-called physical demand-side programs which are closely connected to reliability expectations.

In short Table 3 gives a general summary of potential advantages of demand-side programs.

Verbong et al. (2013) suggest the method of Strategic Niche Management (SNM) to research smart grids which could bring new results and solutions while the new energy system shows more similarities to niche products/services such as it is easy to distinguish from the traditional electric power supply chains and requires special use.

The management of networks in emerging business nets is under-researched (e.g. Möller and Svahn 2009; Planko et al. 2017). In this case we should use a mixed approach instead of B2B and B2C contexts separately. For managing smart grids a certain mixture of B2B and B2C angles could give more efficient, operable and useful solutions due to increased roles of end-users (both business and individual ones).

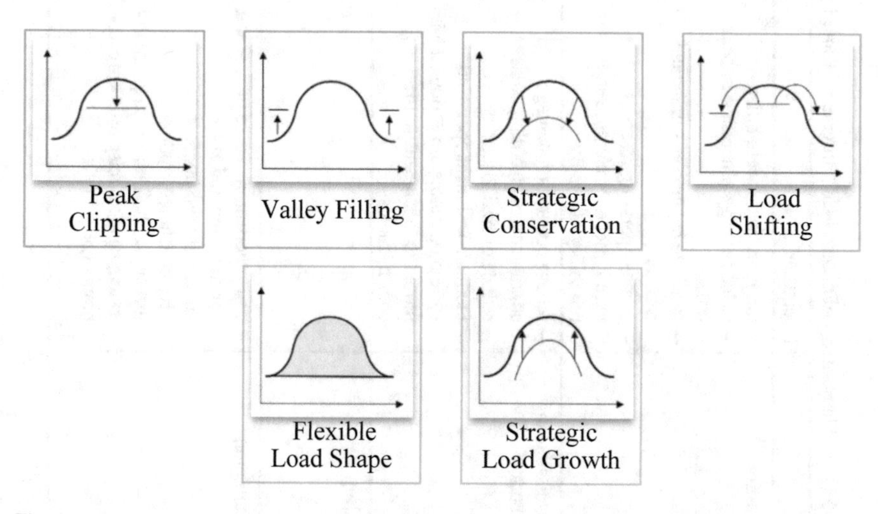

Fig. 2 Demand side management techniques and strategies (Own study based on Gellings 1985; Logenthiran et al. 2011; Alagoz et al. 2012)

Table 3 Possible positives of demand-side programs in smart grids (Own study based on Siano 2016; Han and Piette 2008; Conchado and Linares 2010)

	Operation	Development	Market
Generation	– improve system reliability – reduce energy generation in peak times – facilitate balance of supply and demand – reduce operating reserves requirements or increase short-term reliability of supply	– avoid investment in peaking units – reduce capacity reserves requirements or increase long-term reliability of supply – allow more penetration of intermittent renewable sources	– reduce cost of energy and possibly emissions – reducing system costs – prevents the exercise of market power by electric power producers
Transmission and distribution	– relieve congestion manage contingencies, avoiding outages – reduce overall losses – facilitate technical operation – system operators are endowed with more flexible means to meet contingencies	– defer investment in network reinforcement or increase long-term network reliability	
Retailing			– reduce risk of imbalances – reduce price volatility – new portfolio – more consumer choice
Consumption	– consumers more aware of cost and consumption and even environmental impacts – give consumers options to maximize their utility: opportunity to reduce electricity bills or receive payments	– take investment decisions with greater awareness of consumption and cost	– increase demand elasticity – participant bill savings – bills savings for other customers – generally customers have more options for electricity costs management – two-way communications interval meters that allow customer utility bills revealing the actual energy usage pattern

Planko et al. (2017) investigated network effectiveness in emerging business fields with 36 (B2B) members and five pilot projects. In the system-building networks the level of knowledge and information share together with common aims should be determined. These elements however may change or develop in the changeable environment. Finally they have the view that in the emerging smart grid sector, it is still a challenge to identify the proper performance indicators because experience with the new technology and the fast-changing environment lack which increase uncertainty.

Nordin et al. (2017) investigated empirically network management in emergent high-tech business contexts. In their view network management is "a set of activities undertaken by a focal firm, to define and realize the development of a network to support the emergence of a new business field and the focal firm's future position within it" (2017:3). They emphasize the management responsibility of the focal actor and suggest developing the following capacities:

(a) Network construction capability: looking for and getting to know the actors of network (both individuals and organizations) (casting); open communications and trust development (jamming); those activities which change thinking and stereotypes of actors (frame breaking).
(b) Context handling capability: "the capability to make sense of and influence the emerging business field by a future-driven and boundaryless outlook" (2017:8).
(c) Network position consolidation capability: management of information flow within the network (harvesting), the focal actor's leveraging the enhanced acknowledgement of … and their increased credibility in the emerging industry" (2017:10) (Upgrading).

In the study of U.S. Department of Energy (www.smartgrid.gov 2014) the experts have the view that utilities, regulators, vendors, and third party providers should more closely work together furthermore change their management practices so as to implement flexible and sustainable smart grids. They add that the involved professional stake-holders need efficient strategies and advanced technologies for internal communication and cooperation and the increased consumer data-management.

2.3 *The Planned Role of Economic Incentives and Smart Meters in Smart Grids*

The increased roles of end-users as co-providers have become more obvious when we think of the fact that 80% of the electric-vehicle recharging will occur at home (Lo Schiavo et al. 2013). So it is worthy to plan this activity not just from the technological and the supply chain aspects but also its human angles. The right behaviour, knowledge, openness and cooperative approach should be guaranteed by the end-users. Then disciplined, motivated electricity-buyers will use properly the batteries of EVs and will solve scheduled charging or uploading electricity back to the grid (Yu et al. 2014).

Before going into details let's review the relevant basic terms. Within the complex smart grid the cornerstone is a microgrid, which "is a localized grouping of electricity generations, energy storages, and loads. In the normal operation, it is connected to a traditional power grid (macrogrid)" (Fang et al. 2012:8). The microgrids are connected with the power flow and the information flow.

The appliances of the advanced metering infrastructure (AMI) make possible a safe two-way communication between centres of power utilities and individual end-buyers (Kahrobaee et al. 2013). The AMI can be well used for economic incentives such as saving on electricity bill (Zhou et al. 2016). The home energy management system (HEMS) means a complex optimal system which plays a vital role in monitoring, managing electricity generation and besides these it records electricity consumption and storage (Han et al. 2011; Son and Moon 2010). As you see in Fig. 3 HEMS has the following functions: monitoring, logging, control, alarm and management. Here we can notice the schedulable and non-schedulable devices as well, where EV belongs to schedulable appliances. The special roles and possibilities of EVs in smart grid were discussed in the previous subchapter. The home energy management system handles smart meters and the communication and networking system towards electric power supply chain (Asare et al. 2014). We can say that HEMS is an important, integrated part of future smart grids. Certain benefits of smart grids—such as cost- effectiveness, flexibility, provision of differentiated services and user friendly, advanced smart power technology—can be achieved through HEMS (Zhou et al. 2016). HEMS offers various possibilities and requires advanced knowledge of

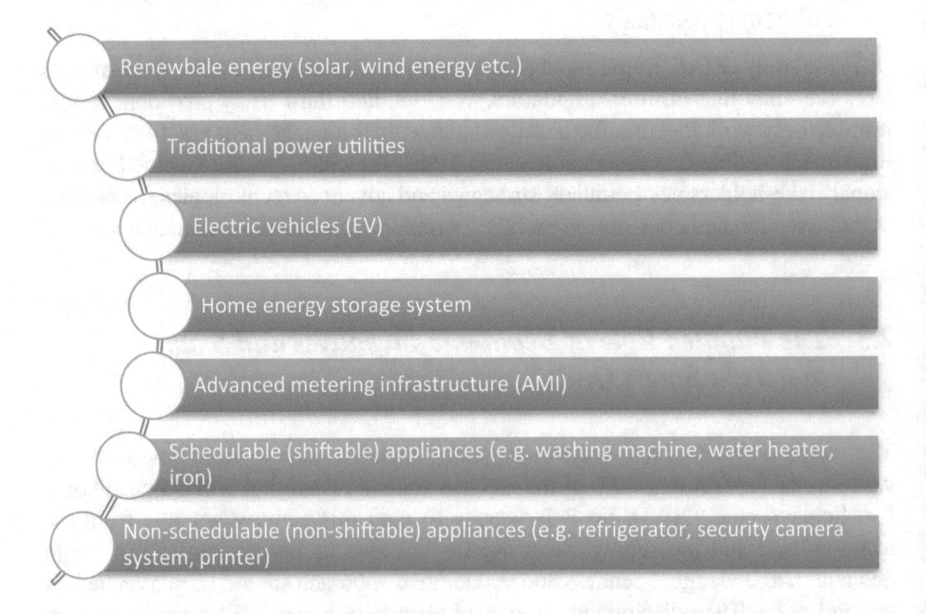

Fig. 3 Building elements of home energy management system (HEMS) (Own study based on Ahmad et al 2017; Zhou et al. 2016)

new technologies but will the majority of individual end-users be aware of all these? When will they learn them and how can they practice the proper using without serious consequences?

A smart meter is generally an electrical meter that records consumption, forwards this information back for billing purposes (Federal Energy Regulatory Commission 2010). Smart meter serves as measuring the energy consumption of end-users but as an advanced, modern device it also makes possible a two-way communication scheme (Benzi et al. 2011; Depuru et al. 2011). So smart meters give possibilities for data acquisition, data processing and advanced metering equipment management (Zhou et al. 2016).

If we accept the almost obvious condition that smart meters and smart metering basically belong to the concept of smart grids, the decision makers and administration should urgently handle and maybe solve the negative opinion, experiences and critics against smart meters by the public. For example in the United Kingdom nearly seven million have been installed since 2009 and originally they wanted to build in smart meters in every home by 2020 which ambitious plan had to be eased so now it is just offered for users (Meadows 2017). The Dutch government also had to give up their plans on mandatory installation of smart meters (Kema 2010).

Demand-Side Management (DSM) focuses on those solutions which support buyers and end-users to optimize or minimize their electricity consumption. If consumers use DSM successfully, they can reduce their energy bills and achieve more efficient and sustainable energy use. In the literature scholars discuss different price-based demand response strategies in which pricing play the dominant role. Within price-based demand response authors distinguish 3 types of pricing: (a) time-of-use pricing: price changes usually daily; (b) real-time pricing: the price may change even in every hour based on actual whole-sale price; and (c) critical peak pricing: specially calculated price for the peak times (Zhou et al. 2016). Siano (2014) calls the last two pricing methods as dynamic pricing and time-of-use pricing belongs to static pricing in his terminology.

The incentive-based demand response serves more complex aims as to motivate end-users to react to higher price so demand will decrease accordingly (Zhou et al. 2016). First tool of the incentive-based demand response is the capacity service program (CASP) which handles certain failures e.g. generator failure within the electric power supply chain. The second tool—the emergency demand response (MDR)—is a little similar to the previous one because it also focuses on failures but rather at system-level (e.g. blackout or system failure). The third is called demand side bidding (DSB) where buyers send their bids to the public utilities depending on current whole-sale price. The fourth is the interruptible load (IL) that means certain electricity volumes which can be interrupted or reduced in case of emergency. The last one is the direct load control (DLC). Like the previous tool (IL) this also implies refuse of load, now due to maintaining reliability (Zhou et al. 2016). According to calculation by Hubert and Grijalva (2012) these programs can bring 9.8–86.6% savings to the utilities and end-users but in each case the starting point of these theories is that individual and business buyers behave as homo economicus, namely absolutely rationally, having all necessary knowledge and information. Most publications on management challenges of smart grids represent this approach (e.g. Zhou et al. 2016; Planko et al. 2017; Siano 2014 ; Tushar et al. 2015) and neglect the views, suggestions of behavioural economics. (For more discussion about behavioural economics see Sect. 4.1.) The theory of Loganthiran et al. (2012) belongs to the few exceptions. Their concept focuses on demand side, a dynamic pricing method and a mechanism how the buyers may react to price change. However, Loganthiran et al. (2012) make clear that the flexible loads of energy will decrease end-users' comfort. Another exception is the paper of Fang et al. (2012) who suggest to research EV owners' behaviour.

On the other hand if we accept any pricing methods explained above we should make sure that the chosen price mechanism will be able to shepherd the end-users in the right direction (e.g. Bai 2016). Similarly to Geelen et al. (2013) we also have doubts that a good dynamic pricing system may maintain or guarantee safe operation. We have the view that more effective, proper communication of professional and individual buyers is highly needed so trough education and trust management the organizers of smart grids can expect the right approach, behaviour and cooperation from the end-users. Faruqui et al. (2010) found that a certain pricing scheme was more successful if it was connected to a prepaid electricity program.

Besides pricing incentives the so called distributed energy resource (DER) may serve efficiency and stability of the system. This is a type of decentralised energy resource in form of small energy loads which are based on current energy demand. The spread of this method is also supported by the tendency of shift from large energy plants to small-scale generators (Gelazanskas and Gamage 2014). Among these smaller power plants there are more renewable ones which bring uncertainty into the energy system (Gelazanskas and Gamage 2014).

The proposed demand side management strategy can be a proper scenario to handle complex management requirements of smart grids while its building element are the system identification and model design to achieve the desired load (Gelazanskas and Gamage 2014). The complicated approach represented by this strategy becomes obvious if we identify the following involved aspects: weather, price (selling and buying), time (e.g. outside temperature, wind, humidity) and calendar, desired and actual load.

3 The Largest Projects and the Geographical Possibilities of Smart Grids

3.1 An Overview of the Most Important Projects and Programs

Actually, as the necessary technical knowledge already exists, the real challenge is to put all these in a reliable system (Fang et al. 2012), that's why pilot projects and legal directives can directly promote the application of smart grids.

USA: Acts of Congress: Energy Independence and Security Act of 2007, American Recovery and Reinvestment Act of 2009 ($3.4 billion in funding for the SG Investment Grant Program and $615 million for the SG Demonstration Program); U.S. federal government—policy for smart grids (Fang et al. 2012). These huge programs include installation of over 65 million Advanced "Smart" Meters, Customer Interface Systems, Distribution and Substation Automation, Volt/VAR Optimization Systems, over 1000 Synchrophasors, Dynamic Line Rating, Cyber Security Projects, Advanced Distribution Management Systems, Energy Storage Systems, and Renewable Energy Integration Projects (www.smartgrid.gov 2014).

Hawaii also has set an aim to receive 70% of energy needs from renewable sources by 2030. For this end the Hawaiian Electric Co. (HECO) launched a pilot project which involves tips for costumers how to decrease their electricity use (Jeff 2012).

China consumes the most electricity and this consumption is going to be tripled by 2035. The Chinese energy use increases yearly by 1.5% and "despite slowing energy demand growth, China still consumes around one quarter of world energy in 2040" (BP Energy Outlook 2018). The renewable energy sources have been given increased priority that's why this source has grown by 9.5% (BP Energy Outlook 2018). In 2011 China's national utility, the State Grid Corporation of China, plans to invest $250 billion in electric power infrastructure development and between 2016–2020 additional $240 billion. Their plans focus on increase of energy effectiveness, decreasing carbon emissions and securing Chinese end-users more information on their electricity consumption (Hart 2011).

The strong Chinese interest toward smart grids originates both from the essential GDP growth and the increased number of urban habitants. This second group may play increased roles in future smart grids, especially if the predictions come true and by 2020 5 million EVs will run in China (Xu et al. 2000).

China firstly gave priority to development of smart grids technology in the 12th Five-Year Plan (2011–2015) and they plan to increase the proportion of renewables (mainly wind and solar energy) to 15% in consumption by 2020 (Xu et al. 2014). In the diverse Five-Year Plans China will develop each activity of electric power supply chains such as generation, transmission, substation, distribution, electricity utilization and dispatching (Xu et al. 2014).

European Union (EU): In 2017 totally 950 projects have been realized in 50 countries. From these, 540 projects have focused on R&D. The largest investments have been in Germany (140 projects), the United Kingdom, France and Spain (Smart grid projects outlook 2017:3).

Only a few small members are exceptions to this trend such as Luxembourg and Denmark. In Luxembourg, under the umbrella project of European H2020 Project (ATENA), Creos Luxembourg—the only owner and operator of the Luxemburg electricity networks and natural gas pipeline—plans to install 250,000 electricity smart meters and 800 public charging stations for electric cars and hybrid plug-in cars as preparing for smart grid application (The official portal of the Grand Duchy of Luxembourg 2015; www.esas.eu 2017).

Besides EU funding the role of private investment is very high in Great Britain and Luxembourg (83 and 98% of the total national investment respectively). Private funding represents an important share in Belgium, Denmark (mainly Copenhagen and its surrounding), Spain (the Basque region), France (Paris and its surrounding), Italy and the Netherlands (Smart grid projects outlook 2017:26–27). The national financing in the EU shows a little different picture because not just Denmark and Germany represent themselves (which member states have received high amount of EU funds) but also other countries such as Greece, France, Austria, Poland and Finland (Smart grid projects outlook 2017:29).

In smart grid financing the national rate is the highest in Denmark (36% by ForskEL, total amount about EUR 17.5 million—Smart grid projects outlook 2017) where the main aim is 'to grant funding for research, development and demonstration projects needed to utilise environmentally friendly electricity generation technologies, including the development of an environmentally friendly and secure electricity system' (Energinet.dk 2016).

The earliest business deployment of smart grid was the so called Italian Enel's Telegestore Project which included installation of more than 32 million smart meters between 2001 and 2006 (openei.org 2012).

India: The annual growth of Indian GDP is 8% which means an expected increase in energy need of 300% (of the current level). In India two-third of the high electricity demand is planned to be generated in smart grids (indiasmartgrid.org 2018). India is also active in installation of smart meters but here the planned deployments would be fulfilled above certain level of energy consumption. India's National Electricity Mobility Mission (NEMM) Plan 2020 involves 6–7 million EVs (ISGF Annual Report 2016–17). The Smart grid Mission (Ministry of Power, Government of India) is just completing the 'All India Metering Status Survey 2018' where data-collection was finished at the end of 2017 (nsgm.gov.in 2018). India's ranking in a list for US businesses has been gradually increasing for ears, in 2017 got the 4th place (Smart Grid Top Market Report).

Japan: In a study for U.S. exporters Japan landed the 7th place in 2017 concerning smart grids facilities and solutions (Smart Grid Top Market Report 2017). This is a really big and increasing market while—despite the start of re-using Japanese nuclear power plants—Japan is convinced to reform its energy sector so as to achieve higher energy stability, more efficient operation and more consumer options. Within the 4th Basic Energy Plan they also plan not only termination of the traditional monopoly situation of the energy industry but also to go on with liberalisation of electricity markets (ITA Smart Grid Top Markets Report 2016). In 2015 Japan invested the most in smart meters and plan to install smart meters almost in each home (totally about 80 million) (ITA Smart Grid Top Markets Report 2016).

3.2 *The Chances of Smart Grids from Geographical Approach*

Some states of the USA are active in distributed energy resource (DERs) programs. In 2018 Arizona completes the "Value-of-solar" project which involves establishment of net energy metering (NEM) and time-of-use (TOU) or demand-based rate in case of new retail customers. There is a "Residential Rate Reform" in California which means new minimum billing each residential buyer to opt-out rates. In Hawaii the "Grid Modernization Strategy" includes ambitious investments in network infrastructure upgrades (Scottmadden Energy Industry Update 2018). In order to achieve V2G and G2V supply certain infrastructure is needed. These US states promote or investigate energy storage: Arizona, California, Hawaii, Maryland, Massachusetts, Nevada, New Jersey, New Mexico, New York, North Carolina, Oregon, and Washington (Scottmadden Energy Industry Update 2018).

So called smart grid roadmaps (roadmap, route map, strategy, etc.) show the direction in more member states of the European Union. These roadmaps—usually organized by the competent national authorities or in some cases by private organisations—show that development of smart grid has a high priority in these countries: Denmark, Germany, Ireland, France, Austria, Slovenia, Sweden and the United Kingdom (Smart grid projects outlook 2017).

The geographical distribution of most smart grid projects in Europe is generally around large urban agglomerations; more concretely between Belgium, the south of the Netherlands and the west of Germany. Certain European capitals attract projects and investments, like London, Copenhagen, Madrid, Rome and Paris (Smart grid projects outlook 2017).

When we have a look at the global map and remember the regions of mentioned projects we may miss Russia. According to statistics the sales of new EVs were as many as 95 pieces and till the beginning of 2018 totally 1771 electric cars have been registered (rusautonews.com 2018). The charging stations can be found dominantly in the Moscow region (80 points) but Rosseti, the Russian power company announced to build 1000 charging stations by the end of 2018 (eurasianetwork.eu 2017). Local people claim that due to huge distances in Russia the diesel cars are still the most popular.

4 Greening Questions and Possibilities in Smart Grid Management

The India Smart Grid Forum is aware that for successful smart grids several efforts and activities are desirable, such as dynamic expansion of the number of EVs, more innovation in transport and energy sector, improvement of energy security and the quality of air in cities (ISGF Annual Report 2016–17).

If we wish to integrate green energy into daily life we should discuss sincerely the management tasks of smart grids together with relevant, unsolved challenges. Demand-side management is a key point in smart grids because the end-users have increased tasks and therefore responsibilities as well. When they consume electricity in smart and effective way (e.g. V2G or smart metering) they contribute to greening of their electric power supply chains. (The functions and benefits of demand-side management are discussed previously in this chapter.)

So it is time to start to educate both individual and professional end-users and to make them understand their new roles and possibilities in near future smart grids. Engineers and experts are busy with investigation of potential technical and IT problems and solution scenarios but they seem to neglect the increasing human participation in complex smart grids.

As we see in Fig. 3 earlier home energy management system (HEMS) gives a good natural base for spread of renewable energy at homes which confirms that smart grids are directly able to provide greening and sustainable environment.

Smart homes and an advanced home energy management system could guarantee for consumers the relevant positives of smart grids and demand-response programs. Wilson et al. (2015) see that smart home technologies have to be adapted to existing home environment and many people think their home as "places loaded with emotion, meaning and memories" (Baillie and Benyon 2008:227). Wilson et al. (2015) also call attention to that producers know very little about their smart home buyers and in our opinion the better understanding and serving of these consumers will be essential when smart homes are to become one of the main scenes of cooperation with end-users.

4.1 The Questions of Flexibility and Resilience in Smart Grids from Economic and Behavioural Approach

If we think over the increased role and responsibility of end-users in future smart grid, we also should know how to deal with the problem of free-riders in smart grid. Free-riders enjoy the full-scale of public services—e.g. lighting tower, street-lighting, heating or electricity—but they refuse to give their required contribution to it.

Planko et al. (2017) empirically found that pilot network participants, who represented each level of the supply chain, started with big enthusiasm but more of them were disappointed by the end. The main problem was caused by different motivating levels which resulted in worse and better performances and finally conflicts within the supply chain. Here the solution was that less ambitious members left the supply chain which was pleasant for the staying partners as well. But the supply chain became smaller. In a test smart grid it is an operable solution but this cannot be an answer in practice. Or shall we exclude these members from energy consumption, similarly to those buyers who have not paid their electricity bills for a very long time?

It is probably technically possible but a deep discussion among experts is needed and after that correct communication toward end-users would be desirable.

As demonstrated above, both individual and professional end-users will have increased functions in smart grids. Are they aware of it? Are they smart enough to actively participate in new electric power supply chains? Do they want this kind of entirely new cooperation? Goulden et al. (2014) investigated these questions and found that individual consumers (in the UK) would not like this system and do not want to be part of it while they experience close monitoring and rough reduction in their freedom to use electricity as they wish.

Behavioural economics combines the elements of psychology and neoclassical economics, but its roots date back to the 18th century. Amos Tversky and Daniel Kahneman (received the Nobel Prize in Economics in 2002) were the first researchers who built their statements on observing the behaviour of people in their decisions in risky situations. They came to the conclusion that, in precarious situations, people are deciding on the expected utility. Furthermore, after the theory of bounded rationality by Herbert Simon (received the Nobel Prize in Economics in 1978) there has been an increasing number of internationally and scientifically accepted models that deal with irrational aspects of business decision making and irrational behaviour. Based on these, should we agree with the mainstream general view on smart grids that both individual and professional consumers always behave rationally, so the task of experts is simply to introduce some operational scenarios and methods? We have strong doubts about that. Similarly to us, more scholars urge empirical surveys of behavioural angles of modern electric power supply chain (e.g. Van Der Schoor and Scholtens 2015; Fang et al. 2012).

A large-scale research on consumer behaviour was completed which was commissioned by the U.S. Department of Energy (www.smartgrid.gov 2014). In this study we cannot see the empirical results just the 'lessons learned', however, the authors call attention to the education of both consumers and stakeholders so as to achieve smooth consumer engagement. They see that in case of time-based rate programs end-users will need and should have the relevant expertise.

Geelen et al. (2013) propose a change in end-users' behaviour using Roger's model of the adoption of innovations. This process of adoption and awareness would be supported by communication programs in different media. They also suggest giving positive feed-back and other benefits for effective energy-users (e.g. they wash at nights). The community-based model uses the motivating power of peers and cooperation (Gardner and Stern 1996; Wilson and Dowlatabadi 2007). This approach could be used well in social media.

All these recommendations belong to social innovation which "is the process of developing and deploying effective solutions to challenging and often systemic social and environmental issues in support of social progress" (www.gsb.stanford.edu 2018). "Social innovation is not the prerogative or privilege of any organizational form or legal structure. Solutions often require the active collaboration of constituents across government, business, and the non-profit world" (Soule, Malhotra, Clavier www.gsb.stanford.edu 2018).

5 Conclusions

As a summary we see that a smart grid is more than a supply chain using advanced technological solutions, but a complicated business network as well that's why end-users and other involved actors will have increased roles and larger impact. Smart homes and EVs could be two of the key elements which may attract end-users to wish to cooperate in smart grids. However the development and rather the spread of both are at early stage, we are before the real boom.

With regard of management of grid-to-vehicle (G2V) and vehicle-to-grid (V2G) supply we see think that in order to achieve reliable operation it should be investigated whether—not just individual energy-users but also—the business users should give up their freedom for full charging at any time (while they may have a special agreement with their public service provider?). Can a firm accept such condition which reduces its flexibility and maybe profitability? Should business electric users establish preventive measures—both technological and economic—or will the (national) public service provider guarantee the necessary electric power?

Some relevant pilot projects and narrower range of tests concerning demand side management (e.g. Planko et al. 2017; or Morgan 2012) strategies and the advanced metering infrastructure (AMI) practices have been completed, but certain difficulties seem to remain. Firstly, if earlier functions of participants in the electric power supply chains change, these require fundamental improvements of buyers' behaviour and cooperative ability as well. Secondly, if the pricing system and other economic incentives cannot achieve flattening and resiliency, security problems may arise in electricity service, especially during peak-times. That's why experts should prepare scenarios not only for IT and technological failures and threats but in case of human mistakes as well. Thirdly, already a wide range of participants is involved in electric power networks but this tendency will strengthen further in the future. So in the modern smart grids the various forms of ownership—and consequently different operating principles, mechanisms, interests and priorities—may result in operational risks and uncertainties. These trends can effect directly or indirectly global and/or national standardisation development as well.

The Chinese ambitious smart grid projects serve double purposes. China wants not only greening and modern electric power supply chains but also to get a stable, leading role in the standardization of the global sector (Hart 2011). With regard to the EU it seems that size, total population and electricity consumption of the given country are the decisive factors in interest and development of smart grids (Smart grid projects outlook 2017).

Globally several major smart grid standardization roadmaps have been established in the USA, the European countries, China, Japan, and South-Korea (Fang et al. 2012). And these are those regions which nowadays do their best and most to achieve smart grids which still exist only in theory (Verbong et al. 2013).

Similarly to the logic of voluntary unemployment (people choose not to work due to low wage or increased value of free time or waiting for better working conditions) we think that assumption of purely rational behaviour cannot give grounded answer how the members of smart grid communities will behave, more concretely how they may choose for example between current pleasant energy prices and more freedom or flexibility for energy-use, actually giving up something from their life-style.

Finally we suggest the following points for consideration and future surveys:

(a) What do end-users mean by efficient energy-use; what temperature in their flat, what range of flexibility of energy use? (Geelen et al. 2013)
(b) What do we know about the proper or 'wrong' use of appliances by individual users or generally those equipment which will be vital in smart grids (Geelen et al. 2013)
(c) What are the price-considerations of buyers and required or acceptable pricing-methods?
(d) What does happen if the pricing system and smart meters cannot operate properly and fulfil balancing?
(e) What does happen to free-riders in a smart grid?
(f) What will be the main performance parameters and minimum requirements?
(g) How and under what conditions will the public provider guarantee stable electricity supply?
(h) What will be the responsibility of each actor in the smart grid?

References

8 things to know about electric cars in Russia (19 Aug 2017). Available via https://eurasianetwork.eu/2017/08/19/7-things-to-know-about-electric-cars-in-russia/. Accessed 1 Aug 2018

Ahmad A, Khan A, Javaid N, Hussain HM, Abdul W, Almogren A, Alamri A, Niaz IA (2017) An optimized home energy management system with integrated renewable energy and storage resources. Energies 10:549 https://doi.org/10.3390/en10040549

Alagoz BB, Kaygusuz A, Karabiber A (2012) A user-mode distributed energy management architecture for smart grid applications. Energy 44(1):167–177

Albert Chiu, Ali Ipakchi, Angela Chuang, Bin Qiu, Brent Hodges, Dick Brooks et al (2009) Framework for integrated demand response (DR) and distributed energy resources (DER) models. http://www.neopanora.com/. Accessed 9 Aug 2018

All India Metering Status Survey (2018) The Smart Grid Mission (Ministry of Power, Government India). Available via http://www.nsgm.gov.in/content/all-india-metering-status-survey. Accessed 3 Aug 2018

America's Electric Cooperatives: 2017 fact sheet. Available via https://www.electric.coop/electric-cooperative-fact-sheet/. Accessed 7 Aug 2018

Amin SM, Giacomoni AM (2012) Smart grid, safe grid. IEEE Power Energ Mag 10(1):33–40

Asare B, Kling WL, Ribeiro PF (2014) Home energy management systems: evolution, trends and frameworks. In: Proceedings of the universities power engineering conference, pp 1–5

Baillie L, Benyon D (2008) Place and technology in the home. Comput Support Coop Work 17:227–256

Bajor P, (2007). The bullwhip-effect in the electricity supply. In Proceedings papers of business sciences: symposium for young researchers (FIKUSZ), p 19–25

Balch O (2015) Energy co-ops: why the UK has nothing on Germany and Denmark. The Guardian, International Edition, 02.10.2015. Available via https://www.theguardian.com/public-leaders-network/2015/oct/02/energy-cooperatives-uk-germany-denmark-community. Accessed 7 Aug 2018

Benzi F, Anglani N, Bassi E, Frosini L (2011) Electricity smart meters interfacing the households. IEEE Trans Ind Electron 58(10):4487–4494

Bp Energy Outlook (2018) Edition, Available via https://www.bp.com/content/dam/bpcountry/de_ch/PDF/Energy-Outlook-2018-edition-Booklet.pdf. Accessed 12 Aug 2018

Burger B (2016) Germany's electricity export surplus brings record revenue of over Two Billion Euros. Available via https://www.ise.fraunhofer.de/en/press-media/news/2016/germanys-electricity-exports-surplus-brings-record-revenue-of-over-two-billion-euros.html. Accessed 7 Aug 2018

Burger B (2018) Power generation in Germany—assessment of 2017. Fraunhofer Institute for Solar Energy Systems ISE. Available via www.ise.fraunhofer.de www.energy-charts.de. Accessed 7 Aug 2018

Chiu A, Ipakchi A, Chuang A, Qiu B, Hodges B, Brooks D, et al (2009) Framework for integrated demand response (DR) and distributed energy resources (DER) models. http://www.neopanora.com/. Accessed 9 Aug 2018

Community Renewable Electricity Generation: Potential sector growth to 2020—Final Report (2014) Report to Department of Energy and Climate Change

Conchado A, Linares P (2010) The economic impact of demand-response programs on power systems. A survey of the state of the art. Economics for Energy. ISSN 2172/8437

Cooperatives—European Commission. Available via https://ec.europa.eu/growth/sectors/social-economy/cooperatives_en. Accessed 7 Aug 2018

De Gennaro M, Paffumi E, Scholz H, Martini G (2014) GIS-driven analysis of e-mobility in urban areas: an evaluation of the impact on the electric energy grid. Appl Energy 214:94–116

Defining Social Innovation (2018) Stanford graduate school of business. Available via https://www.gsb.stanford.edu/faculty-research/centers-initiatives/csi/defining-social-innovation. Accessed 6 Aug 2018

Depuru SSSR, Wang L, Devabhaktuni V, Gudi N (2011) Smart meters for power grid—challenges, issues, advantages and status. In: Proceedings of the power systems conference and exposition (PSCE), pp 1–7

Electromobility: deployment of the national infrastructure of 800 public charging stations by 2020. In: The official portal of the Grand Duchy of Luxembourg. Available via http://luxembourg.public.lu/en/le-grand-duche-se-presente/index.html. Accessed 2 Aug 2018

Energinet.dk (2016) ForskEL Call 2017

ESAS installs 100,000 smart electricity meters for Creos in Luxembourg. Available via https://www.esas.eu/en/news/esas-installs-100000-smart-electricity-meters-creos-luxembour. Accessed 2 Aug 2018

Experiences from the Consumer behaviour Studies on Engaging Customers (2014) U.S. Department of Energy Available via https://www.smartgrid.gov/recovery_act/overview/consumer_behavior_studies.html Accessed 1 Aug 2018

Fang X, Misra S, Xue G, Yang D (2012) Smart grid—the new and improved power grid: a survey. IEEE Commun Surv Tutorials 14(4):944–980

Faruqui A, Sergici S, Sharif A (2010) The impact of informational feedback on energy consumption—a survey of the experimental evidence. Energy 35:1598–1608

Federal Energy Regulatory Commission (2010) Assessment of demand response and advanced metering. Staff Report http://www.ferc.gov/legal/staff-reports/2010-dr-report.pdf

Gardner GT, Stern PC (1996) Environmental problems and human behavior. Pearson Custom Publishing, Boston, MA

Geelen D, Reinders A, Keyson D (2013) Empowering the end-user in smart grids: recommendations for the design of products and services. Energy Policy 61:151–161

Gelazanskas L, Gamage KAA (2014) Demand side management in smart grid: a review and proposals for future direction. Sustain Cities Soc 11:22–30
Gellings CW (1985) The concept of demand-side management for electric utilities. Proc IEEE 73(10):1468–1470
Global EV Outlook (2017) OECD/IEA (International Energy Agency). Website: www.iea.org
Goulden M, Bedwell Ben, Rennick-Egglestone Stefan, Rodden Tom, Spence Alexa (2014) Smart grids, smart users? The role of the user in demand side management. Energy Res Soc Sci 2:21–29
Hadley SW, Tsvetkova AA (2009) Potential impacts of plug-in hybrid electric vehicles on regional power generation. Electricity J 22(10):56–68
Han J, Piette MA (2008) Solutions for summer electric power shortages: demand response and its applications in air conditioning and refrigeration systems. In: Refrigeration, air conditioning, & electric power machinery, vol 29, issue 1, pp 1–4
Han J, Choi CS, Park WK, Lee I (2011) Green home energy management system through comparison of energy usage between the same kinds of home appliances. In: Proceedings of the 15 h IEEE international symposium on consumer electronics (ISCE), pp 1–4
Hannan MA, Hoque MM, Mohamed A, Ayob A (2017) Review of energy storage systems for electric vehicle applications: issues and challenges. Renew Sustain Energy Rev 69:771–789
Hart M (2011) China pours money into smart grid technology. Center for American Progress. Available via https://www.americanprogress.org/issues/green/reports/2011/10/24/10473/china-pours-money-into-smart-grid-technology/. Accessed 3 Aug 2018
Hubert T, Grijalva S (2012) Modelling for residential electricity optimization in dynamic pricing environments. IEEE Trans Smart Grid 3(4):2224–2231
Hu Q, Li F, Fang X, Bai L (2016) A framework of residential demand aggregation with financial incentives. IEEE Trans Smart Grid 9(1): 497–505
ISGF Annual Report 2016–17 (2017) India smart grid forum. Available via http://www.indiasmartgrid.org/reports/ISGF%20ANNUAL%20REPORT%202016%20-17.pdf. Accessed 3 Aug 2018
ITA Smart Grid Top Markets Report (2016) U.S. Department of Commerce, International Trade Administration. Available via www.trade.gov/topmarkets. Accessed 3 Aug 2018
Jeff (2012) Balancing Hawaiian wind power with demand response. 2 Feb 2012 Green Tech Media. Available via https://www.greentechmedia.com/articles/read/balancing-hawaiian-wind-power-with-demand-response. Accessed 2 Aug 2018
Kahrobaee MS, Rajabzadeh RA, Kiat SL, Asgarpoor S (2013) A multi-agent modelling and investigation of smart homes with power generation, storage, and trading features. IEEE Trans Smart Grid 4(2):659–668
Kema (2010) Smart meters in the Netherlands. Revised Financial Analysis and Policy Advice, Arnhem
Kempton W, Udo V, Huber K, Komara K, Letendre S, Baker S, Brunner D, Pearre N (2009) A test of vehicle-to-grid (V2G) for energy storage and frequency regulation in the PJM system. Mid-Atlantic Grid Interactive Cars Consortium
Lo Schiavo L, Delfanti M, Fumagalli E, Olivieri V (2013) Changing the regulation for regulating the change: innovation-driven regulatory developments for smart grids, smart metering and e-mobility in Italy. Energy Policy 57:506–517
Loganthiran T, Srinivasan D, Shun TZ (2012) Demand side management in smart grid using heuristic optimisation. IEEE Trans Smart Grid 3(3):1244–1252
Logenthiran T, Srinivasan D, Shun TZ (2011) Multi-agent system for demand side management in smart grid. In: IEEE PEDS 2011, Singapore, 5–8 Dec 2011
Meadows S (2017) Six reasons to say no to a smart meter. The Telegraph (on-line). Available via https://www.telegraph.co.uk/money/consumer-affairs/six-reasons-say-no-smart-meter/. Accessed 30 July 2018
Mohsenian-Rad A-H, Wong VWS, Jatskevich J, Schober R, Leon-Garcia A (2010) Autonomous demand-side management based on game-theoretic energy consumption scheduling for the future smart grid. IEEE Trans Smart Grid 1(3):320–331

Möller K, Svahn S (2009) How to influence the birth of new business fields—network perspective. Ind Mark Manage 38:450–458. https://doi.org/10.1016/j.indmarman.2008.02.009

Morgan T (2012) Smart grids and electric vehicles: made for each other? International Transport Forum, Discussion Paper, No. 2012-2

New electric cars sales have increased by 29% in 2017 in Russia (28 Jan 2018). Available via http://rusautonews.com/2018/01/28/new-electric-cars-sales-have-increased-by-29-in-2017-in-russia. Accessed 2 Aug 2018

Nordin F, Ravald A, Möller K, Mohr JJ (2017) Network management in emergent high-tech business contexts: critical capabilities and activities. Industrial Marketing Management, Available online 16 Oct 2017, https://doi.org/10.1016/j.indmarman.2017.09.024

Palensky P, Dietrich D (2011) Demand side management: demand response, intelligent energy systems, and smart loads. IEEE Trans Ind Inf 7(3):381–388

Parsons GR, Hidrue MK, Kempton W, Gardner MP (2014) Willingness to pay for vehicle-to-grid (V2G) electric vehicles and their contract terms. Energy Econ 42:313–324

Planko J, Chappin MMH, Cramer JM, Hekkert MP (2017) Managing strategic system-building networks in emerging business fields: a case study of the Dutch smart grid sector. Ind Mark Manage 67:37–51

Putrus G, Lacey G, Bentley E (2015) Towards the integration of electric vehicles into the smart grid. Springer, Berlin, pp 345–366

RECALIBRTAION The Scottmadden Energy Industry Update (2018) vol 18, issue 1

Roe C, Evangelos F, Meisel J, Meliopoulos AP, Overbye T (2009) Power system level impacts of PHEVs. In: 42nd Hawaii international conference on system sciences, pp 1–10

Schneider K, Gerkensmeyer C, Kintner-Meyer M, Fletcher R (2008) Impact assessment of plug-in hybrid vehicles on pacific northwest distribution systems. Power & Energy Society General Meeting, pp 1–6

Siano P (2014) Demand response and smart grids—a survey. Renew Sustain Energy Rev 30:461–478

Smart grid projects outlook (2017)

Smart Grid Top Markets Report (2017) U.S. Department of Commerce, International Trade Administration. Available via https://www.trade.gov/topmarkets/pdf/Smart_Grid_Top_Markets_Report.pdf. Accessed 3 Aug 2018

Son YS, Moon KD (2010) Home energy management system based on power line communication. In: Proceedings of the 28th international conference on consumer electronics (ICCE)

Sortomme E, Hindi MM, MacPherson SDJ, Venkata SS (2011) Coordinated charging of plug-in hybrid electric vehicles to minimize distribution system losses. IEEE Trans Smart Grid 2(1):198–205

Technology Roadmap—Smart Grids (2011) International energy agency

Telegestore (Smart Grid Project). Available via https://openei.org/wiki/Telegestore_(Smart_Grid_Project). Accessed 1 Aug 2018

The Dutch energy transition (2018) Rabobank. Available via https://www.rabobank.com/en/images/infographic_the_dutch_energy_transition.pdf. Accessed 7 Aug 2018

Tushar W, Chai Bo, Yuen Chau, Smith David B, Wood Kristin L, Yang Zaiyue, Vincent Poor H (2015) Three-party energy management with distributed energy resources in smart grid. IEEE Trans Ind Electron 62(4):2487–2498

Van Der Schoor T, Scholtens B (2015) Power to the people: local community initiatives and the transition to sustainable energy. Renew Sustain Energy Rev 43:666–675. https://doi.org/10.1016/j.rser.2014.10.089

Verbong GPJ, Beemsterboer Sjouke, Sengers Frans (2013) Smart grids or smart users? Involving users in developing a low carbon electricity economy. Energy Policy 52:117–125

What Smart Grid can do for India? (2018) India smart grid knowledge portal. Available via http://www.indiasmartgrid.org/sgg4.php. Accessed 3 Aug 2018

Wilson C, Dowlatabadi H (2007) Models of decision making and residential energy use. Annu Rev Environ Resour 32:169–203

Wilson C, Hargreaves Tom, Hauxwell-Baldwin Richard (2015) Smart homes and their users: a systematic analysis and key challenges. Pers Ubiquit Comput 19:463–476

Xu D, Wang M, Wu C, Chan K (2000) Evolution of smart grid in China. McKinsey on Smart Grid. Available via https://www.mckinsey.com/~/media/McKinsey/dotcom/client_service/EPNG/PDFs/McK%20on%20smart%20grids/MoSG_China_VF.ashx. Accessed 3 Aug 2018

Xu Zhao, Xue Yusheng, Wong Kit Po (2014) Recent advancements on smart grids in China. Electric Power Compon Syst 42(3–4):251–261. https://doi.org/10.1080/15325008.2013.862327

Yu R, Ding J, Zhong W, Liu Y, Xie S (2014) PHEV charging and discharging cooperation in V2G networks: a coalition game approach. IEEE Internet Things J 1(6):578–589

Zhou B, Li W, Chan KW, Cao Y, Kuang Y, Liu X, Wang X (2016) Smart home energy management systems: concept, configurations, and scheduling strategies. Renew Sustain Energy Rev 61(2016):30–40

Printed in the United States
By Bookmasters